Genetic Engineering

Editor

Ananda M. Chakrabarty, Ph.D.

Member
General Electric Research and Development Center
Schenectady, New York

CRC Press, Inc.
Boca Raton, Florida

Library of Congress Cataloging in Publication Data
Main entry under title:

Genetic engineering.

Bibliography: p.
Includes index.
1. Genetic engineering. I. Chakrabarty,
Ananda M., 1938—   [DNLM: 1. Genetic interven-
tion. QH442 G326]
QH442.G44   575.1   78-13781
ISBN 0-8493-5259-2

©1978 by CRC Press, Inc.
Second Printing, 1979
Third Printing, 1981
Fourth Printing, 1983

International Standard Book Number 0-8493-5259-2

Library of Congress Card Number 78-13781
Printed in the United States

PREFACE

Over the past several years, there has been tremendous progress in our understanding of the molecular basis of gene action. With the advent of the recombinant DNA techniques, it has now been possible to introduce various segments of eukaryotic and prokaryotic genes, both in bacteria as well as in cultured mammalian cells. The improved techniques of the chemical synthesis of deoxyoligonucleotides and their cloning in bacterial cells have resulted in the production of a human hormone by such cells, and there will undoubtedly be reports of success in the microbial production of other biologically active animal proteins in the future. The initial fears of accidental release of harmful microorganisms harboring recombinant DNA and creating enormous biohazard problems are gradually subsiding. This is mainly due to recent evidences that rule out the possibility of deliberate introduction of pathogenic characteristics to nonpathogenic microorganisms.

The primary objective of this book is to familarize the reader with the basic concepts of genetic engineering research. A detailed description of the general procedures, the nature of restriction enzymes and vectors, and the development of alternate host-vector systems have been included. The roles of various translocatable elements and the techniques of cloning DNA copies of mammalian messenger RNA have also been emphasized. The futuristic roles of recombinant DNA techniques in genetic therapy, the social, moral, and ethical questions, have been discussed in order to give the readers a wider perspective of this area of research. It is hoped that the book will not only help in the understanding of the development of genetic engineering techniques, but will also stimulate the reader's interest in the realization of the enormous potential this area of research holds for mankind.

A. M. Chakrabarty

THE EDITOR

Ananda M. Chakrabarty was born in Sainthia, India, on April 4, 1938, and received his Ph.D. in Biochemistry from Calcutta University in 1965. He later joined the faculty of the University of Illinois, where he was a research associate and a visiting lecturer in the Biochemistry Division. He joined the staff of the General Electric Company, Corporate Research and Development, in 1971, and has pursued research in genetics of hydrocarbon biodegradation. He also teaches at the State University of New York at Albany, where he is an Adjunct Professor of Biological Sciences. Dr. Chakrabarty received an IR-100 award for constructing multi-plasmid *Pseudomonas* strains for oil spill cleanup and was selected Industrial Research Scientist of the year in 1975. He participated in the Asilomar as well as La Jolla Recombinant DNA conferences for developing guidelines involving recombinant DNA research and acted as a workshop chairman in the National Academy of Sciences forum on recombinant DNA.

Dr. Chakrabarty is a member of the Society of Biological Chemists, American Society for Microbiology, Society of Industrial Microbiology, and American Association for the Advancement of Science.

CONTRIBUTORS

Frederick M. Ausubel, Ph.D.
Assistant Professor
Department of Biology
Harvard University
Cambridge, Massachusetts

John F. Brown, Jr., Ph.D.
Manager—Life Science Branch
General Electric Corporate Research and
  Development Center
Schenectady, New York

Ananda M. Chakrabarty, Ph.D.
Member
General Electric Research and
  Development Center
Schenectady, New York

Robert B. Helling, Ph.D.
Associate Professor
Division of Biological Sciences
University of Michigan
Ann Arbor, Michigan

Edward Loechler
Graduate Student
Biochemistry
Brandeis University
Waltham, Massachusetts

Margaret I. Lomax, Ph.D.
Assistant Research Scientist
Department of Cellular
and Molecular Biology
University of Michigan
Ann Arbor, Michigan

Tracy McLellan, Ph.D.
Member
Recombinant DNA Group
Science for the People
Cambridge, Massachusetts

Kenneth Murray, Ph.D.
Professor
Department of Molecular Biology
University of Edinburgh
Edinburgh, Scotland

Noreen E. Murray, Ph.D.
Senior Lecturer
Department of Molecular Biology
University of Edinburgh
Edinburgh, Scotland

Bob Park, M. A.
Member
Recombinant DNA Group
Science for the People
Cambridge, Massachusetts

Richard O. Roblin, Ph.D.
Head, Molecular Biology
of Tumor Cells Group
Cancer Biology Program
NCI—Fredrick Cancer Research Center
Fredrick, Maryland

Winston Salser, Ph.D.
Professor
Molecular Biology Institute
University of California at Los Angeles
California

David Shore
Graduate Student
Department of Biochemistry
Stanford University Medical School
Stanford, California

Peter Starlinger, Ph.D.
Institut für Genetik
der Universität zu Köln
Köln, West Germany

J. Gregor Sutcliffe, Ph.D.
Postdoctoral Fellow
Scripps Clinic and Research Foundation
Department of Cellular
and Developmental
Biology
La Jolla, California

Scott Thacher
Graduate Student
Harvard University
Cambridge, Massachusetts

Gary A. Wilson, Ph.D.
Associate Professor
Department of Microbiology
University of Rochester Medical Center
Rochester, New York

Philip A. Youderian
Graduate Student
Biology
Massachusetts Institute of Technology
Cambridge, Massachusetts

Frank E. Young, M.D., Ph.D.
Professor and Chairman
Department of Microbiology
University of Rochester Medical Center
Rochester, New York

TABLE OF CONTENTS

Chapter 1

THE MOLECULAR CLONING OF GENES — GENERAL PROCEDURES

R.B. Helling and M.I. Lomax

TABLE OF CONTENTS

I. INTRODUCTION

Our understanding of the structure, organization, and function of the genes of *Escherichia coli* and of some viruses has progressed far beyond our understanding of the genes of higher organisms. This is largely because bacterial and viral genomes are relatively simple and techniques for precise genetic manipulation of small segments of DNA are only available for bacteria and viruses. The ability to maintain selected genes, regardless of source, within *E. coli* would allow an application of the sophisticated genetic methods available for *E. coli* to any organism.[1-4a] Hitherto, this has been impossible for many reasons.

Several barriers prevent the exchange of genetic information between different species. By definition, mating is restricted in higher organisms to members of the same species. Transformation (the ability to introduce DNA molecules from the environment into the cell) is a well-known phenomenon in several bacterial species; however, many other bacteria, including *E. coli*, do not normally undergo transformation. Even when it can occur, the foreign DNA must come from the same bacterial strain or it will be degraded by the host cell's site-specific restriction endonuclease.[5] These nucleases recognize specific nucleotide sequences in DNA and cleave the DNA at or near these recognition sites. Cells containing such a nuclease also contain a methylating enzyme (restriction-methylase) which modifies newly synthesized DNA strands by methylating within the nuclease-sensitive site. Methylation of either DNA strand protects against nuclease activity. Since foreign DNA is generally not methylated in the same sequence, it is cleaved on contact with the nuclease. The EcoRI nuclease appears to be sequestered in the periplasmic space outside the cell membrane, where it can digest foreign DNA before entry.[6] No restriction and modification system has been demonstrated in eukaryotes.

Further difficulties must be overcome before foreign DNA can be maintained stably in a bacterial cell. Bits of DNA taken into competent cells during transformation must be incorporated into a "replicon" in order to be replicated. A replicon is any DNA molecule capable of self-replication, such as a chromosome, virus, or plasmid. This incorporation is almost always dependent on genetic exchange (recombination) with a homologous part of the replicon; nonhomologous DNA is not exchanged into a replicon.[7,8] Furthermore, genetic exchange requires the activity of a set of specific recombination enzymes.

In the last few years, these barriers to genetic exchange have been overcome. New procedures allow scientists to add nonhomologous DNA to small replicons (plasmids or viruses) in vitro and transfer the expanded replicons into bacterial cells. The added DNA is perpetuated during cellular growth as part of the replicon.[9] These procedures are referred to as recombinant DNA technology or the molecular cloning of genes.

The molecular cloning of genes is the result of the discovery of: (1) a subset of restriction enzymes which cleave DNA at specific sites, leaving single-stranded, complementary ends ("sticky ends") through which DNA from diverse sources can be joined,[10-12] and (2) a transformation procedure for *E. coli*.[13] Molecular cloning involves the following steps, which are diagrammed in Figure 1.

1. The desired DNA segment is dissected from the total DNA of the donor organism and freed of other contaminating DNA segments. Alternatively, cells receiving the desired DNA are separated from other cells at a later step.
2. The DNA segment is joined to the replicon through their complementary ends.
3. The replicon, including added DNA, is transferred into a cell.
4. Cells which have received and maintain the replicon are isolated or, in the case of a virus, the replicon itself is identified by the appearance of viral plaques.
5. The presence of the desired additional DNA is verified.

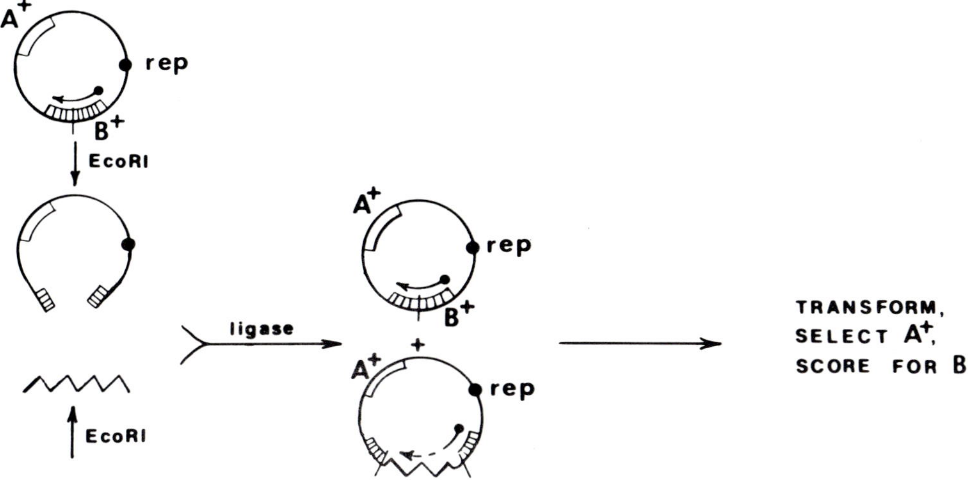

FIGURE 1.   Molecular cloning of genes using a plasmid as cloning vehicle. The replicator locus is designated by *rep*. Gene A identifies and selects cells containing the plasmid. It corresponds to the tetra-cycline-resistance determinant of pSC101, the colicin-resistance determinant of colE1 and the ampicillin-resistance determinant of pSF2124. Gene B represents a gene inactivated by the cloning process as, for example, a gene for colicin production in colE1 and pSF2124. The function of the equivalent gene in pSC101 has not been identified. As shown, transcription initiated at the gene B promotor can continue across the inserted DNA. This allows the expression of foreign genes that either lack a promotor or have one that is not recognized by the host cell.

The advantages of cloning segments of DNA are

1.   Discrete fragments of DNA can be isolated from complex mixtures of DNA molecules which would be difficult to fractionate by conventional physical or chemical methods.
2.   The cloned genes and their products may be made in large yield and high purity.
3.   Genes from any organism can be studied in the simplicity of the *E. coli* cell using the techniques available to the *E. coli* geneticist.
4.   Genetic systems allowing the manipulation of small, defined segments of DNA can be established in cells of mammals and other organisms.

This chapter will describe the general procedures involved in molecular cloning. Where more than one technique is available to accomplish a particular step, an attempt will be made to compare the relative advantages and disadvantages of each.

II. GENERAL PROCEDURES

A. Cutting and Joining DNA Segments
1. Cutting DNA

Unless care is taken, most DNA is sheared randomly to a molecular weight of 1 to 40×10^6 during preparation. By subjecting carefully prepared DNA to controlled shearing, large DNA fragments of a defined size suitable for cloning can be obtained.[14,15] For most purposes, however, it is advantageous to cleave DNA at defined points, thereby generating a unique set of discrete fragments, rather than at random. A large number of site-specific restriction endonucleases are available for this purpose (see K. Murray, this volume).[16,16a] Several enzymes are especially useful for cloning fragments of DNA because they produce the largest cleavage fragments and also make staggered

TABLE 1

DNA Sequence Specificity of Restriction Nucleases Making Staggered Cuts

Enzyme	Specificity	Source
BamHI	5′ G↓GATCC 3′	Bacillus amyloliquifaciens H
BglII	5′ A↓GATCT 3′	Bacillus globiggi
EcoRI	5′ G↓AATTC 3′	Escherichia coli RY13
HindIII	5′ A↓AGCTT 3′	Hemophilus influenzae Rd
SalI	5′ G↓TCGAC	Streptomyces albus G
XmaI	5′ C↓CCGGG 3′	Xanthomonas malvacearum

Note: Only one of the complementary DNA strands is shown. In each case, the recognition site is symmetrical and the opposite strand is identical in this site. ↓ designates points of cleavage. The complementary strand is also subject to cleavage as shown for BamHI:

$$5' \ldots \ldots G{\downarrow}GATCC \ldots \ldots 3'$$
$$3' \ldots \ldots CCTAG{\uparrow}G \ldots \ldots 5'$$

$$5' \ldots \ldots G \qquad \blacktriangledown \quad GATCC \ldots \ldots 3'$$
$$3' \ldots \ldots CCTAG \quad + \qquad G \ldots \ldots 5'$$

cuts, which are of importance in joining molecules. The recognition sequences of these enzymes are presented in Table 1.

Each recognition sequence consists of six base pairs. If the bases were distributed randomly, subsequent digestion of large DNA molecules with 50% G-C base pairs should produce fragments averaging 4096 base pairs, 4.1 kilobases (kb) in length. In practice, the average fragment size varies because (1) the G-C composition of the DNA is greater or less than 50%, (2) the neighboring base sequence can affect specificity,[17-19] or (3) the cleavage sequence is distributed nonrandomly. BamHI and SalI give the largest cleavage products, with an average size of 6 and 8 kb, respectively. Other restriction enzymes, recognizning shorter sequences, produce smaller products as expected (K. Murray, this volume).

2. Joining DNA Molecules Through Hydrogen Bonds Between Complementary Single-stranded Ends

Molecules may be joined through hydrogen bonds, but ultimately the union must be secured by covalent bonds. The ligation may be performed in vitro; otherwise, separate molecules or hydrogen-bonded joint molecules must be ligated by the cell after uptake.[9,14,15] The recovery of successfully transformed replicons is more efficient if ligation is done prior to transformation rather than in vivo.

In practice, DNA segments are usually joined through their single-stranded cohesive ends and covalently sealed by DNA ligase. Complementary ends can be formed in two distinct ways, either through the addition of defined single-stranded sequences (homopolymeric tails) or by cleavage with restriction enzymes making staggered cuts and generating cohesive termini (sticky ends).

The first procedure was developed by Lobban and Kaiser[21] and Jackson et al.[22] (Figure 2).   The enzyme terminal transferase from calf thymus catalyzes formation of single-stranded homopolymeric tails on double-stranded DNA if the reaction is carried out in the presence of a single type of nucleoside triphosphate.[23] Accordingly, poly dA tails can be added to one DNA preparation and poly dT tails can be added to another. The two types of DNA can then be joined through their complementary ends. If the

TERMINAL TRANSFERASE PROCEDURE

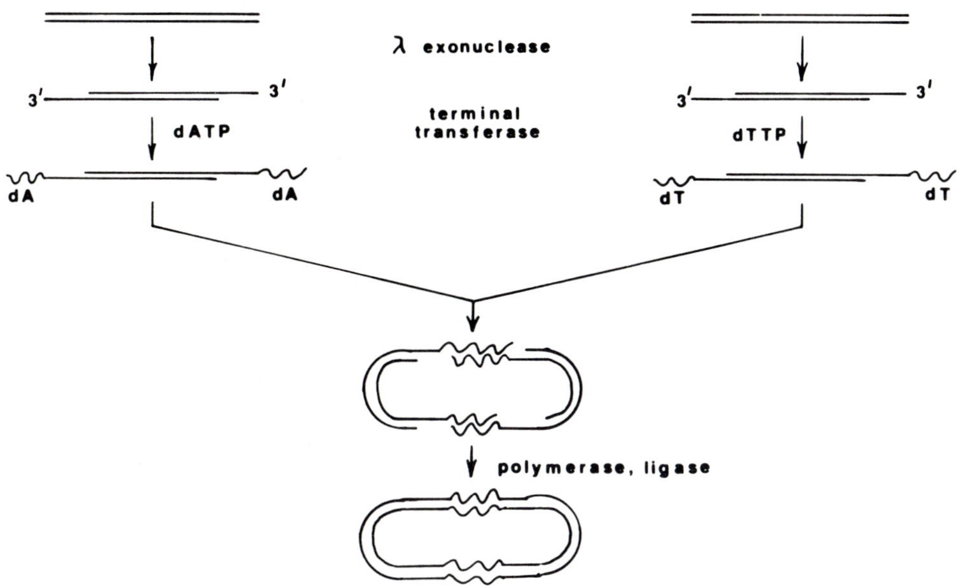

FIGURE 2.   Coupling DNA segments by means of poly dA and poly dT complementary single-stranded ends.

sequences are 40 nucleotides or longer, the joint is quite stable, and the hybrid molecule can be used in transformation without in vitro ligation.[21,23a]

Terminal transferase adds deoxyribonucleotides sequentially to the 3'-OH end of a DNA strand. Under usual conditions utilizing Mg^{++} as cofactor, the substrate must have a 3' single-strand extension containing three or more bases for the addition of nucleotides to occur.[23] This extension is produced by chewing back the 5'-ends of double-stranded DNA with λ-exonuclease.[21,22] (Most restriction enzymes producing staggered cuts generate DNA segments which terminate in 5' rather than 3' single-stranded ends.) Recently, it has been found that by using Co^{++} instead of Mg^{++} as cofactor, flush-ended DNA or DNA with 5' single-stranded ends serves as a primer for terminal transferase,[24] obviating the need for exonuclease digestion.

The second procedure for forming cohesive ends involves treating the native duplex DNA with an appropriate restriction endonuclease.   Each nuclease listed in Table 1 cleaves a symmetrical nucleotide sequence (palindrome). As the sequence is both unique and overlapping, DNA fragments produced by any one of the enzymes can be joined through their identical and complementary ends. DNA ligase can subsequently be used to seal the union permanently.

The hydrogen-bonded joint formed between EcoRI-generated cohesive ends has a melting temperature (T_m) of 5 to 6°C.[11] The T_m is a function of the base composition of the hybrid region, increasing with G-C content. Thus, the joints formed through XmaI-generated ends are expected to be more stable (higher T_m) than those produced by EcoRI-generated ends. Of course, the more base pairs in the overlap, the more stable the pairing.

3. Joining DNA Molecules Through Covalent Bonds

The enzyme DNA ligase catalyzes the covalent linkage of the 3'-hydroxyl terminus of one strand of DNA to the 5'-phosphate end of a second if both strands are paired on the same template.[25] Thus, the enzyme can seal single-strand breaks such as those present in hydrogen-bonded cohesive ends. Two types of DNA ligase have been well characterized[25] and are used extensively in cloning DNA molecules. DNA ligase from *E. coli* requires NAD as cofactor. T4 ligase, like ligases from higher organisms, requires ATP as cofactor. In addition to repairing single-strand breaks, the T4 ligase can join RNA and DNA strands aligned on a complementary DNA template. T4 DNA ligase is available commercially, as are terminal transferase and many restriction endonucleases. A third enzyme, RNA ligase, joins single-stranded DNA molecules together, RNA molecules together, or RNA molecules to single-stranded DNA molecules. [26,26a]

An additional and unexpected property of the T4 ligase was discovered by Sgaramella.[27] The enzyme can join "flush"-ended (fully base-paired) DNA duplexes, although at a rate considerably slower than when it seals single-stranded nicks. Flush-end joining is of considerable importance because it allows joining of DNA molecules which lack cohesive single-stranded ends. Thus, DNA segments obtained either by shear, by using two different restriction enzymes, or by using restriction enzymes which fail to make staggered cuts can be joined without adding single-stranded "tails." When necessary, DNA polymerase can be used to fill out nonflush-ended duplexes preparatory to joining by ligase.

4. Formation of Circles vs. Multimers

An end union can be intramolecular (circle formation) or intermolecular (joining of different DNA molecules). The relative rates of the two types of joining depend on both the DNA concentration and segment length.[28,29] Intermolecular joining is favored by a high concentration of DNA molecules. However, the efficiency of transformation by circular DNA is far greater than by linear DNA.[30] For this reason, after segments of DNA are joined at high DNA concentrations, the mixture is often diluted to a concentration at which circularization occurs preferentially. The rigidity of double-stranded DNA impedes the circularization of fragments the size of a random coil segment or smaller.[28] (The random coil segment is roughly 260 nucleotide pairs; however, the size is a function of ionic strength.) On the other hand, the rate of circularization is expected to decrease as the length of the DNA molecule increases because the two ends are less likely to encounter each other. As a result of this bias and an apparent size-dependent bias in transformation,[4,31] long DNA segments are cloned less efficiently than short segments.

5. Comparison of Joining Procedures

Joining DNA segments generated by restriction enzymes which form cohesive ends is the simplest procedure currently in use. Furthermore, hybrid molecules made in this way and cloned in an appropriate host cell can be dissected later by the same enzyme to regenerate the original DNA segments. This property is extremely useful in identifying the cloned DNA segment and purifying it for further study or manipulation.

On the other hand, the ability to clone intact genes depends on whether a restriction target appears within the gene. If it does, it is still possible to clone the gene in longer molecules produced either by incomplete nuclease digestion or by digestion in the presence of distamycin.[32] Alternatively, a different restriction enzyme could be used.

Terminal joining catalyzed by T4 ligase is simple and applicable to longer DNA molecules than are usually available after treatment with a restriction enzyme. Furthermore, if cleavage occurs at random rather than at specific sequences, each gene will appear intact on some fragments and therefore can be cloned. Any two double-stranded DNA molecules can be joined irrespective of the base sequences of their ends. Terminal joining is of particular importance in cloning defined, chemically synthesized DNA sequences where it is undesirable to add additional base pairs, e.g., in adding promotors or specific restriction enzyme cleavage sequences[33,34] or in joining fragments produced by different restriction enzymes.[35] A disadvantage of flush-ended joining is the inability to separate the joined fragments after cloning to obtain molecules identical to those originally joined.

The terminal transferase process is useful because: (1) very long segments of DNA can be joined, (2) DNA molecules can be joined irrespective of the base sequences of their ends, and (3) joining is restricted to molecules from different sources. A relatively high yield of transformants can be obtained even without ligation in vitro if the cohesive ends are sufficiently long. A disadvantage is that fragments identical to those cloned from the original donor organism cannot be easily recovered. It is possible to recover the starting material, but with altered ends;[36] however, in special cases, the donor DNA can be recovered without alteration.[16,35] Single-strand breaks incurred during DNA preparation can cause a further problem. Terminal transferase adds nucleotides at these positions as well as at the true ends of DNA molecules. If the DNA is then cloned, deletion, addition, or replacement at the break points may occur.

6. Adaptors

Short, defined nucleotide sequences have been chemically synthesized as adaptors for adding other DNA to an appropriate cloning vector. For example, after blunt-end ligation to the "other" DNA, the EcoRI target sequence serves as an adaptor for joining it to EcoRI ends of a cloning vehicle. In this way, a synthetic *lacO* locus flanked by two EcoRI sites was inserted in the plasmid pMB9.[33] The adjacent EcoRI sites allow the *lacO* sequence to be removed easily from the plasmid when desired. Chemically synthesized *lacO* regions have also been cloned by slightly different procedures.[34,37] It is likely that a variety of adaptors offering promotors, operators, ribosome-binding sites, new restriction sites, etc., flanked by EcoRI or other termini for cloning will be available in the near future.

B. Cloning Vehicles

1. Phage Vectors

The use of bacterial viruses as vectors for accepting and transmitting cloned genes is discussed in detail elsewhere in this book.[38] A derivative of λ has been certified by the National Institutes of Health (NIH), Bethesda, Maryland for use in experiments requiring stringent biological containment (λ.1 of Table 2),[39] and other λ derivatives elegantly designed for cloning genes have been constructed.[38,40,41] Attempts are underway to develop alternative virus-bacterial host systems for recombinant DNA experiments (see F. Young, this volume).

Figure 3 outlines procedures for cloning genes in λ derivatives. DNA to be cloned is joined to λ DNA in vitro. The recombinant DNA is inserted into bacterial cells where functional viral particles are formed. Successful packaging in λ heads requires the λ DNA ends and a total DNA length corresponding to 75 to 109% of the size of the wild-type chromosome.[42-44] Current λ vectors lack nonessential DNA to increase the space available for added DNA. Insertion vectors are available which can accept up to 14 kb of DNA, while replacement vectors accept up to 19 kb (Table 2).

2. Plasmid Vectors

The discovery that DNA fragments produced by EcoRI endonuclease contained sticky ends immediately suggested a means for joining DNA from diverse sources. The obvious choice of a replicon to which foreign genes could be added was the coliphage λ. It was a relatively small virus and had been extensively studied genetically and biochemically. However, EcoRI cleaves λ at five points. Multiple cleavage posed the problem of how new DNA could be added to the λ genome while simultaneously retaining

TABLE 2

Molecular Cloning Vehicles

Plasmids

Vehicle	Size (kb[a])	Copy number[b]	Known genes[c]	Useful restriction sites[d]	Ref.
pSC101	8.8	1—3	*tet*	BamHI (tet), EcoRI, HindIII, SalI (tet)	1, 3, 9, 14, 20, 47, 104
colE1	6.5	20	*col,imm*	EcoRI (col), HaeII (6), HaeIII (14), Hinc II, Pst I (2), SalI, SmaI	**3, 45, 49, 50, 169, 170**
pMB9	5.4	20	*col,imm,tet*	EcoRI, SalI, HindIII	3, 23a, 33, 172
pSF2124	11.3	10	*amp,col,imm*	BamHI, EclI (3), EcoRI (col). HindII (2), **PstI(4). SmaI**	3, 20, 53, 136, 148, 173, 174
pBR322	4.0	20⁺	*amp,tet*	BamHI (tet), EcoRI, HindIII (tet), PstI (amp), SalI (tet)	175
PCR1	13.4	10⁺	*kan,imm*	EcoRI, HindIII	3, 176
λdv1	8.2	80	λ(cIII-Q)	BamHI, EcoRI, HindIII	132, 142, 177
colE1-cos λ-*guaA*	33	?	*imm,guaA*,λ(R-J)	BamHI (λE), EcoGI (2), SmaI (3)	146, 178—180
pPL10	6.8	20	*bac*	EcoRI (2), HindIII	55
pCD1	7.5	10—20?	*col,imm,thy,tet*	BamHI, BglII(2). EcoRI (2)	181
RP1 (RP4)	55	1—3	*car,kan,neo,tet*	EcoRI	3, 59, 61—63, 142
RK2	58	1—3	*amp,kan,tet*	BamHI, BglII, EcoRI, HindIII, PstI, SalI (2), SmaI	3, 58, 62

Viruses

Vehicle	Size (kb)	Insert maximum[e]	Useful restriction sites	Ref.
λ.1 (λgt WES·λC)	40.6 (effective)[f]	11	EcoRI (2)	39
λ.2 (λgt WES·λB)	34.2 (effective)[f]	17.5	EcoRI (2), SstI (2)	40
λ.3 (641)	38	14	EcoRI	41
λ.4 (791)	33 (effective)[f]	19	EcoRI (2)	41
SV40a	1.67	3.3	BamHI	76, 78, 163
SVGT-1	3	2	HincII (2), Hpa (2), HindIII (3)	78

[a]   kb (kilobase pairs) = 1000 base pairs = $1.54 \times 10^{-6} \times$ molecular weight.
[b]   Plasmid copies per chromosome during exponential growth.
[c]   Abbreviations refer to colicin E1 production (*col*): immunity to colicin E1 (*imm*), bacteriocin production (*bac*); resistance to ampicillin (*amp*), carbenicillin (*car*); kanamycin (*kan*); neomycin (*neo*); tetracycline (*tet*); and thymidylate synthetase activity (*thy*).
[d]   In parentheses: gene cleaved if known; number of restriction sites if more than one.
[e]   Largest piece which can be added and still allow packaging in the virion.
[f]   Effective size refers to portion of original virus retained after cloning.

essential λ genes in their proper relationship. Ultimately, the problem was resolved easily by the deletion of certain EcoRI restriction sites and nonessential genes.[42-44]

The potential difficulty of using λ as a cloning vehicle suggested to Boyer it might be simpler to find a plasmid which could serve as a cloning vehicle. Plasmids are extra-chromosomal elements, i.e., small, covalently closed circular duplex DNA molecules found in the cytoplasm of the cell and capable of autonomous replication. In principal, the plasmid cloning vector only needs necessary replication functions (*rep*) which would not be inactivated by inserting DNA at a unique EcoRI site in the plasmid. It is

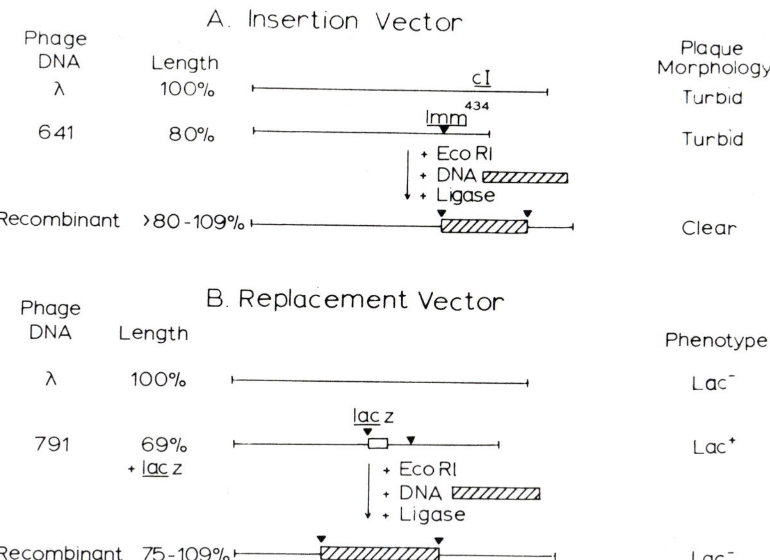

FIGURE 3.   Molecular cloning of genes using the virus λ as a cloning vehicle. The addition of DNA at the single available site in an insertion vector splits a gene. Lack of the activity encoded by that gene is apparent by the plaque morphology. In this case, a clear plaque appears instead of a turbid plaque. A nonessential segment of DNA is replaced by foreign DNA when a replacement vector is the cloning vehicle. If the nonessential segment contains a gene whose expression is readily identifiable, the λ derivative containing new DNA is easily recognized. In this case, the loss of the *lacZ* gene results in a different plaque color when assayed appropriately. A variety of useful replacement and insertion vectors, several of which have been approved for use in experiments requiring EK2 containment, have been described.[41,183,184]

also desirable that the plasmid contain a gene by which plasmid-carrying cells could be selected from a large population of cells, most of which lacked the plasmid. Several such plasmids were identified and one, pSClOl, was used for the first molecular cloning experiments.[9] Most subsequent gene cloning in plasmids has been done with pSClOl, a second plasmid (colE1),[45] or one of their derivatives.

a. Plasmid Biology

Like viruses, plasmids appear to be associated with virtually all bacterial genera.[46] All known genetic transfer systems involving bacterial conjugation are dependent on plasmid-borne genes. A resistance to antibiotics and heavy metals, the production of bacteriocins and toxins, and unusual metabolic capabilities such as degradation of complex organic compounds frequently result from the presence of a plasmid. However, they are not required for survival or growth of the cell under most conditions because truly essential genes are in the chromosome.

Plasmids range in size from about 1 to 300 kb in contrast to the much larger chromosome of most bacteria (3000 to 5000 kb). As expected, plasmids with complex attributes, such as ability to direct gene transfer in conjugation, are generally 60 kb or larger. The larger plasmids are present in only about one to three copies per bacterial chromosome, i.e., control of their replication is *stringent*. A small plasmid (pSClOl) derived from a *rep*-containing fragment of a larger plasmid (R6-5) is also subject to stringent control (Table 2).[47] Replication of pSClOl, like that of the *E. coli* host chromosome, is dependent on continued protein synthesis. However, pSClOl replication is independent of DNA polymerase I (*polA*) function.[48]

Other naturally occurring small plasmids, such as colE1, RSF1030, and CloDF13, exhibit *relaxed* control of replication, i.e., are present in multiple copies per chromosome during exponential growth (Table 2).[49] Replication of these plasmids is dependent on polymerase I activity but independent of continued protein synthesis. Thus, 1000 to 3000 copies of the colE1 plasmid per cell (about 40 to 50% of the total DNA) can be attained by adding chloramphenicol to prevent protein synthesis and replication of the chromosome.[50] An equivalent mass of a colE1 plasmid containing additional cloned DNA can be achieved in the same way. The number of copies decreases proportionately to the increased size of the plasmid.[45] The chloramphenicol procedure is an important means of obtaining large amounts of a DNA segment cloned in a plasmid.

pSC101 and colE1 can exist as separate plasmids in the same cell. However, two plasmids with the same *rep* functions are *incompatible* and cannot both be maintained. When pSC101 and colE1 are fused to form a single plasmid, multiple copies are maintained under the control of the colE1-derived *rep* locus.[51] Replication procedes from the pSC101-derived *rep* locus only if the plasmid number per cell becomes low. Nevertheless, the conjugate plasmid is incompatible with both colE1 and pSC101.[51] These results show that at least in the case of pSC101, replication is subject to repression by a product of the *rep* region. The repressor level is low enough to allow replication to proceed only when few copies of the plasmid are present. The results also suggest that some naturally occurring plasmids may contain more than one set of *rep* genes, even though only one set is functional.[51,51a,51b]

b. Cloning Genes in a Plasmid

A cloning vehicle should be as small as possible to reduce the possibility of including undesirable genes and prevent self-transmission by conjugation to another cell. The yield and purity of cloned genes and the efficiency of ligation and transformation are also greater with smaller DNA molecules. The minimum size of a plasmid vector is determined only by the requirements for *rep* functions, a gene-insertion site, and a gene to which selection can be applied in transformation.

Some of the properties of an ideal plasmid cloning vehicle are depicted in Figure 1 In addition to the essential features described above, two other attributes are desirable. One is an insertion site in a gene whose function is easily identifiable. As seen in Figure 1, plasmids containing cloned DNA inserted in gene B *lack* a functional gene B product (insertional inactivation). If the product is easily identifiable, cells containing a plasmid with cloned DNA can be picked out relatively easily from cells containing only an unmodified plasmid. It may be possible to select them directly.

A second desirable property of the plasmid cloning vehicle is a site for inducing transcription across the inserted fragment. Foreign genes may not be transcribed if their normal punctuation signals are not recognized properly. If so, an adjacent promoter in the cloning vehicle might allow transcription. Such plasmid-initiated transcription should be inducible (or repressible) because uncontrolled transcription could be inhibitory or lethal to the cell.[45]

ColE1 and its derivatives possess both of these attributes. The single EcoRI site in colE1 is in a gene affecting colicin production. Cells containing a plasmid with DNA inserted at the EcoRI site are identified easily because they are incapable of producing colicin (a protein which kills other *E. coli* cells lacking the plasmid).[45] Furthermore, transcription of the added DNA can be initiated by mitomycin C,[52] and presumably by ultraviolet irradiation or other treatments inducing colicin from colE1.

The availability of cloning vectors with single insertion sites for different restriction enzymes in genes subject to insertional inactivation (e.g., *amp*, *kan*, *tet*) allows a choice of restriction enzymes for cleaving plasmid and donor DNA. Also, the construction

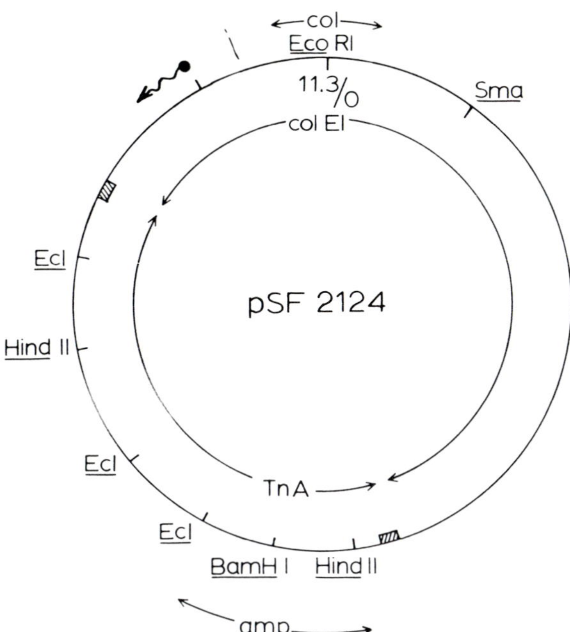

FIGURE 4. Map of the plasmid pSF2124. The TnA sequence, containing the *amp* gene and bounded by inverted repeat sequences was inserted within colE1 by recombination in vivo.[53] Locations of restriction-enzyme cleavage sites were determined in several laboratories.[6,139,148,166,170] Replication probably proceeds counterclockwise from a position 1.25 kb counterclockwise from the EcoRI site because pSF2124 is a derivative of colE1.[185] The plasmid contains no HindIII sites,[139] but does contain PstI, HaeII, and HaeIII sites.[185]

and use of short adaptors with defined promoters such as *lac*, *trp*, or λ allow greater flexibility in controlling transcription across cloned genes.

Figure 4 is a map of pSF2124, a plasmid currently used for cloning genes.[53] The plasmid is a colE1 derivative which contains a gene (*amp*) for ampicillin resistance. Genes can be inserted at the single EcoRI site in vitro. A bacterial cell can be transformed with the modified plasmid, and all nontransformed cells killed with ampicillin. Transformants containing cloned DNA can be easily distinguished because they fail to produce colicin, unlike cells with an unmodified plasmid. pBR322 is a very small plasmid certain to be useful in molecular cloning.[54] This plasmid contains two genes (*amp* and *tet*) which can be used for selection and in which cloned genes can be identified. The single Pst site is within the *amp* gene, and the sole Bam, Hind III, and Sal sites are within the *tet* gene. A single EcoRI site is present elsewhere.

Plasmid cloning vectors are being developed for use in hosts other than *E. coli*. pPL10 is a small plasmid of *Bacillus* which exhibits relaxed replication (Table 2).[55] It has no readily selectable genes but could be modified by adding such a gene. pCD1, a derivative of *E. coli* plasmid pMB9 (a colE1 plasmid) containing a *thy⁺* gene from *Bacillus* virus φ3T, is being tested as a possible cloning vehicle for *Bacillus*.[56] However, reports conflict as to whether the plasmid can be maintained in *Bacillus*.[56,57] Neither colE1 nor pSC101 appear to be maintained in bacteria not closely related to *E. coli*.[57,58] Several plasmids from *Staphylococcus aureus* have been transformed into *B. subtilis* and may be useful as cloning vehicles.[58a]

A group of related plasmids ("P plasmids") found in *Pseudomonas* is of considerable importance due to their wide host range. These plasmids can be maintained in most Gram-negative bacteria, unlike other known plasmids. For this reason, they show great promise as general cloning vehicles for use in a variety of bacteria.[59] For the same reason, it is prudent to use variants unable to conjugate to prevent an unrestricted dissemination of cloned genes. P plasmids have been shown to pick up chromosomal genes and transfer them to unrelated bacteria by "normal" recombination and mating.[60] Several laboratories are modifying P plasmids such as RPl and RK2 (Table 1) to reduce their size.[61-64] This will eliminate undesirable drug-resistance genes and genes required for conjugation. Furthermore, the smaller the plasmid, the greater the purity of the cloned genes, and the less likely the presence of other undesirable genes.

C. Purifying Plasmid DNA

Several simple procedures are available for obtaining plasmid DNA free from contaminating chromosomal DNA. If protein synthesis is not required to initiate plasmid DNA replication, chloramphenicol is added before the cells reach stationary phase to selectively inhibit chromosome duplication. Prior to harvesting, the plasmid-containing cells are killed by treatment with chloroform or diethylpyrocarbonate for several minutes to reduce possible biohazards in handling large numbers of cells.[65]

A convenient method for lysing cells and selectively obtaining plasmid DNA is based on the high salt-cleared lysate procedure developed by Hirt for extracting viral DNA from animal cells.[66-68] The cells are collected and resuspended in a small volume of an isotonic solution (a buffer containing 25% sucrose). Lysis is subsequently effected gently so that the chromosome remains attached to the cell envelope. The process usually involves treatment with EDTA and lysozyme to break down the peptidoglycan layer of the cell wall and the addition of the anionic detergent sodium dodecylsulfate (SDS) to disrupt membranes. This lysis procedure is applicable to many bacterial genera. Variant procedures involving proteolytic enzymes, ribonuclease, lysostaphin, or nonionic detergents may be useful in some cases, e.g., to lyse cells of *Staphylococcus* or *Pseudomonas*.[63,69]

Following lysis, NaCl is added to the gently opened cells to a 1 M concentration to selectively precipitate the large DNA of the chromosome. After centrifugation, the supernatant solution containing plasmid DNA is removed from the pellet containing chromosomal DNA and cell debris. If the volume is large, the plasmid DNA can be concentrated and further purified by precipitation with polyethylene glycol.[68,71,72] More commonly, phenol is used to remove proteins and part of the RNA. The DNA is subsequently precipitated with ethanol and resuspended in a small amount of buffer. At this stage, the plasmid may be used in transformation or studied by agarose-gel electrophoresis. However, it is contaminated with RNA and chromosomal DNA fragments which should be removed before the plasmid DNA is used in cloning.

Plasmid DNA is usually purified from a cleared lysate by CsCl density-gradient centrifugation in the presence of ethidium bromide or propidium iodide (dye-buoyant density procedure).[73] SDS forms a precipitate in a CsCl solution; therefore, it is either removed prior to CsCl addition or replaced by a nonionic detergent for lysis.[74] Supercoiled plasmid DNA binds less dye than open circular or linear DNA and bands at a denser position in the gradient. Consequently, the supercoiled DNA molecules in this band can be recovered free of contaminating chromosomal (linear) DNA, RNA, or protein.

3. Possible Vectors for Eukaryotic Cells

Several genes have been joined to a defective Simian virus 40 (SV40) chromosome,[3,75-79] including λ genes, an *E. coli* suppressor gene coding for tRNA^{Tyr}su$^+$III, and *Drosophila* genes. The hybrid molecules replicate in monkey kidney cells growing in culture if SV40 is present to provide missing functions. Defective polyoma virus could probably serve as a cloning vehicle for mouse cells in an analogous way. The defective nature of the viruses used provides a substantial degree of containment of the hybrid molecules.

Other potential cloning vehicles for higher cells include mitochondrial and chloroplast DNA. Their utility is complicated by their normal location in an organelle. Small yeast plasmids (6 kb) is not associated with nucleus or mitochondria may be useful for cloning genes in yeast and other organisms.[80,81]

C. Transformation

Although transformation occurs under normal growth conditions in a few bacteria, such as *Streptococcus pneumoniae, Bacillus, Hemophilus, Neisseria,* and *Acinetobacter,* it does not occur normally in most bacteria. The discovery that Ca^{++} treatment allowed *E. coli* cells to incorporate DNA from the environment may have been critical to the use of plasmids as cloning vectors. This procedure is simpler than alternative procedures available for use with phage vectors.[82,83] The Ca^{++} procedure has been used successfully with *E. coli,*[13,30,84,85] *Salmonella,*[57,86,87] *Enterobacter,*[87] *Staphylococcus,*[88] and *Pseudomonas,*[89,90] and is presumably effective with many other bacteria.

Ca^{++}-mediated transformation involves washing cells in Mg^{++} and adding DNA to cells suspended in Ca^{++} at 4°C. After binding the DNA, the cells are given a short heat pulse at 37 to 42°C and diluted with broth. Following gene expression, the cells are plated on media selective for transformants.[1,13,86] The treatment has no obvious effect on growth or viability of *E. coli* cells.[87]

Taketo has systematically studied Ca^{++}-induced transformation of *E. coli* by viral nucleic acid.[87,91,92] He found the procedure allowed uptake of either RNA or single- or double-stranded DNA. Transformation by duplex viral DNA was inhibited by heterologous duplex DNA but not by RNA or single-stranded DNA. The optimal Ca^{++} concentration was 70 to 80 mM. No cation could substitute for Ca^{++}, although Mg^{++} (6 mM) increased the number of Ca^{++}-induced transformants about twofold. Zn^{++} was severely inhibitory.

The transformability of cells in exponential growth phase was gradually lost after they entered the stationary phase. Although the type of growth medium used prior to treatment had little effect, the efficiency of transformation dropped precipitously at growth temperatures below about 32°C. However, a low temperature was obligatory for the Ca^{++}-mediated binding of DNA; no transformation occurred if the Ca^{++} treatment was performed at 37°C.

The proportion of unsaturated fatty acids in membrane phospholipids decreases as the temperature of growth increases.[92a] As a result, membranes formed at higher temperatures are less fluid and more rigid. Taketo proposed that Ca^{++} accelerates the stacking of liquid crystals which are formed after chilling the envelope phospholipids by its binding to phosphate groups.[92] DNA penetrates the membrane through spaces opened by shrinkage of the lipid phase.

Ca^{++} is of general use in inducing transformability, although the conditions under which this occurs differ significantly among different bacterial groups. These differences may reflect, at least in part, differences in the postuptake processing of DNA. Several studies have shown that in *E. coli* the *recBC* nuclease severely inhibits transformation, probably through its exonucleolytic activity.[84,85,91,93,94] Exonuclease I activity is also detrimental.[84] The transformation of genes which must recombine into the chromosome can only be detected in mutant recipients (*recB recC sbcB*) lacking both nucleases. Obviously mutant cells lacking any restriction endonuclease must be used as recipients to prevent the degradation of incoming foreign DNA. Transformability of *endA* strains lacking endonuclease I is also increased over that in wild type cells.[91,95,95a] Likewise, some mutants altered in the structure of the cell envelope are occasionally transformed at a higher rate.[86,96]

Circular DNA molecules or molecules which readily circularize through cohesive ends (such as λ) are resistant to degradation by exonucleases and consequently transform at far higher efficiencies than noncircular DNA[9,30,93-95] The efficiency of transformation decreases with the increasing size of the molecule transformed.[31]

Transformation efficiency is very low — about 1 transformant per 10^5 molecules of circular pSCl0l or colE1 DNA or per 2×10^4 λ molecules.[1] Kretschmer et al.[97] showed that most DNA molecules are potentially capable of transforming cells successfully, but that only rare bacterial cells are competent for transformation. They found that a cell successfully transformed for one plasmid had a high probability of being transformed for a second plasmid present in the medium. These results suggest that it might be possible to improve the efficiency of transformation by isolating a suitable mutant of *E. coli*. Alternatively, another bacterium exhibiting higher transformation frequency could be used.

D. Isolation and Purification of DNA to be Cloned

Most DNA purification procedures include cell lysis by a detergent, removal of proteins with phenol and/or proteolytic enzymes, removal of RNA with ribonuclease, and ethanol precipitation. The detergent SDS from many sources is contaminated with material which destroys biological activity of DNA. Matheson, Coleman, and Bell® SDS is satisfactory as supplied. Detergent which inactivates DNA or exhibits color must be passed through a column of activated carbon or reprecipitated from ethanol before use. Shaking with organic solvents, as used in the Marmur procedure,[98] causes extensive breakage of DNA[99] and is generally avoided.

DNA from cytoplasmic organelles is isolated from purified organelles to avoid contamination by chromosomal DNA. Supercoiled DNA, such as found in organelles or replicative intermediates of many viruses, is purified by dye buoyant-density centrifugation.[73] In contrast to total or organellar DNA purification, the isolation of specific genes from bulk DNA is generally a task of considerable difficulty.

1. Isolation of Specific Genes from Bulk DNA by Physical Differences

Some genes differ substantially in base composition from that of bulk DNA. If the gene is present in multiple copies per genome (repeated sequences), the corresponding DNA may form a "satellite" band distinct from main-band DNA in a CsCl or Cs_2SO_4 density gradient. The density difference may be enhanced by Ag^+ or Hg^+. The first eukaryotic genes cloned in a prokaryotic cell (ribosomal RNA genes of *Xenopus laevis*) were isolated in this way.[100]

DNA that is not physically connected with a chromosome (such as amplified rRNA genes of *Tetrahymena*[101,102]) or intact DNA from tiny chromosomes (such as in yeast[103]) can sometimes be separated from the much larger bulk of chromosomal DNA by agarose-gel electrophoresis or rate-zonal centrifugation in a sucrose gradient.

Restriction endonuclease-generated DNA fragments can also be separated according to size by agarose-gel electrophoresis. Below 6 to 10×10^6 daltons, a linear relationship between the logarithms of the molecular weights of DNA molecules and their relative electrophoretic mobilities is approximated when a low electric field strength (1.5 V/ cm gel) is applied.[104] This relationship may deviate noticeably from linearity when the mobilities of molecules differing greatly in molecular weight are compared, particularly during high voltage electrophoresis. The mobilities of DNA molecules also vary according to conformation; therefore, various conformers of a plasmid (supercoiled, open circular, and noncircular) can be separated.[28,29,104,105] Base composition has relatively little influence on separation of double-stranded DNA.[19,104,106] However, complementary single strands of DNA can be separated, presumably because they differ in base composition.[105] Coelectrophoresis of DNA fragment(s) of unknown molecular weight with DNA fragments whose molecular weight has been accurately determined (Table 3) allows a size estimation of the unknown fragment.

If the donor genome is simple (e.g., a virus), it may be possible to identify the band

TABLE 3

DNA Standards for Gel Electrophoresis

DNA	Restriction fragment		Molecular weight (M daltons)	Length (kb[a])	Ref.
λ	None		30.8	47.5	19
λ	EcoRI	·A	13.7	21.1	19, 104
		·B	4.74	7.3	
		·C	3.73	5.7	
		·D	3.48	5.4	
		·E	3.02	4.7	
		·F	2.13	3.3	
P22	EcoRI	·A	12.96	20.0	104, 182
		·B	6.10	9.4	
		·C	4.77	7.3	
		·D	2.65	4.1	
		·E	1.59	2.4	
		·F	0.75	1.16	
		·G	0.72	1.11	
		·H	0.56	0.86	
pSF2124	EcoRI		7.3	11.3	53
PM2	HpaII		6.68	10.3	105
pSC101	EcoRI		5.8	8.9	9
colE1	EcoRI		4.2	6.5	45
φX174	Pst		3.51	5.4	105
SV40	EcoRI		3.33	5.1	28
φX174RF	HincII	·R1	0.681	1.049	186, 187
		·R2	0.501	0.775	
		·R3	0.396	0.609	
		·R4	0.322	0.495	
		·R5	0.255	0.392	
		·R6a	0.224	0.345	
		·R6b	0.222	0.341	
		·R6c	0.218	0.335	
		·R7a	0.193	0.297	
		·R7b	0.189	0.291	
		·R8	0.137	0.211	
		·R9	0.105	0.162	
		·R10	0.051	0.079	
φX174RF	HaeIII	·Z1	0.877	1.342	186,
		·Z2	0.704	1.078	
		·Z3	0.570	0.872	
		·Z4	0.396	0.606	
		·Z5	0.202	0.310	
		·Z6a	0.177	0.271	
		·Z6b	0.181	0.278	
		·Z7	0.153	0.234	
		·Z8	0.107	0.194	
		·Z9	0.077	0.118	
		·Z10	0.046	0.072	

[a]kb = kilobase pairs.

containing the desired gene through knowledge of the fragment order in the genome.[1,104] If the gene is from a bacterium exhibiting a high efficiency of transformation involving recombination with the chromosome, such as *B. subtilus*, the location of a desired gene within the gel can be determined after electrophoresis by slicing the gel into fractions and transforming with the DNA fragments from each fraction.[107,108]

DNA fragments separated by agarose-gel electrophoresis can also be identified *in situ* if complementary RNA is available. Southern developed a procedure for transferring DNA fragments from gels to nitrocellulose filter strips after alkali denaturation.[109] Denatured DNA bound to the filter hybridizes with complementary RNA. Hybrid bands containing radioactive RNA are visualized by radioautography or fluorography. Alternatively, the filter strips can be cut into sections and counted in a liquid scintillation spectrometer, although this method has less resolving power. The Southern procedure is not reliable with DNA molecules of less than about 10^3 kb which bind poorly to the nitrocellulose.[109,110]

Once the location of the desired DNA fragment is determined in a portion of the gel, it can be eluted from the remainder of the gel and used in cloning. To be sure, if restriction fragments of cellular DNA are fractionated according to size, the eluted DNA from a single fraction will still consist of a mixture of fragments. Nevertheless, it is highly purified relative to the initial preparation.

Several methods have been developed for recovering DNA fragments from the agarose gel. The simplest involves cutting the band from the gel, mashing in buffer to extract the DNA, and removing the agarose by centrifugation. Another simple procedure is the "freeze-squeeze" method.[111] Gel slices are placed between sheets of parafilm and are frozen. The liquid is extruded from the gel by squeezing the parafilm between thumb and forefinger, while the remaining agarose is removed by centrifugation. Agarose will dissolve in saturated KI solutions; the DNA can be recovered by chromatography on hydroxylapatite or benzoylated-napthoylated DEAE cellulose.[28] Tiny agarose particles sometimes prevent the recovered DNA from serving as an enzyme substrate. The only reliable procedure for obtaining DNA totally free of agarose is centrifugation in a KI density gradient.[112]

Polyacrylamide-gel electrophoresis is generally used for separating and purifying DNA molecules smaller then 1.5 kb but is less satisfactory than agarose-gel electrophoresis for larger DNA molecules.

2. Isolation of Specific Genes by Complementary RNA

Identification and purification of most unique genes depends on the availability of its respective RNA transcript. The gene can be identified unambiguously by hybridization with the RNA because recognition depends on gene structure rather than function. The purification of mRNAs from many unique genes has been achieved. The RNA can be used in two distinct ways to isolate DNA from specific genes. It can either serve as a template for in vitro synthesis of a DNA copy or be used as a probe to find and isolate homologous fragments of DNA synthesized in vivo.

a. RNA-directed Synthesis of DNA

RNA-dependent DNA polymerase (a reverse transcriptase and usually from avian myeloblastosis virus) catalyzes the synthesis of a DNA strand (cDNA) complementary to the RNA. The enzyme generates a 3′-terminal "hook," or hairpin structure with a short double-stranded region, which make the cDNA self-priming for the subsequent synthesis of a second DNA strand by *E. coli* DNA polymerase I.[113-115] The RNA strand is then removed by alkaline hydrolysis, and the hairpin loop connecting the two complete DNA strands is cut by S1 nuclease. The resulting double-stranded structure can be joined to a cloning vehicle either by adding homopolymeric single-strand tails with terminal transferase or by flush-end ligation. Apparently, the RNA-cDNA hybrid molecule can also be joined to a cloning vehicle in vitro and subsequently converted to double-stranded DNA in vivo.[116]

A disadvantage of procedures based on reverse transcription is that nontranscribed DNA flanking the structural gene and containing initiation and termination signals will not be synthesized. However, once the transcribed region is cloned, it can be used as a probe to isolate fragments of cellular DNA containing nearby sequences. Enzymatically synthesized DNA from rabbit α- and β-globin mRNA and chick ovalbumin mRNA has been cloned in plasmids of *E. coli*.[114,116-120]

b. Chromatography Involving Nucleic Acid Hybridization

When bound to an inert matrix, either RNA or DNA can hybridize with a complementary strand in solution and thus be used for affinity chromatography. The most useful affinity columns employ nucleic acids bound to cellulose.[121] A high binding efficiency results from recycling the column effluent through the column.[122] The trapping of nonspecifc complexes formed through the hybridization of repetitive sequences is minimized by denaturing the recycled material between each passage.[122,123] Bound RNA can be used to purify complementary DNA, which can be eluted and used in turn to isolate either the other DNA strand or more RNA. Duplex DNA can be obtained by annealing the separately purified complementary strands.

Eukaryotic DNA segments capable of hybridizing with specific mRNAs can be purified by taking advantage of the 3′-poly A tail on most eukaryotic mRNAs.[124-126] The tail is available for hybridization to column-bound poly U or poly dT sequences. Single-stranded DNA is made available for hybridization with RNA by a controlled treatment of the DNA with λ-exonuclease to digest the 5′-ends and expose the beginning of the coding strand of a gene. Alternatively, the other end of the coding strand can be exposed by treatment with *E. coli* exonuclease III, which digests DNA from the 3′-end. The hybridization of mRNA with the middle of a gene is achieved by making gaps in one strand of duplex DNA with pancreatic deoxyribonuclease and exonuclease III.

An alternative chromatographic procedure depends on the high affinity of organic mercurial compounds for mercaptans.[125-127] DNA hybridized with mercuriated RNA binds to sulfhydryl-agarose and is removed from nonhybridized DNA. Extremely little nonspecific binding of DNA is observed, and the column can be used repeatedly. A high degree of purification of a specific DNA segment hybridized with mercuriated RNA can be achieved by binding it to sulfhydryl-agarose, eluting with mercaptoethanol, binding to poly U-agarose, and eluting with ribonuclease.[125]

It should be possible to isolate DNA that is partially hybridized with RNA by direct physical methods. Under conditions favoring stability of the DNA-RNA hybrid, RNA hybridizes with its complementary DNA strand, displacing the homologous strand of duplex DNA over a localized region.[128] Molecules with such hybridized regions differ in density from fully double-stranded DNA and, presumably, can be separated by density-gradient centrifugation.

E. Stability of Plasmids with Cloned Fragments

Transformation with a plasmid or virus containing an added DNA segment does not guarantee that the DNA will be maintained or the cell will maintain itself in competition with other cells lacking that segment. The inserted piece may interfere with replication of the plasmid, reduce the "fitness" of the host cell, or be subject to mutation or recombination at an unusual rate.

Few randomly added genes will ever contribute to the well-being of a cell. Most genes (particularly if expressed to form RNA and protein) will prove detrimental to the cell since the energy expended in replication and expression is wasted, and in some cases, because the gene products are directly inhibitory to the cell. For example, the uptake of a gene which produced an unregulated ribonuclease would be lethal to a cell.

Generally, cloned DNA is maintained through many generations of duplication without change, as is shown by sequencing both β-globin gene segments in pMB9 and segments of sea urchin histone H_2A and H_3 genes in pSC101[114,129,130] and by restriction enzyme patterns of globin, histone, ribosomal RNA, and many other genes. However, a series of observations support previous expectations that some plasmids and viruses with cloned fragments can be maintained with difficulty or not at all.

Virtually all foreign DNA segments cloned in λ which have been studied confer a selective disadvantage to the virus, relative to wild-type λ.[131] λdv containing cloned *lac, bio,* or *gua* DNA is highly unstable, and cells lacking the plasmid appear at a high frequency.[132] The *bio*-11 segment from λdv *bio*-11 cloned in λdv provides sufficient biotin to satisfy the nutritional requirement. However, *bio*-11 cloned in colE1 either does not provide sufficient biotin, is unstable, or is lethal to the host cell.[132]

Furthermore, the EcoRI fragment containing both the *trp* genes and the ϕ80 N gene from ϕ80*trp* can not be cloned successfully in colE1 unless the adjacent EcoRI fragment from ϕ80*trp* is cloned with it.[45] Presumably, the second fragment (which contains the ϕ80 immunity region) is necessary to repress N gene expression. A high level of N gene product may be lethal to the cell. Analogously, the λ fragment containing the λ immunity region can only be cloned in pSC101 under conditions conducive to lysogeny by wild-type λ.[133] Several DNA fragments from *Drosophila melanogaster* successfully cloned in pSC101 cannot be cloned in colE1 derivatives.[133a] Two sets of genes have been successfully cloned in one orientation relative to plasmid genes. However, clones do not appear with the DNA in the opposite orientation. These are the *X. laevis* oocyte-type 5S RNA genes cloned in pSC101[134] and the rabbit β-globin gene cloned in pMB9.[135]

trpR Cells containing the *E. coli trp* operon cloned in colE1 are at a large competitive disadvantage with *trp*R$^+$ cells containing the same plasmid with *trp*R$^+$ or *trp*R cells either containing colE1 alone or lacking any plasmid.[45] The copy number of pSF1010 is altered by in vivo insertion of the TnA sequence in one region of the plasmid, but not in another region.[136]

The structure of some plasmids is subject to change during growth. Bedbrook and Ausubel found that plasmids frequently form dimers and larger multimers.[137] The conversion results from the classical recombination of homologous plasmid molecules, since it is *recA*$^+$-dependent but only partially dependent on the *recB*$^+$, *recC*$^+$, and *recF*$^+$ genes.

ColE1 containing a kanamycin-resistance gene (*kan*) undergoes frequent deletions of part of the *kan* DNA; however, the colE1 moiety remains unaltered.[138] Derivatives of pSF2124 containing yeast rDNA sequences are also unstable, and variants containing less rDNA accumulate during growth.[139] Similarly, Helling and Lomax's laboratory has found alterations in pSC101-derivatives containing *Xenopus* rDNA by partially digesting the plasmid DNA with EcoRI-endonuclease, separating the products by agarose-gel electrophoresis, and transforming with DNA from gel regions that did not correspond to EcoRI fragments in the original plasmid. The resulting transformants showed altered gel patterns corresponding to changes in the DNA derived from *Xenopus*. None of the 20 examined showed a change in the pSC101 portion. Presumably, the altered DNA resulted from recombination in the region of the *Xenopus* rDNA containing tandem repeated sequences.

recA⁻ Mutant cells are often used as hosts to avoid changes in the plasmid or virus resulting from recombination between homologous DNA segments. However, *recA* cells grow poorly, are less viable than the wild-type,[140] may have a reduced transformation frequency, and cannot be induced for colicin production.[52] Furthermore, plasmids containing transposable elements, such as the short DNA sequences (insertion sequences) commonly bracketing antibiotic-resistance genes, are capable of recombination with nonhomologous DNA even in the absence of classical recombination.[141,142]

The validity of many studies of genes cloned in foreign cells depends on the integrity of those genes before and after transformation. This integrity is essential if the gene expression is studied or recovery of gene products identical to those in the original cell is desired. Present experience suggests that the vast majority of added DNA sequences are faithfully duplicated in the host cell; however, some (particularly reiterated sequences) may undergo structural change at a significant rate. Therefore, it is essential in many studies to verify that the cloned gene is a faithful copy of the native one.

F. Gene Banks

If one is only interested in a single cloned DNA segment, the desired cell can be isolated and a stock culture easily maintained. All other transformed cells can be discarded. However, this wastes other cloned segments from the source DNA. Therefore, "banks" of hundreds or thousands of cloned fragments from the genomes of *E. coli*, *Saccharomyces cerevisiae*, mammals, and other organisms are being established. Investigators can screen such a bank in search of a cell with any desired cloned DNA, thus obviating the need to carry out the entire cloning procedure repeatedly.

In principal, entire genomes can be established in banks as sets of discrete fragments. Complete sets of chloroplast, mitochondrial, and viral DNAs can be maintained provided no segment is severely disruptive to the cloning vehicle or cell in which it is housed. There is no certainty of attaining a bank with all segments of a cellular genome, due to the large number of different DNA segments into which any cell genome is fragmented by conventional procedures.

One can estimate the probability P that a given unique DNA sequence is present in a collection of transformed colonies as follows:[15]

$$P = 1 - [(1 - \frac{X}{L}) f]^N$$

where X = length of desired segment of DNA; L = average length of DNA fragment cloned; f = fraction of genome represented by an average fragment; and N = number of transformant colonies containing cloned DNA. This equation assumes that each DNA segment is transformed and maintained with equal efficiency.

Clarke and Carbon calculated that a bank size (N) of 720 transformed colonies was sufficient to give a probability of 0.90 that one *E. coli* gene was present in a hybrid plasmid with an average molecular weight of 8.5×10^6.[15] A bank of twice this size gave a probability of 99%. Obviously, the size of the necessary bank increases as the genome becomes more complex. The bank may be a population containing many DNA fragments cloned randomly in different cells. A serious disadvantage of this method is that transformed cells containing recombinant DNA molecules grow at different rates; therefore, the process of culturing the cells to obtain a sufficient amount of DNA could lead to selective loss of certain recombinants. Certainly, the distribution of cloned fragments in the purified DNA would not represent the fragment distribution in the original organism. Alternatively, the storage bank may consist of many individual transformed cells grown and maintained in pure culture.

G. Selection, Identification, and Characterization of Desired Clones
1. Direct Identification of Clones with Specific Genes
a. Genetic Methods

Recombinant plasmids or viruses carrying specific *E. coli* genes are selected directly by using appropriate mutants as recipients for transformation. For example, recombinant vectors containing *trp, gal, lop, lacZ, ara, thyA,* or other genes from *E. coli* have been constructed.[15,45,143-146] If the recipient strain is F⁺, the recombinant plasmid can be transferred easily to other recipients. This provides a rapid and simple procedure for identifying a recombinant plasmid containing a specific gene among a large bank of plasmids with randomly cloned genes.[15]

Individual colonies in the bank are replicated to auxotrophic cells on appropriate selective media. From the positions of growth on the replica plate, colonies on the master containing a plasmid complementing the mutant recipient gene can be identified. This procedure is extremely useful when it can be applied. However, the NIH Guidelines prohibit use of a sexually competent donor strain for cloning uncharacterized genes from unrelated organisms.[146a]

Although selective methods are very powerful, they are generally limited to genes for which bacterial mutations exist and to genes which can be expressed in bacterial hosts. Genes from many bacteria, both Gram-positive and Gram-negative, are expressed to give the expected phenotype when transplanted to *E. coli* or other bacteria.[56,57,60,147-149] Segments of yeast DNA that complement mutant genes of *E. coli* have been cloned in phage lambda and in a plasmid.[150,151] However, the expression of the desired gene can vary with the cloning vector and the host cell.[1,132,149] Consequently, direct selection may fail to produce a desired clone even when the desired gene comes from a related bacterium.

b. In Situ Hybridization

Colonies or plaques containing specific genes can be identified *in situ* if a labeled RNA probe complementary to that gene can be obtained. In the colony-hybridization procedure developed by Grunstein and Hogness,[152] transformed colonies on nitrocellulose filters are treated with alkali to denature the plasmid DNA, which remains bound to the filter. After washing to remove cell debris, the filters can be hybridized with the appropriate RNA probe. Colonies which hybridize with RNA are located by radioautography or fluorography. Similar procedures have been developed for λ and SV40 plaques.[153-154] The colony-hybridization technique has been used extensively, and a simpler version of the original procedure has been developed.[155]

c. In Situ Immunoassay

Clones containing specific genes can be identified by *in situ* immunoassay if the antibody for the gene product is available.[156] The technique is applicable both to plaques and colonies; therefore, either viral or plasmid cloning vectors can be used. This method depends upon the presence of the gene product, rather than its activity. Thus, it should be useful for identifying genes for structural proteins, eukaryotic proteins for which no *E. coli* mutants exist, enzyme subunits, etc.

2. Indirect Enriching Methods for Clones and Specific Genes

In many cases, it may not be possible to identify specific clones by any of the previously described procedures. Consequently, indirect methods must be employed to enrich for specific recombinant plasmids or viruses or to identify specific recombinant clones. In such experiments, usually the entire genome of an organism or organelle is cloned as separate fragments, and the heterogeneous population of plasmids or viruses is fractionated by the desired means.

a. Fractionation by Density Differences

Recombinant DNA can be fractionated either before or after transformation by centrifugation to equilibrium in CsCl-density gradients.[156a,156b] The DNA on the heavy or light side of the main DNA band can be used to transform cells to enrich for either heavier or lighter classes of recombinant DNA. Obviously, this technique is useful only if the cloned DNA has a buoyant density significantly different from the plasmid or viral vector.

b. Fractionation by Size

It may be useful to prefractionate the recombinant DNA molecules by size before the transformation procedure. Fractionation is usually done by a sucrose gradient centrifugation.[157] Agarose-gel electrophoresis can also be used. Glover et al. fractionated a ligated mixture of pSC101 and *Drosophila* DNA to obtain only the dimeric molecules and eliminate the multimeric forms.[158] The same procedure can be used after transformation to enrich for rare recombinant molecules containing large fragments of foreign DNA.

c. Sib Selection

Sib selection[159] facilitates the isolation of clones which must be identified by a tedious or expensive assay procedure, e.g., one based on in vitro protein synthesis, enzyme activity, or hybridization with a complementary nucleic acid. The heterogeneous population of transformed cells is divided into pools, each containing many different clones. Each is tested by the assay procedure to locate the pool containing the desired gene. That pool is subdivided and retested until the individual clone is identified. Transformant cells containing cloned histone genes from unfractionated sea urchin DNA were successfully isolated in this manner.[160]

3. Determination of DNA Fragment Number, Size, and Order

As discussed previously, agarose-gel electrophoresis is the best available procedure for quickly characterizing the number and size of fragments in a population of DNA molecules restricted to discrete size classes above about 1.5 kb. Polyacrylamide-gel electrophoresis becomes more important as the fragment size diminishes. These procedures are especially useful in determining the fragment composition of recombinant DNA molecules constructed through a ligation of the sticky ends produced by a restriction endonuclease. Cleavage by the same nuclease regenerates the original DNA fragments which are easily identified on gels.

There are many advantages to gel electrophoresis compared with other available analytical procedures. It is simple, relatively inexpensive, and fast. The resolution of molecules closely similar in size can be achieved (Figure 5), and multiple samples can be processed simultaneously. DNA bands can be visualized by staining with ethidium bromide, which binds to DNA and fluoresces under ultraviolet irradiation; consequently, very small amounts of DNA can be analyzed (as little as 10 ng of DNA per band).[104] DNA bands can be recovered from the gel and used for subsequent studies, e.g., analysis with different restriction nucleases, molecular hybridization, nucleotide sequencing, or transformation.

The relative amounts and sizes of DNA molecules can also be determined by electron microscopy (EM).[9,161,162] Although more laborious and requiring an electron microscope, the frequency of different size classes of DNA is more accurately determined by EM than by other procedures, and size estimates are more reliable.

The order of restriction endonuclease-generated DNA segments can be determined by a variety of procedures involving EM and gel electrophoresis. For example, the map of EcoRI segments of λ DNA was developed from the EM of heteroduplexes of

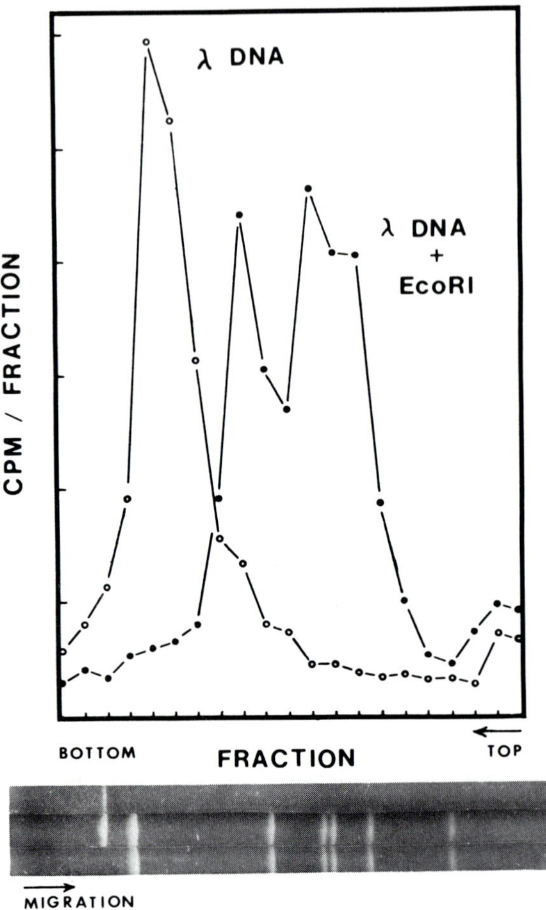

FIGURE 5. Separation of DNA molecules by sucrose-gradient centrifugation and agarose-gel electrophoresis. The upper part of the figure depicts the separation of 180 ng of ^{32}P-labeled intact λ DNA and the six DNA fragments produced by digestion of 80 ng of λ DNA withhthe EcoRI restriction endonuclease. The DNA was centrifuged for 2 hr at 55,000 r/min in a Spinco® SW56 swinging bucket rotor containing a 5 to 20% sucrose gradient. The lower part of the figure shows the positions of intact λ DNA (upper gel), intact λDNA plus EcoRI-digested λ DNA (middle gel), and of EcoRI-digested λ DNA (lower gel). 220 ng of DNA was added to the lower gel. A voltage gradient of 1.5 V/cm was applied to the gels.[104] The superiority of gel electrophoresis for separation of DNA molecules of less than 10×10^6 to 15×10^6 in molecular weight is obvious. However, sucrose-gradient centrifugation is useful for the separation of molecules of much higher molecular weight than could be resolved by present electrophoretic procedures.

fragment strands with strands of the intact chromosome.[19,161] This map was confirmed by comparing the EcoRI fragment electrophoretic patterns of a variety of λ derivatives differing by deletion and/or addition of DNA from other genomes.[1,104]

The most general procedures for identifying adjacent fragments and determining their original order in the chromosome are the comparison of the fragment electrophoretic patterns produced by two or more restriction endonucleases, singly and in combination. This has been done, e.g., with the tumor virus SV40;[163,164] an alternative procedure is to compare the DNA electrophoretic patterns completely or partially digested by a single restriction and endonuclease.[1,32,165]

Smith and Birnstiel devised a simple procedure for mapping the order of DNA fragments generated by a restriction nuclease.[166] The 5′-ends of the DNA strands are labeled enzymatically with ^{32}P by using polynucleotide kinase. The end-labeled DNA is treated with a restriction nuclease which cleaves the DNA at only one point. The resulting two fragments, each radioactive at one end, are separated by agarose-gel electrophoresis, purified, and incompletely digested with a different restriction endonuclease that produces many fragments. The partially digested fragments are separated according to size by agarose-gel electrophoresis and visualized by radioautography. Although a complex mixture of products results, only those fragments containing one of the original ends, and hence radioactive, will be visualized. A single gel should have radioactive bands corresponding to the single end fragment, the end fragment joined to the second fragment, etc. The difference in size from band to band corresponds to the size of each additional fragment. Thus, the sequence of restriction sites and their distance from the end of the molecule can be determined from one partial digest. Once the two original pieces have been purified, samples of each can be digested by many different site-specific nucleases, quickly generating a complete map of the fragments. This procedure was used to sequence the endonuclease-generated fragments of the cloned repeat unit of sea urchin histone DNA.[166]

Newly developed procedures for visualizing RNA-DNA hybrids by EM allow the precise localization of individual genes.[128,167,168] EM and gel electrophoresis complement each other in ordering DNA fragments.

ACKNOWLEDGMENTS

We thank our many colleagues who made their papers and unpublished results available to us prior to publication. We also thank Ron Lomax for drawing several figures and Ms. Connie Mele for excellent assistance in the preparation of this manuscript. Our work is supported by Grant GM21287 from the National Institutes of Health, Bethesda, Maryland.

REFERENCES

1. **Helling, R.B.,** Eukaryotic genes in prokaryotic cells, *Stadler Genet. Symp.,* 7, 15, 1975.
2. **Cohen, S.N.,** The manipulation of genes, *Sci. Am.,* 233, 24, 1975.
3. **Sinsheimer, R. L.,** Recombinant DNA, *Annu. Rev. Biochem.,* 46, 415, 1977.
4. **Collins, J.,** Gene cloning with small plasmids, *Curr. Top. Microbiol. Immunol.,* 78, 122, 1977.
4a. **Curtiss, R.,** Genetic manipulation of microorganisms: potential benefits and biohazards, *Annu. Rev. Microbiol.,* 30, 507, 1976.
5. **Murray, K.,** this volume.
6. **Tanaka, T. and Weisblum, B.,** Construction of a colicin El-R factor composite plasmid *in vitro:* means for amplification of deoxyribonucleic acid, *J. Bacteriol.,* 121, 354, 1975.
7. **Bodmer, W.F. and Ganesan, A.T.,** Biochemical and genetic studies of integration and recombination in *Bacillus subtilis* transformation, *Genetics,* 50, 717, 1964.
8. **Young, F.E. and Wilson, G.A.,** Transformation of bacteria by heterologous DNA, in preparation.
9. **Cohen, S.N., Chang, A.C.Y., Boyer, H.W., and Helling, R.B.,** Construction of biologically functional bacterial plasmids *in vitro, Proc. Natl. Acad. Sci. U.S.A.,* 70, 3240, 1973.

10. **Sgaramella, V.,** Enzymatic oligomerization of bacteriophage P22 DNA and of linear simian virus 40 DNA, *Proc. Natl. Acad. Sci. U.S.A.,* 69, 3389, 1972.

11. **Mertz, J.E. and Davis, R.W.,** Cleavage of DNA by RI restriction endonuclease generates cohesive ends, *Proc. Natl. Acad. Sci. U.S.A.,* 69, 3370, 1972.

12. **Hedgpeth, J., Goodman, H.M., and Boyer, H.W.,** DNA nucleotide sequence restricted by the RI endonuclease, *Proc. Natl. Acad. Sci. U.S.A.,* 69, 3448, 1972.

13. **Mandel, M. and Higa, A.,** Calcium-dependent bacteriophage DNA infection, *J. Mol. Biol.,* 53, 159, 1970.

14. **Wensink, P.C., Finnegan, D.J., Donelson, J.E., and Hogness, D.S.,** A system for mapping DNA sequences in the chromosomes of *Drosophila melanogaster, Cell,* 3, 315, 1974.

15. **Clarke, L. and Carbon, J.,** A colony bank containing synthetic colEl hybrid plasmids representative of the entire *E. coli* genome, *Cell,* 9, 91, 1976.

16. **Roberts, R.J.,** Restriction endonucleases, *Crit. Rev. Biochem.,* 4(2), 123, 1976.

16a. **Roberts, R.J.,** The role of restriction endonucleases in genetic engineering, Tenth Miles International Symposium, in press.

17. **Polisky, B., Greene, P., Garfin, D.E., McCarthy, B.J., Goodman, H.M., and Boyer, H.W.,** Specificity of substrate recognition by the *Eco*RI restriction endonuclease, *Proc. Natl. Acad. Sci. U.S.A.,* 72, 3310, 1975.

18. **Murray, K., Hughes, S.G., Brown, J.S., and Bruce, S.A.,** Isolation and characterization of two sequence-specific endonucleases from *Anabaena variabilis, Biochem. J.,* 159, 317, 1976.

19. **Thomas, M. and Davis, R.W.,** Studies on the cleavage of bacteriophage lambda DNA with *Eco*RI restriction endonuclease, *J. Mol. Biol.,* 91, 315, 1975.

20. **Hamer, D.H. and Thomas, C.A., Jr.,** Molecular cloning of DNA fragments produced by restriction endonucleases *Sal*I and *Bam*I, *Proc. Natl. Acad. Sci. U.S.A.,* 73, 1537, 1976.

21. **Lobban, P.E. and Kaiser, A.D.,** Enzymatic end to end joining of DNA molecules, *J. Mol. Biol.,* 78, 453, 1973.

22. **Jackson, D.A., Symons, R.H., and Berg, P.,** Biochemical method for inserting new genetic information into DNA of simian virus 40: circular SV40 DNA molecules containing lambda phage genes and the galactose operon of *Escherichia coli, Proc. Natl. Acad. Sci. U.S.A.,* 69, 2904, 1972.

23. **Bollum, F.J.,** Terminal deoxynucleotidyl transferase, in *The Enzymes,* Boyer, P.D., Ed., Academic Press, New York, 10, 1974, 145.

23a. **Cockburn, A.F., Newkirk, M.J., and Firtel, R.A.,** Organization of the ribosomal RNA gene of *Dictyostelium discoideum*: mapping of the non-transcribed spacer region, *Cell,* 9, 605, 1976.

24. **Roychoudhury, R., Jay, E., and Wu, R.,** Terminal labelling and addition of homopolymer tracts to duplex DNA fragments by terminal deoxynucleotidyl transferase, *Nucl. Acids Res.,* 3, 101, 1976.

25. **Lehman, I.R.,** DNA ligase: structure, mechanism, and function, *Science,* 186, 790, 1974.

26. **Walker, G.C., Uhlenbec, O.C., Bedows, E., and Gumport, R.I.,** T4 induced RNA ligase joins single-stranded oligoribonucleotides, *Proc. Natl. Acad. Sci. U.S.A.,* 72, 122, 1975.

26a. **Sugino, A., Goodman, H.M., Heyneker, H.L., Shine, J., Boyer, H.W., and Coyzarelli, N.R.,** Interaction of bacteriophage T4 RNA and DNA ligases in joining of duplex DNA at base-paired ends, *J. Biol. Chem.,* 252, 2987, 1977.

27. **Sgaramella, V., van de Sande, J.H., and Khorana, H.G.,** Studies on polynucleotides. C. A novel joining reaction catalysed by the T4-polynucleotide ligase, *Proc. Natl. Acad. Sci. U.S.A.,* 67, 1468, 1970.

28. **Dugaiczyk, A., Boyer, H.W., and Goodman, H.M.,** Ligation of *Eco*RI endonuclease-generated DNA fragments into linear and circular structures, *J. Mol. Biol.,* 96, 171, 1975.

29. **de Vries, F.A.J., Collins, C.J., and Jackson, D.A.,** Joining of simian virus 40 DNA molecules at endonuclease R. *Eco*RI sites by polynucleotide ligase and analysis of the products by agarose-gel electrophoresis, *Biochem. Biophys. Acta,* 435, 213, 1976.

30. **Cohen, S.N., Chang, A.C.Y., and Hsu, L.,** Nonchromosomal antibiotic resistance in bacteria: genetic transformation of *Escherichia coli* by R-factor DNA, *Proc. Natl. Acad. Sci. U.S.A.,* 69, 2110, 1972.

31. **Collins, C.J., Jackson, D.A., and de Vries, F.A.J.,** Biochemical construction of specific chimeric plasmids from ColEl DNA and unfractionated *E. coli* DNA, *Proc. Natl. Acad. Sci. U.S.A.,* 73, 3838, 1976.

32. **Nosikov, V.V., Braga, E.A., Karlishev, A.V., Zhuze, A.L., and Polyanovsky, O.L.,** Protection of particular cleavage sites of restriction endonucleases by distamycin A and actinomycin D, *Nucl. Acids Res.,* 3, 2293, 1976.

33. **Heyneker, H.L., Shine, J., Goodman, H.M., Boyer, H.W., Rosenberg, J., Dickerson, R.E., Narang, S.A., Itakura, K., Lin, S.-Y., and Riggs, A.D.,** Synthetic *lac* operator DNA is functional *in vivo, Nature,* 263, 748, 1976.

34. **Marians, K.J., Wu, R., Stawinski, J., Hozumi, T., and Narang, S.D.,** Cloned synthetic lac operator DNA is biologically active, *Nature,* 263, 744, 1976.

35. **Bachman, K., Ptashne, M., and Gilbert, W.,** Construction of plasmids carrying the CI gene of bacteriophage λ, *Proc. Natl. Acad. Sci. U.S.A.,* 73, 4174, 1976.

36. **Hofstetter, H., Schamböck, A., Van den Berg, J., and Weissmann, C.,** Specific excision of the inserted DNA segment from hybrid plasmids constructed by the poly(dA)·poly(dT) method, *Biochim. Biophys. Acta,* 454, 587, 1976.

37. **Sadler, J. R., Tecklenburg, M., Betz, J. L., Goeddel, D. V., Yansura, D. G., and Caruthers, M. H.,** Cloning of chemically synthesized lactose operators, *Gene,* 1, 305, 1977.

38. **Murray, N.,** this volume.

39. **Enquist, L., Tiemeier, D., Leder, P., Weisberg, R., and Sternberg, N.,** Safer derivatives of bacteriophage λgt · λC for use in cloning of recombinant DNA molecules, *Nature,* 259, 596, 1976.

40. **Tiemeier, D., Enquist, L., and Leder, P.,** Improved derivative of a phage λ EK2 vector for cloning recombinant DNA, *Nature,* 263, 526, 1976.

41. **Murray, N.E., Brammer, W.J., and Murray, K.,** Lambdoid phages that simplify the recovery of *in vitro* recombinants, *Mol. Gen. Genet.,* 150, 53, 1977.

42. **Rambach, A. and Tiollais, T.,** Bacteriophage having EcoRI endonuclease sites only in the non-essential region of the genome, *Proc. Natl. Acad. Sci., U.S.A.,* 71, 3927, 1974.

43. **Murray, N.E. and Murray, K.,** Manipulation of restriction targets in phage to form receptor chromosomes for DNA fragments, *Nature,* 251, 476, 1974.

44. **Thomas, M., Cameron, J.R., and Davis, R.W.,** Viable molecular hybrids of bacteriophage lambda and eukaryotic DNA, *Proc. Natl. Acad. Sci. U.S.A.,* 71, 4579, 1974.

45. **Hershfield, V., Boyer, H.W., Yanofsky, C., Lovett, M.A., and Helinski, D.R.,** Plasmid ColEl as a molecular vehicle for cloning and amplification of DNA, *Proc. Natl. Acad. Sci. U.S.A.,* 71, 3455, 1974.

46. **Falkow, S.F.,** *Infectious Multiple Drug Resistance,* Pion, London, 1975, 58.

47. **Cohen, S.N. and Chang, A.C.Y.,** Recircularization and autonomous replication of a sheared R-factor DNA segment in *Escherichia coli* transformants, *Proc. Natl. Acad. Sci. U.S.A.,* 70, 1293, 1973.

48. **Timmis, K., Cabello, F., and Cohen, S.N.,** Utilization of two distinct modes of replication by a hybrid plasmid constructed *in vitro* from separate replicons, *Proc. Natl. Acad. Sci. U.S.A.,* 71, 4556, 1974.

49. **Helinski, D.R.,** Plasmid DNA replication, *Fed. Proc.,* 35, 2026, 1976.

50. **Clewell, D.B.,** Nature of ColEl plasmid replication in *Escherichia coli* in the presence of chloramphenicol, *J. Bacteriol.,* 110, 667, 1972.

51. **Cabello, F., Timmis, K., and Cohen, S.N.,** Replication control in a composite plasmid constructed by *in vitro* linkage of two distinct replicons, *Nature,* 259, 285, 1976.

51a. **Rownd, R.H., Perlman, D., and Goto, N.,** Structure and replication of R-factor DNA in *Proteus mirabilis,* in *Microbiology 1974,* Schlessinger, D., Ed., American Society for Microbiology, Washington, D.C., 1975, 76.

51b. **Crossa, J.H., Luttrapp, L.K., Heffron, F., and Falkow, S.,** Two replication sites on R-plasmid DNA, *Mol. Gen. Genet.,* 140, 39, 1975.

52. **Selker, E., Brown, K., and Yanofsky, C.,** Mitomycin C-induced expression of *trpA* of *Salmonella typhimurium* inserted into the plasmid *ColEl, J. Bacteriol.,* 129, 388, 1977.

53. **So, M., Gill, R., and Falkow, S.,** The generation of a ColEl-Ap^R cloning vehicle which allows detection of inserted DNA, *Mol. Gen. Genet.,* 142, 239, 1975.

54. **Boyer, H.W.,** personal communication.

55. **Lovett, P.S., Duvall, E.J., and Keggins, K.M.,** *Bacillus pumilis* plasmid pPLl0: properties and insertion into *Bacillus subtilis* by transformation, *J. Bacteriol.,* 127, 817, 1976.

56. **Young, F.E., Duncan, C., and Wilson, G.A.,** Development of the *Bacillus subtilis* model system for recombinant molecule technology, in *Genetic Modification: Impact of Molecules on Genetic Research,* Beers, R.F. and Bassett, E.G., Eds., Raven Press, New York, 1977.

57. **Ehrlich, S.D., Burstyn-Pettegrew, H., Stroynowski, I., and Lederberg, J.,** Expression of the thymidylate synthetase gene of the *Bacillus subtilis* bacteriophage phi-3-T in *Escherichia coli, Proc. Natl. Acad. Sci. U.S.A.,* 4145, 1976.

58. **Helinski, D.R., Hershfield, V., Figurski, D., and Meyer, R.J.,** Construction and properties of plasmid cloning vehicles, Tenth Miles International Symposium, 1977.

58a. **Ehrlich, S.D.,** Replication and expression of plasmids from *Staphylococcus aureus* in *Bacillus subtilus, Proc. Natl. Acad. Sci. U.S.A.,* 74, 1680, 1977.

59. **Jacob, A.E., Cresswell, J.M., Hedges, R.W., Coetzee, J.N., and Beringer, J.E.,** Properties of plasmids constructed by the *in vitro* insertion of DNA from *Rhizobium leguminosarum* or *Proteus mirabilis* into RP4, *Mol. Gen. Genet.,* 147, 315, 1976.

60. **Olsen, R.H. and Gonzales, C.,** *Escherichia coli* gene transfer to unrelated bacteria by a histidine operon-RP1 drug resistance plasmid complex, *Biochem. Biophys. Res. Commun.,* 59, 377, 1974.

61. **Jacob, A.E. and Grinter, N.J.,** Plasmid RP4 as a vector replicon in genetic engineering, *Nature,* 255, 503, 1975.

62. **Meyer, R.. Figurski, D., and Helinski, D.R.,** Molecular vehicle properties of the broad host range plasmid RK2, *Science,* 190, 1226, 1975.
63. **Chakrabarty, A.M., Mylroie, J.R., Friello, D.A., and Vacca, J.G.,** Transformation of *Pseudomonas putida* and *Escherichia coli* with plasmid-linked drug-resistance factor DNA, *Proc. Natl. Acad. Sci. U.S.A.,* 72, 3647, 1975.
64. **Shipley, P.L. and Olsen, R.H.,** Isolation of a non-transmissible antibiotic resistance plasmid by transductional shortening of R factor RP1, *J. Bacteriol.,* 123, 20, 1975.
65. **Weissman, C. and Boll, W.,** Reduction of possible hazards in the preparation of recombinant plasmid DNA, *Nature,* 261, 428, 1976.
66. **Hirt, B.,** Selective extraction of polyoma DNA from infected mouse cell cultures, *J. Mol. Biol.,* 26, 365, 1969.
67. **Guerry, P., LeBlanc, D.J., and Falkow, S.,** General method for the isolation of plasmid deoxyribonucleic acid, *J. Bacteriol.,* 116, 1064, 1973.
68. **Meyers, J.A., Sanchez, D., Elwell, L.P., and Falkow, S.,** Simple agarose gel electrophoretic method for the identification and characterization of plasmid deoxyribonucleic acid, *J. Bacteriol.,* 127, 1529, 1976.
69. **Lindberg, M., Sjostrom, J.E., and Johansson, T.,** Transformation of chromosomal and plasmid characters in *Staphylococcus aureus, J. Bacteriol.,* 109, 884, 1972.
71. **Humphreys, G.O., Willshaw, G.A., and Anderson, E.S.,** A simple method for the preparation of large quantities of pure plasmid DNA, *Biochim. Biophys. Acta,* 383, 457, 1975.
72. **Johnston, J. B. and Gunsalus, J. C.,** Isolation of metabolic plasmid DNA from *P. putida, Biochem. Biophys. Res. Comm.,* 75, 13, 1977.
73. **Radloff, R., Bauer, W., and Vinograd, J.,** A dye-bouyant-density method for the detection and isolation of closed circular duplex DNA: the closed circular DNA in HeLa cells, *Proc. Natl. Acad. Sci. U.S.A.,* 57, 1514, 1967.
74. **Clewell, D.B. and Helinski, D.R.,** Supercoiled circular DNA-protein complex in *Escherichia coli:* purification and induced conversion to an open circular form, *Proc. Natl. Acad. Sci. U.S.A.,* 62, 1159, 1969.
75. **Ganem, D., Nussbaum, A.L., Davoli, D., and Fareed, G.C.,** Propagation of bacteriophage λ-DNA in monkey cells after covalent linkage to a defective simian virus 40 genome, *Cell,* 7, 349, 1976.
76. **Nussbaum, A.L., Davoli, D., Ganem, D., and Fareed, G.C.,** Construction and propagation of a defective simian virus 40 genome bearing an operator from bacteriophage λ, *Proc. Natl. Acad. Sci. U.S.A.,* 73, 1068, 1976.
77. **Ganem, D., Nussbaum, A.L., Davoli, D., and Fareed, G.C.,** Isolation, propagation, and characterization of replication requirements of reiteration mutants of simian virus 40, *J. Mol. Biol.,* 101, 57, 1976.
78. **Goff, S.P. and Berg, P.,** Construction of hybrid viruses containing SV40 and lambda phage DNA segments and their propagation in cultured monkey cells, *Cell,* in press.
79. **Hamer, D. H., Davoli, D., Thomas, C. A., Jr., and Fareed, G. C.,** Simian virus 40 carrying an *Escherichia coli* suppressor gene, *J. Mol. Biol.,* 112, 155, 1977.
80. **Hollenberg, C.P., Degelmann, A., Kustermann-Kuhn, B., and Royer, H.D.,** Characterization of 2-μm DNA of *Saccharomyces cerevisceae* by restriction fragment analyses and integration in an *Escherichia coli* plasmid, *Proc. Natl. Acad. Sci. U.S.A.,* 73, 2072, 1976.
81. **Guerineau, M., Grandchamp, C., and Slonimski, P.P.,** Circular DNA of yeast episome with two inverted repeats: structural analysis by a restriction enzyme and electron microscopy, *Proc. Natl. Acad. Sci. U.S.A.,* 73, 3030, 1976.
82. **Radding, C.M. and Kaiser, A.D.,** Gene transfer by broken molecules of λDNA: activity of the left half-molecule, *J. Mol. Biol.,* 7, 225, 1963.
83. **Wackernagel, W.,** An improved spheroplast assay for λ-DNA and the influence of the bacterial genotype on the transfection rate, *Virology,* 48, 94, 1972.
84. **Cosloy, S.D. and Oishi, M.,** Genetic transformation in *Escherichia coli* K12, *Proc. Natl. Acad. Sci. U.S.A.,* 70, 84, 1973.
85. **Wackernagel, W.,** Genetic transformation in *E. coli:* the inhibitory role of the *recBC* DNAse, *Biochem. Biophys. Res. Commun.,* 51, 306, 1973.
86. **Lederberg, E.M. and Cohen, S.N.,** Transformation of *Salmonella typhimurium* by plasmid deoxyribonucleic acid, *J. Bacteriol.,* 119, 1072, 1974.
87. **Taketo, A.,** Sensitivity of *Escherichia coli* to viral nucleic acid. V. Competence of calcium-treated cells, *J. Biochem.,* 72, 973, 1972.
88. **Sjöström, J.E., Lindberg, M., and Philipson, L.,** Transfection of *Staphylococcus aureus* with bacteriophage deoxyribonucleic acid, *J. Bacteriol.,* 109, 285, 1972.
89. **Chakrabarty, A.,** this volume.
90. **Chakrabarty, A. M.,** Molecular cloning in Pseudomonas, in *Microbiology 1976,* Schlessing, D., Ed., American Society for Microbiology, Washington, D.C., 1977, 579.

91. **Taketo, A. and Kuno, S.,** Sensitivity of *Escherichia coli* to viral nucleic acid. VII. Further studies on Ca^{++}-induced competence, *J. Biochem.,* 75, 59, 1974.

92. **Taketo, A.,** Sensitivity of *Escherichia coli* to viral nucleic acid. VIII. Idiosyncracy of Ca^{++}-dependent competence for DNA, *J. Biochem.,* 75, 895, 1974.

92a. **Miller, J. B. and Koshland, D. E., Jr.,** Membrane fluidity and chemotaxis: effects of temperature and membrane lipid composition on the swimming behavior of *Salmonella typhimurium* and *Escherichia coli, J. Mol. Biol.,* 111, 183, 1977.

93. **Sgaramella, V., Ehrlich, S. D., Bursztyn, H., and Lederberg, J.,** Enhancement of transfecting activity of bacteriophage P22 DNA upon exonucleolytic erosion, *J. Mol. Biol.,* 105, 587, 1976.

94. **Ehrlich, S. D., Sgaramella, V., and Lederberg, J.,** Transfection of restrictionless *Escherichia coli* by bacteriophage T7 DNA: effect of *in vitro* erosion of DNA by λ exonuclease, *J. Mol. Biol.,* 105, 603, 1976.

95. **Benzinger, R., Enquist, L. W., and Skalka, A.,** Transfection of *Escherichia coli* spheroplasts. V. Activity of *recBC* nuclease in *rec⁺* and *rec⁻* Spheroplasts measured with different forms of bacteriophage DNA, *J. Virol.,* 15, 861, 1975.

95a. **Kahmann, R., Kamp, D., and Zipser, D.,** Transfection of *Escherichia coli* by Mu DNA, *Mol. Gen. Genet.,* 149, 323, 1976.

96. **Bursztyn, H., Sgaramella, V., Ciferri, O., and Lederberg, J.,** Transfectability of rough strains of *Salmonella typhimurium, J. Bacteriol.,* 124, 1630, 1975.

97. **Kretschmer, P.J., Chang, A.C.Y., and Cohen, S.N.,** Indirect selection of bacterial plasmids lacking identifiable phenotypic properties, *J. Bacteriol.,* 124, 225, 1975.

98. **Marmur, J.,** A procedure for the isolation of deoxyribonucleic acid from microorganisms, *J. Mol. Biol.,* 3, 208, 1961.

99. **Kelly, M.S. and Pritchard, R.H.,** Unstable linkage between genetic markers in transformation, *J. Bacteriol.,* 89, 1314, 1965.

100. **Morrow, J.F., Cohen, S.N., Chang, A.C.Y., Boyer, H.W., Goodman, H.M., and Helling, R.B.,** Replication and transcription of eukaryotic DNA in *Escherichia coli, Proc. Natl. Acad. Sci. U.S.A.,* 71, 1743, 1974.

101. **Gall, J.G.,** Free ribosomal RNA genes in the macronucleus of *Tetrahymena, Proc. Natl. Acad. Sci. U.S.A.,* 71, 3078, 1974.

102. **Yao, M.-C., Krimmel, A.R., and Gorovsky, M.A.,** A small number of cistrons for ribosomal RNA in the germinal nucleus of a eukaryote, *Tetrahymena pyriformis, Proc. Natl. Acad. Sci. U.S.A.,* 71, 3082, 1974.

103. **Petes, T.D., Newlon, C.S., Biejers, B., and Fangman, W.L.,** Yeast chromosomal DNA: size, structure, and replication, in *Chromosome Structure and Function,* Cold Spring Harbor Laboratory, New York, 38, 1974, 9.

104. **Helling, R.B., Goodman, H.M., and Boyer, H.W.,** Analysis of endonuclease R. *Eco*RI fragments of DNA from lambdoid bacteriophage and other viruses by agarose-gel electrophoresis, *J. Virol.,* 14, 1235, 1974.

105. **Johnson, P. H. and Grossman, L. I.,** Electrophoresis of DNA in agarose gels. Optimizing separations of conformational isomers of double and single stranded DNAs, *Biochemistry,* 16, 4217, 1977.

106. **McDonell, N.W., Simon, M.N., and Studier, F.W.,** Analysis of restriction fragments of T7 DNA and determination of molecular weights by electrophoresis in neutral and alkaline gels, *J. Mol. Biol.,* 110, 119, 1977.

107. **Dean, D.H. and Halvorson, H.O.,** Isolation of polygenic fragments of the *Bacillus* chromosome: transformation activity of endo R. *Eco*RI-fragmented DNA, in *Microbiology 1976,* Schlessinger, D., Ed., American Society for Microbiology, Washington, D.C., 1976.

108. **Harris-Warrick, R.M., Elkana, Y., Erlich, S.D., and Lederberg, J.,** Electrophoretic separation of *Bacillus subtilis* genes, *Proc. Natl. Acad. Sci. U.S.A.,* 72, 2207, 1975.

109. **Southern, E.M.,** Detection of specific sequences among DNA fragments separated by gel electrophoresis, *J. Mol. Biol.,* 98, 503, 1975.

110. **Lomax, M.I., Helling, R.B., Hecker, L.I., Schwartzbach, S.D., and Barnett, W.E.,** Cloned ribosomal RNA genes from chloroplasts of *Euglena gracilis, Science,* 196, 202, 1977.

111. **Thuring, R.W.J., Sanders, J.P.M., and Borst, P.,** A freeze-squeeze method for recovering long DNA from agarose gels, *Anal. Biochem.,* 66, 213, 1975.

112. **Blin, N., Gabain, A.V., and Bujard, H.,** Isolation of large molecular weight DNA from agarose gels for further digestion by restriction enzymes, *FEBS Lett.,* 53, 84, 1975.

113. **Efstratiadis, A., Kafatos, F.C., Maxam, A.M., and Maniatis, T.,** Enzymatic *in vitro* synthesis of globin genes, *Cell,* 7, 279, 1976.

114. **Higuchi, R., Paddock, G.V., Wall, R., and Salser, W.,** A general method for cloning eukaryotic structural gene sequences, *Proc. Natl. Acad. Sci. U.S.A.,* 73, 3146, 1976.

115. **Monahan, J.J., McReynolds, L.A., and O'Malley, B.W.,** The ovalbumin gene: *in vitro* enzymatic synthesis and characterization, *J. Biol. Chem.,* 251, 7355, 1976.

116. **Wood, K.O. and Lee, J.C.**, Integration of synthetic globin genes into an *E. coli* plasmid, *Nucl. Acids Res.*, 3, 1961, 1976.

117. **Maniatis, T., Kee, S.G., Efstratiadis, A., and Kafatos, F.C.**, Amplification and characterization of a β-globin gene synthesized *in vitro*, *Cell*, 8, 163, 1976.

118. **Rougeon, F., Kourilsky, P., and Mach, B.**, Insertion of a rabbit β-globin gene sequence into an *E. coli* plasmid, *Nucl. Acids Res.*, 2, 2365, 1975.

119. **Rabbits, T.H.**, Bacterial cloning of plasmids carrying copies of rabbit globin messenger RNA, *Nature*, 260, 221, 1976.

120. **McReynolds, L.A., Monahan, J.J., Bendure, D.W., Woo, S.L.C., Paddock, G.V., Salser, W., Dorson, J., Moses, R.E., and O'Malley, B.W.**, The ovalbumin gene: insertion of ovalbumin gene sequences in chimeric bacterial plasmids, *J. Biol. Chem.*, 252, 1840, 1977.

121. **Shih, T.Y. and Martin, M.A.**, Chemical linkage of nucleic acids to neutral and phosphorylated cellulose powders and isolation of specific sequences by affinity chromatography, *Biochemistry*, 13, 3411, 1974.

122. **Shih, T.Y. and Martin, M.A.**, A general method of gene isolation, *Proc. Natl. Acad. Sci. U.S.A.*, 70, 1697, 1973.

123. **Woo, S.L.C., Smith, R.G., Means, A.R., and O'Malley, B.W.**, The ovalbumin gene. Partial purification of the coding strand, *J. Biol. Chem.*, 251, 3868, 1976.

124. **Venetianer, P. and Leder, P.**, Enzymatic synthesis of solid phase-bound DNA sequences corresponding to specific mammalian genes, *Proc. Natl. Acad. Sci. U.S.A.*, 71, 3892, 1974.

125. **Yenikolopov, G.N., Ryskov, A.P., Nitta, N., and Georgiev, G.P.**, Isolation of native DNA fragments containing structural genes at the beginning, in the middle, or at the end of the coding strand, *Nucl. Acids Res.*, 3, 2645, 1976.

126. **Georgiev, G.P., Ilyin, Y.V., Ryskov, A.P., Tchurikov, N.A., Yenikolopov, G.N., Gvozdev, V.A., and Ananiev, E.V.**, Isolation of eukaryotic DNA fragments containing structural genes and the adjacent sequences, *Science*, 195, 394, 1977.

127. **Dale, R.M.K. and Ward, D.C.**, Mercurated polynucleotides: new probes for hybridization and selective polymer fractionation, *Biochemistry*, 14, 2458, 1975.

128. **Thomas, M., White, R.L., and Davis, R.W.**, Hybridization of RNA to double-stranded DNA: formation of R-loops, *Proc. Natl. Acad. Sci. U.S.A.*, 73, 2294, 1976.

129. **Salser, W., Browne, J., Clarke, P., Heindell, H., Higuchi, R., Paddock, G., Roberts, J., Studnicka, G., and Zakar, P.**, Determination of globin mRNA sequences and their insertion into bacterial plasmids, *Prog. Nucl. Acids Res.*, 19, 177, 1976.

130. **Sures, I., Maxam, A., Cohn, R.H., and Kedes, L.H.**, Identification and location of the histone H2A and H3 genes by sequence analysis of sea urchin (*S. purpuratus*) DNA cloned in *E. coli*, *Cell*, 9, 495, 1976.

131. **Cameron, J.R. and Davis, R.W.**, The effects of *Escherichia coli* and yeast DNA insertions on the growth of λ bacteriophage, *Science*, 196, 212, 1977.

132. **Mukai, T., Matsubara, K., and Takagi, Y.**, Cloning bacterial genes using plasmid λdv, *Mol. Gen. Genet.*, 146, 269, 1976.

133. **Bernardi, A. and Bernardi, G.**, Cloning of all EcoRI fragments from phage λ in *E. coli*, *Nature*, 264, 89, 1976.

133a. **Meagher, R.B., Tait, R.C., Betlach, M., and Boyer, H.**, Protein expression in *E. coli* minicells by recombinant plasmids, *Cell*, 10, 521, 1977.

134. **Carroll, D. and Brown, D.D.**, Adjacent repeating units of *Xenopus laevis* 5S DNA can be heterogeneous in length, *Cell*, 7, 477, 1976.

135. **Maniatis, T., Kee, S.G., Efstratiadis, A., and Kafatos, F.C.**, Amplification and characterization of a β-globin gene synthesized *in vitro*, *Cell*, 8, 163, 1976.

136. **Rubens, C., Heffron, F., and Falkow, S.**, Transposition of a plasmid deoxyribonucleic acid sequence that mediates ampicillin resistance: independence from host *rec* functions and orientation of insertion, *J. Bacteriol.*, 128, 425, 1976.

137. **Bedbrook, J.R. and Ausubel, F.M.**, Recombination between bacterial plasmids leading to the formation of plasmid multimers, *Cell*, 9, 707, 1976.

138. **Bodsch, W.**, Excision of a DNA-sequence determining Kanamycin resistance from a colE1-Km recombinant plasmid during transformation, in preparation.

139. **Cramer, J.H., Farrelly, F.W., Barnitz, J.T., and Rownd, R.H.**, Construction and restriction endonuclease mapping of hybrid plasmids containing *Saccharomyces cerevisiae* ribosomal DNA, *Mol. Gen. Genet.*, 151, 229, 1977.

140. **Capaldo-Kimball, F. and Barbour, S.D.**, Involvement of recombination genes in growth and viability of *Escherichia coli* K-12, *J. Bacteriol.*, 106, 204, 1971.

141. **Cohen, S.N.**, Transposable genetic elements and plasmid evolution, *Nature*, 263, 731, 1976.

142. Faelen, M., Toussaint, A., Van Montagu, M., Van den Elsacker, S., Engler, G., and Schell, J., *In vivo* genetic engineering: the Mu mediated transposition of chromosomal DNA segments onto transmissible plasmids, in *DNA Insertion Elements, Plasmids, and Episomes*, Bukhari, A.I., Shapiro, J., and Adhya, S., Eds., Cold Spring Harbor Laboratory, Cold Spring Harbor, New York, in press.

143. Berg, D.E., Jackson, D.A., and Mertz, J.E., Isolation of a λdv plasmid carrying the bacterial *gol* operon, *J. Virol.*, 14, 1063, 1974.

144. Cameron, J.R., Panasenko, S.M., Lehman, I.R., and Davis, R.W., *In vitro* construction of bacteriophage λ carrying segments of the *Escherichia coli* chromosome: selection of hybrids containing the gene for DNA ligase, *Proc. Natl. Acad. Sci. U.S.A.*, 72, 3416, 1975.

145. Polisky, B., Bishop, R.J., and Gelfand, D.H., A plasmid cloning vehicle allowing regulated expression of eukaryotic DNA in bacteria, *Proc. Natl. Acad. Sci. U.S.A.*, 73, 3900, 1976.

146. Mukai, T., Matsubara, K., and Takagi, Y., *In vitro* construction of colE1 factor carrying genes for synthesis of guanine or thymine, *Proc. Jpn. Acad.*, 51, 353, 1975.

146a. Recombinant DNA research, *Fed. Regist.*, 41, 27902, 1976.

147. Chang, A.C.Y. and Cohen, S.N., Genome construction between bacterial species *in vitro*: replication and expression of *Staphylococcus* plasmid genes in *Escherichia coli*, *Proc. Natl. Acad. Sci. U.S.A.*, 71, 1030, 1974.

148. Courvalin, P., Weisblum, B., and Davies, J., Aminoglycoside-modifying enzyme of an antibiotic-producing bacterium acts as a determinant of antibiotic resistance in *Escherichia coli*, *Proc. Natl. Acad. Sci. U.S.A.*, 74, 999, 1977.

149. Morgan, E.A. and Kaplan, S., Transcription of *Escherichia coli* ribosomal DNA in *Proteus mirabilis*, *Mol. Gen. Genet.*, 147, 179, 1976.

150. Struhl, K., Cameron, J.R., and Davis, R.W., Functional genetic expression of eukaryotic DNA in *Escherichia coli*, *Proc. Natl. Acad. Sci. U.S.A.*, 73, 1471, 1976.

151. Ratzkin, B. and Carbon, J., Functional expression of cloned yeast DNA in *Escherichia coli*, *Proc. Natl. Acad. Sci. U.S.A.*, 74, 487, 1977.

152. Grunstein, M. and Hogness, D.S., Colony hybridization: a method for the isolation of cloned DNAs that contain a specific gene, *Proc. Natl. Acad. Sci. U.S.A.*, 72, 3961, 1975.

153. Jones, K.W. and Murray, K., A procedure for detection of heterologous DNA sequences in lamdoid phage by *in situ* hybridization, *J. Mol. Biol.*, 96, 455, 1975.

153a. Kramer, R.A., Cameron, J.R., and Davis, R.W., Isolation of bacteriophage λ containing yeast ribosomal RNA genes: screening by *in situ* RNA hybridization to plaques, *Cell*, 8, 227, 1976.

154. Villareal, L.P. and Berg, P., Hybridization *in situ* of SV40 plaques: detection of recombinant SV40 virus carrying specific sequences of non-viral DNA, *Science*, 196, 183, 1977.

155. Benton, W.D. and Davis, R.W., Screening of λgt recombinant clones by hybridization to single plaques *in situ*, *Science*, 196, 180, 1977.

156. Skalka, A. and Shapiro, L., *In situ* immunoassays for gene translation products in phage plaques and bacterial colonies, *Gene*, 1, 65, 1977.

156a. Firtel, R.A., Cockburn, A., Frankel, G., and Hershfield, V., Structural organization of the genome of *Dictyostelium* discoideum: analysis by EcoRI restriction endonucleases, *J. Mol. Biol.*, 102, 831, 1976.

156b. Wellauer, P.K., David, I.B., Brown, D.D., and Reeder, R.H., The molecular basis for length heterogenicity in ribosomal DNA from *Xenopus laevis*, *J. Mol. Biol.*, 105, 461, 1976.

157. Cohen, S. N. and Chang, A. C. Y., A method for selective cloning of eukaryotic DNA fragments in *Escherichia coli* by repeated transformation, *Mol. Gen. Genet.*, 134, 133, 1974.

158. Glover, D. M. White, R. L., Finnegan, D. J., and Hogness, D. S., Characterization of six cloned DNAs from *Drosophila melanogaster*, including one that contains the genes for rRNA, *Cell*, 5, 149, 1975.

159. Cavalli-Sforza, L. L. and Lederberg, J., Isolation of pre-adaptive mutants in bacteria by sib selection, *Genetics*, 41, 367, 1956.

160. Kedes, L. H., Chang, A. C. Y., Houseman, D., and Cohen, S. N., Isolation of histone genes from unfractionated sea urchin DNA by subculture cloning in *E. coli*, *Nature*, 255, 533, 1975.

161. Davis, R. W., Simon, M., and Davidson, N., Electron microscope heteroduplex methods for mapping regions of base sequence homology in nucleic acids, in *Methods in Enzymology*, Moldave, K. and Kaplan, N., Eds., Academic Press, New York, XXIB, 1971, 413.

162. Younghusband, H. B. and Inman, R. B., The electron microscopy of DNA, *Annu. Rev. Biochem.*, 43, 605, 1974.

163. Danna, K. J., Sack, G. H., Jr., and Nathans, D., Studies of simian virus 40 DNA. VII. A cleavage map of the SV40 genome, *J. Mol. Biol.*, 78, 363, 1973.

164. Nathans, D., Adler, S. P., Brockman, W., Danna, K. J., Lel, T. N. H., and Sack, G. H., Jr., Use of restriction endonucleases in analyzing the genome of simian virus 40, *Fed. Proc.*, 33, 1135, 1974.

165. Parker, R. C., Watson, R. M., and Vinograd, J., Mapping of closed circular DNAs by cleavage with restriction endonucleases and calibration by agarose gel electrophoresis, *Proc. Natl. Acad. Sci. U.S.A.*, 74, 851, 1977.

166. **Smith, H. O. and Birnstiel, M. L.,** A simple method for DNA restriction site mapping, *Nucl. Acids Res.,* 3, 2387, 1976.

167. **Angerer, L. M., Davidson, N., Murphy, W., Lynch, D., and Attardi, G.,** An electron microscope study of the relative positions of the 4S and ribosomal RNA genes in HeLa cell mitochondrail DNA, *Cell,* 9, 81, 1976.

168. **Wu, M. and Davidson, N.,** Use of gene 32 protein staining of single-strand polynucleotides for gene mapping by electron microscopy: application to the ϕ80d$_3$*ilv su*7 system, *Proc. Natl. Acad. Sci. U.S.A.,* 72, 4506, 1975.

169. **Oka, A. and Takanami, M.,** Cleavage map of colicin E1 plasmid, *Nature,* 264, 193, 1976.

170. **Betlach, M., Hershfield, V., Chow, L., Brown, W., Goodman, H. M., and Boyer, H. W.,** A restriction endonuclease analysis of the bacterial plasmid controlling the EcoRI restriction and modification of DNA, *Fed. Proc.,* 35, 2037, 1976.

171. **Tanaka, T., Weisblum, B., Schnos, M., and Inman, R. B.,** Construction and characterization of a chimeric plasmid composed of DNA from *Escherichia coli* and *Drosophila melanogaster, Biochemistry,* 14, 2064, 1975.

172. **Rodriguez, R. L., Bolivar, F., Goodman, H. M., Boyer, H. W., and Betlach, M.,** in *Molecular Mechanisms in the Control of Gene Expression, ICN-UCLA Symposia on Molecular and Cellular Biology 5,* Nierlich, N. P., Rutter, W. J., and Fox, C. F., Eds., Academic Press, New York, in press.

173. **So, M., Boyer, H. W., Bettlach, M., and Falkow, S.,** Molecular cloning of an *Escherichia coli* plasmid determinant that encodes for the production of heat-stable enterotoxin, *J. Bacteriol.,* 128, 463, 1976.

174. **Gautier, F., Mayer, H., and Goebel, W.,** Cloning of calf thymus satellite I DNA in *Escherichia coli, Mol. Gen. Genet.,* 149, 23, 1976.

175. **Bolivar, F., Rodriguez, R., Greene, P. J., Betlach, M. C., Heynecker, H. I., Boyer, H. W., Crossa, J., and Falkow, S.,** A new multipurpose cloning system. Construction and characterization of new cloning vehicles. II. A multipurpose cloning system, *Gene,* 2, 95, 1977.

176. **Covey, C., Richardson, D., and Carbon, J.,** A method for the deletion of restriction sites in bacterial plasmid deoxyribonucleic acid, *Mol. Gen. Genet.,* 145, 155, 1976.

177. **Matsubara, K.,** Genetic structure and regulation of a replicon of plasmid λdv, *J. Mol. Biol.,* 102, 427, 1976.

178. **Fukumaki, Y., Shimada, K., and Takagi, Y.,** Specialized transduction of colicin E1 DNA in *Escherichia coli* K-12 by phage lambda, *Proc. Natl. Acad. Sci. U.S.A.,* 73, 3238, 1976.

179. **Shimada, K., Fukumaki, Y., and Takagi, Y.,** *In vivo* genetic engineering: interconversion of DNA fragments of specialized transducing phage genome and colE1 factor carrying genes for guanine synthesis, in *DNA Insertion Elements, Plasmids, and Episomes,* Cold Spring Harbor Laboratory, New York, 1977, in press.

180. **Maeda, S., Shimada, K., and Takagi, Y.,** Molecular nature of an *in vitro* recombinant molecule: colicin E1 factor carrying genes for synthesis of guanine, *Biochem. Biophys. Res. Commun.,* in press,

181. **Young, F. E., Duncan, C., and Wilson, G. A.,** Development of the Bacillus subtilus model system for recombinant molecule technology, in *Genetic Modification: Impact of Molecules on Genetic Research,* Beers, R. F. and Bassett, E. G., Eds., Raven Press, New York, 1977, in press.

182. **Jackson, E. N., Miller, H. I., and Adams, M. L.,** EcoRI restriction endonuclease cleavage site map of bacteriophage P$_{22}$DNA, *J. Mol. Biol.,* 118, 347, 1978.

183. **Blattner, F. R., Williams, B. G., Blechl, A. E., Denniston-Thompson, K., Faber, H. E., Furlong, L. -A., Greenwald, D. J., Keifer, D. O., Moore, D. D., Schumm, J.W., Sheldon, E. L., and Smithies, O.,** Charon phages: safer derivatives of bacteriophage lambda for DNA cloning, *Science,* 196, 161, 1977.

184. **Leder, P., Tiemeir, D., and Enquist, L.,** EK2 derivatives of bacteriophage lambda useful in the cloning of DNA from higher organisms: the λgt WES system, *Science,* 196, 175, 1977.

185. **Armstrong, K. A., Hershfield, V., and Helinski, D. R.,** Gene cloning and containment properties of plasmid colE1 and its derivatives, *Science,* 196, 172, 1977.

186. **Sanger, F., Air, G. M., Barrell, B. G., Brown, N. L., Coulson, A. R., Fiddes, J. C., Hutchison, C. A., III, Slocombe, P. M., and Smith, M.,** Nucleotide sequence of bacteriophage ϕX174 DNA, *Nature,* 265, 687, 1977.

187. **Johnson, P.,** personal communication.

Chapter 2

BACTERIOPHAGE λ AS A VECTOR IN RECOMBINANT DNA RESEARCH — ADVANTAGES AND LIMITATIONS

Noreen E. Murray

TABLE OF CONTENTS

I. INTRODUCTION

For more than a decade, bacteriophage λ has been used as a vector for the study of bacterial genes. Transducing derivatives of phage λ have permitted amplification of incorporated genes and their products and provided excellent substrates for both the genetic and physical analyses of the heterologous DNA.[1] The more recent availability of restriction endonucleases that break DNA within a particular nucleotide sequence (target) to give discrete fragments with short, single-stranded cohesive ends[2-4] immediately suggested that transducing derivatives of phage λ might be made in vitro.[5-7] Genetic and biochemical analyses that previously were possible for only some *Escherichia coli* genes would then become more generally applicable.

The chromosome of phage λ, a linear duplex with a molecular weight of about 31 million,[1] must be encapsidated to produce a mature, infectious virion. This requirement imposes a strict upper limit on the size of the phage chromosome;[8] therefore, to use the λ genome as a receptor for fragments of DNA, it is necessary not only to remove all but one of the restriction targets for the chosen enzyme but also to delete some of the phage chromosome. Furthermore, breakage of the remaining site by the restriction enzyme followed by the insertion of the donor DNA should not destroy any essential phage function; the recombinant molecule must remain capable of propagation in the bacterial cell.

The development of the phage chromosome as a receptor for fragments of DNA seemed assured by the availability of well-characterized mutations (including even large deletions), the means of manipulating restriction targets, and by the previous identification of the essential components of the phage transcriptional circuits and replication system. Indeed, the level of understanding achieved by lambdologists[1] permitted the genetic engineer to predict which regions of the genome could be sacrificed and which features could be used to advantage.

This chapter emphasizes how the genetics of phage λ can be used to facilitate the isolation, recognition, and analysis of those derivatives that have incorporated a fragment of heterologous DNA. It aims at identifying the disadvantages and advantages of λ vectors when compared with various plasmids.

II. VECTORS

The first λ vectors[5-7] were designed to accept fragments of DNA produced by endo R.EcoRI[9]* In part, this reflects the fact that R.EcoRI was one of the first enzymes shown to produce fragments of DNA with short cohesive ends;[2,10] more particularly, it is coupled to the advantages offered when a host strain for phage λ codes for a restriction enzyme. This provides simple biological systems for the selection of mutations leading to loss of restriction targets and quick tests for determining the number of restriction targets within a phage genome.[11,12] The development of other λ vectors has been facilitated by the detailed genetic and physical maps of the λ genome,[13] which permitted the use of deletion and substitution mutations in the predictable manipulation of restriction targets, e.g., for RHindIII.[14]

A. Insertion Vectors

A λ genome deleted for some nonessential DNA and retaining a single target for a restriction enzyme (Figure 1, phage 1) serves as an insertion vector. Recombinant phages may be detected if insertion of a fragment of DNA inactivates a nonessential

---

*   The nomenclature used throughout this chapter is detailed in Reference 9, but endoR.EcoRI will be abbreviated to R.EcoRI, etc.

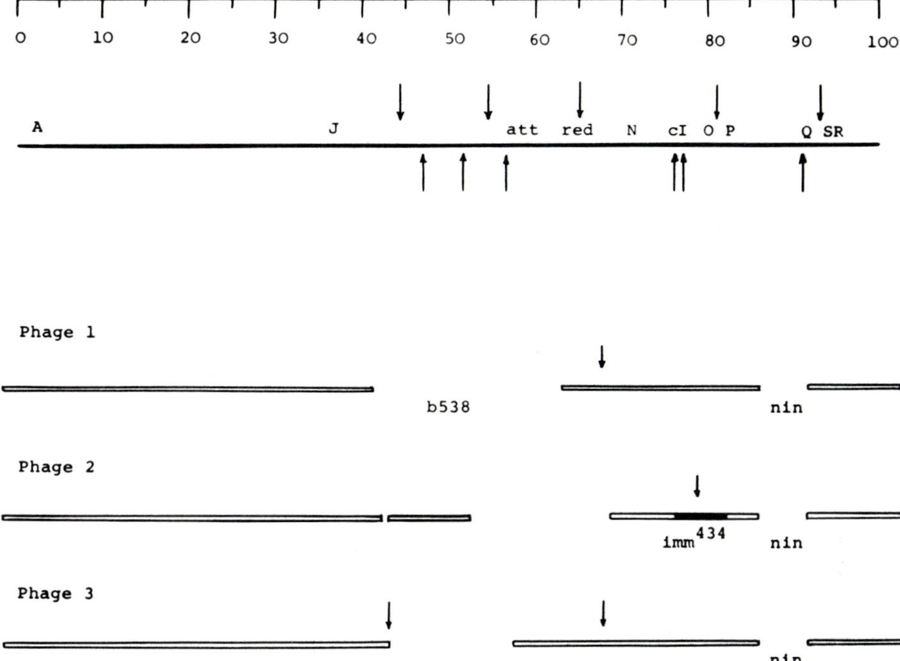

FIGURE 1.   Lambda vectors. The scale at the top of the figure represents the length of wild-type λ DNA. The map immediately below shows some of the important genetic markers of phage λ. Arrows above the genetic map indicate the positions of targets for R.EcoRI, those below the map indicate targets for R.HindIII. Phage 1 is a simple insertion vector; insertion of DNA inactivates the *red* gene. Phage 2 is an alternative insertion vector; insertion of DNA at the restriction target in the *imm*434 region inactivates the repressor gene, resulting in a change from a turbid to a clear plaque morphology. Phage 3 is a replacement vector in which a central, dispensable fragment of DNA is deleted by a restriction enzyme and replaced by donor DNA. The amount of residual vector DNA may be decreased from 72.5 to 65% or less by using alternative restriction targets,[6,14,16,18] and an additional deletion.[16,18]

phage gene.[5] In the vector shown in Figure 1, phage 1, insertion of DNA at the restriction target within a phage gene coding for a recombination function (*redα*) produces a recombination deficient (Red⁻) phage.[5] Red⁻ phages can be detected by their inability to plaque on either a *polA* or a ligase-deficient strain of *E. coli*.[15] Two easier alternatives are now available; one using a derivative of the phage already described (Figure 1, phage 1) is discussed later (Section III.D and IV.A), the other is provided by restriction targets in the gene, *cI*, coding for the phage repressor protein. The latter uses λ*imm*434 phages, since the *cI* gene of this phage contains single targets for the restriction enzymes R.EcoRI and R.HindIII. The presence of donor DNA within the *cI* gene of λ*imm*434 is recognized by a change from the normal turbid plaque morphology to a clear plaque morphology.[16] Recombinant (hybrid) plaques are thus readily detected and frequently comprise over 10% of those recovered. Recovery has been as high as 40%[17] but this probably reflects a particularly favorable state of the donor DNA and the quality of the restriction enzyme used. Vectors (Figure 1, phage 2) in which approximately 20% of the phage genome is deleted have been derived from both recombination proficient (Red⁺) and recombination deficient (Red⁻) phages.[16] These can incorporate the smallest fragments of duplex DNA produced as well as those as large as 9 or 10 kilobases (kb).

B. Replacement Vectors

Replacement vectors are propagated as phages that retain two targets for the restriction enzyme, flanking a replaceable segment of DNA. Treatment with the restriction enzyme deletes the central fragment, leaving the two arms of the vector which comprise all the essential genes of the phage but provide a genome length that is too short to be encapsidated. Plaque-forming phages only result from the incorporation of either the original fragment or an alternative donor fragment of DNA. The minimum size requirement of a phage chromosome is used in this way to increase the efficiency with which recombinant molecules are recovered. In the presence of an excess of donor DNA, the majority of the plaques recovered from a reaction using a replacement vector should contain donor DNA. The requirement for a minimum genome length may even be used to select against the incorporation of small fragments.

The first replacement vector (Figure 1, phage 3) accommodates fragments up to approximately 14 kb.[6] Alternative versions have been constructed to make the necessary space for fragments as large as 18 to 20 kb.[14,16,18,19] Others include a central fragment that imparts a readily recognizable phenotype so that recombinants are easily distinguished from reconstituted vector chromosomes.[16,18] In one such phage, this fragment includes a mutant transfer-RNA gene, *supE*, which is detected by the suppression of an amber mutation in the *lacZ* gene of a bacterial host as either red plaques on lactose-MacConkey agar or as blue plaques on agar containing 5-bromo-3-chloro-2-indolyl-β-D-galactoside (XG). In the former test, immune Lac$^+$ cells in the center of a plaque are red; in the latter, the substituted galactoside is hydrolysed in the agar to give a nondiffusing blue pigment.[20] This latter test is equally applicable to clear and turbid plaques. Alternatively, the central fragment of a vector for the RI system may contain most of the *lacZ* gene of *E. coli*; this is sufficient to complement at least some *lacZ* mutants of *E. coli*.[16,18] Lactose MacConkey or XG plates may be used to detect phages that retain the *lacZ* fragment. Expression of the suppressor or *lacZ* genes, irrespective of their orientation within the λ genome, is implicit in the design of these receptor phages. An even more efficient system would be one in which the bacterial host and medium could be chosen to selectively prevent the propagation of the parental phage; recent attempts to make such a "suicide-selection" have not been successful.

The following approach was used to make replacement vectors for the HindIII system and should be applicable whenever an insertion vector is available.[16] A fragment resulting from the digestion of *E. coli* DNA with R.HindIII was first cloned in an insertion vector (Figure 2, phages 1 and 2). Incorporation of the bacterial DNA increases the length of the phage genome thereby permitting the deletion of further nonessential regions of the λ genome. A replacement vector results when the new deletion removes enough phage DNA to make the chromosome too short to be recovered in the absence of the central fragment (Figure 2, phage 3). Available techniques readily select deletion derivatives of phage λ (see Section VI.B).

C. Vectors for Other Enzyme Systems

The vectors described in Sections A and B have been made for both the EcoRI and HindIII systems. New vectors are needed as other restriction enzymes become available. On the false premise that nucleotides are distributed randomly throughout DNA molecules, a given hexanucleotide sequence is expected to occur once in 4^6 ($\sim$ 4000) nucleotides. This implies that a small plasmid (about 6kb) will frequently lack a particular restriction target, while the λ genome most recently estimated as 49kb[21] will usually contain too many. Considering those hexanucleotide recognition sequences already identified, the variation within the λ genome extends from one target for R.Xho[22] to as many as 18 for R.Pst.[24] The task of removing unwanted restriction targets is anticipated to be more demanding than that of incorporating new ones, since the current in vitro techniques permit the experimenter to splice into vectors a fragment of DNA that

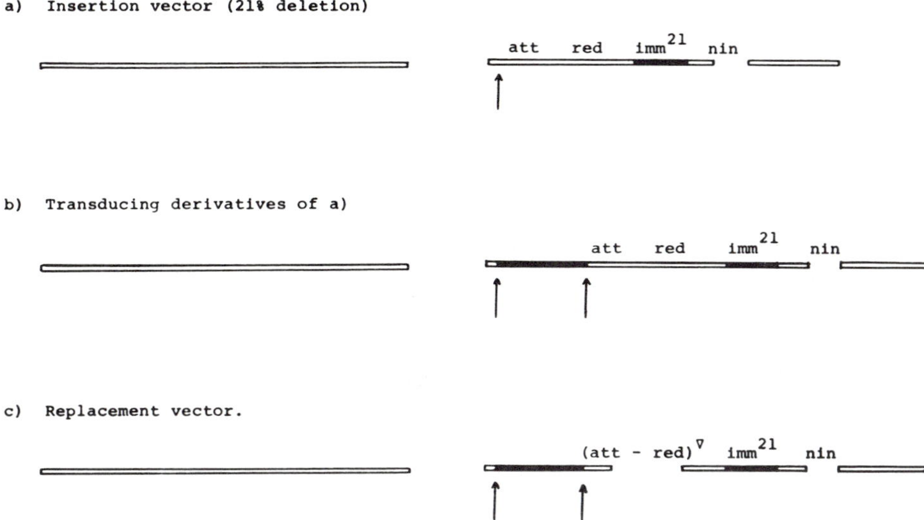

FIGURE 2.   Derivation of a replacement vector from an insertion vector. Phage a is an integration-proficient insertion vector for DNA fragments generated by R.HindIII. Phage b is a transducing derivative of a, including a fragment of *E. coli* DNA. The deletion of phage DNA from phage b produces a replacement vector, since the loss of the central DNA fragment from phage c yields a genome with all essential functions, but a DNA content of only 70% of that of wild-type λ.

includes even a series of different recognition sequences.[25] However, as already mentioned, for phage λ, the five targets for R.EcoRI and six for R.HindIII have been manipulated with relative ease.

A survey of restriction targets located within the λ genome (Figure 3) indicates a number of other useful enzymes for fragmenting DNA prior to cloning in phage λ vectors. The sources and some characteristics of the enzymes are listed in this volume.[26]

Since all but one of the five targets for R.BamI are located in nonessential regions of the genome,[27] (Figure 3), the ideal λ vector for this enzyme system hinges on the removal of the target in the left arm of the chromosome. Efforts to remove this target have proved unsuccessful, but its presence has been tolerated by choosing a suitable way of recovering the recombinant molecules (see Section III.C). Three restriction enzymes, R.BamI, R.BglII, and R.MboI, produce fragments bearing a 5′-terminal tetranucleotide extension, GATC.[28] A vector for the BamI system will thus serve for both R.BglII and R.MboI, although the cloned fragment can only be retrieved when both the vector and the donor DNA are fragmented with R.BamI.

Insertion vectors are available for both R.SalI and R.SstI, and replacement vectors have also been made.[19,29] Phage λ has only two targets for each of these enzymes, and they are both confined to nonessential regions of the genome (see Figure 3). Both R.SalI and R.SstI create fragments with cohesive ends, although their nucleotide sequences have not yet been published.[28] Similarly, insertion vectors for DNA fragments made with R.Xho already exist as this enzyme makes a single staggered break in λ⁺ DNA within a hexanucleotide sequence located about 70% from the left end of the λ chromosome.[22]

The enzymes R.Xma and R.Sma recognize the same hexanucleotide sequence; however, only the former yields fragments of DNA with cohesive ends.[28] The positions of the three targets recognized by these two enzymes[23] are shown in the last line of Figure 3. To make a replacement vector for the Xma system, it is only necessary to remove the rightmost target in the presence of an extra small deletion such as *nin5*. Indeed, this target has been removed by substituting a region of the Φ80 genome.[30] If R.Sma

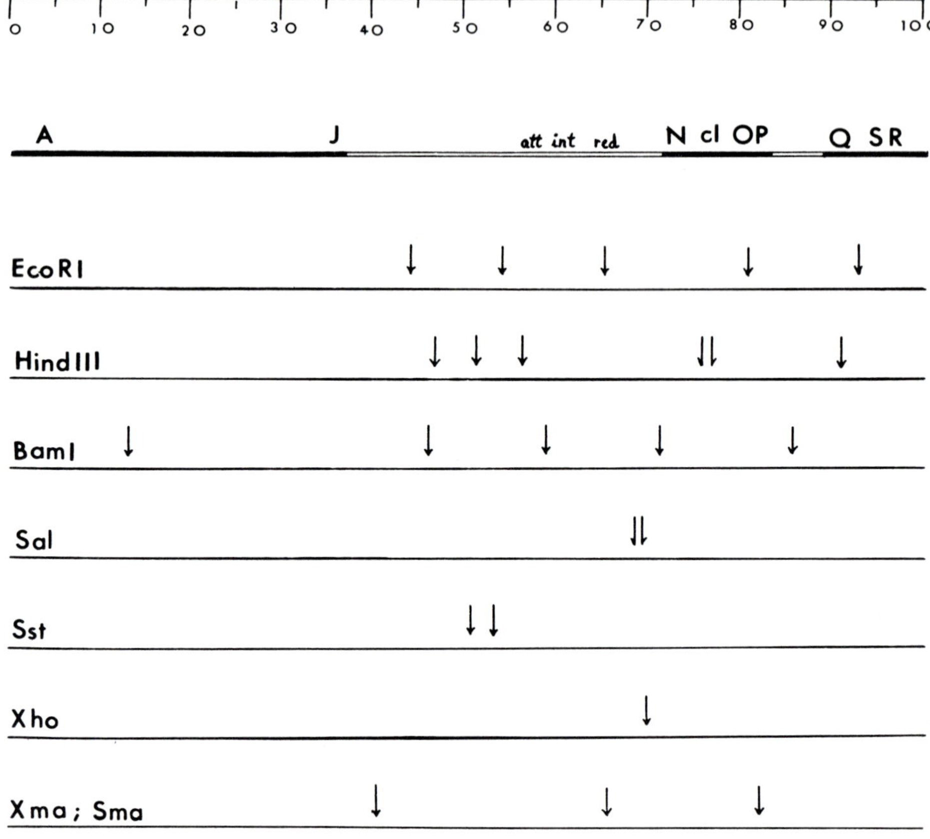

FIGURE 3.   Restriction targets within the λ genome. The scale at the top of the figure represents the length of wild-type λ DNA. The map immediately below shows some of the important genetic markers. Arrows above the other lines indicate the positions of the restriction targets for R.EcoRI, R.HindIII, R.BamI, R.SalI, R.Sst, R.Xho, R.Xma, and R.Sma.

breaks DNA to produce "flush" ends, and flush ends may be joined by T4 polynucleotide ligase,[31] this vector could also be used for any fragments lacking cohesive ends.

It is pertinent to mention that any vector made to accept fragments of DNA generated by R.EcoRI also serves when the more relaxed specificity of this enzyme is achieved (R.EcoRI*); both activities generate fragments having the 5′-terminal tetranucleotide sequence, AATT.[32]

This brief survey shows a variety of enzymes that can already, or will soon, be used to fragment DNA for cloning in λ vectors. In addition, λ vectors are useful for cloning fragments following digestion of DNA with two restriction enzymes. This has particular application for the products of repeated physical fractionations of fragments such as those formed by a sequential digestion with R.EcoRI and R.HindIII. These two fractionations may yield fragments with either identical or different cohesive ends. Suitable vectors for the HindIII and EcoRI systems are then digested with their appropriate enzyme and mixed with donor fragments of DNA which will then be recovered regardless of their terminal sequences.

Replacement vectors are also available for mixed enzyme systems, e.g., R.SstI with R.EcoRI, R.BamI, or R.HindIII.[19] Another alternative is the use of small adaptor fragments. New λ vectors are expected in which the targets for two, and preferably more, restriction enzymes have been manipulated to provide a single genome able to serve as a receptor for fragments generated by any one of two or more enzyme specificities.

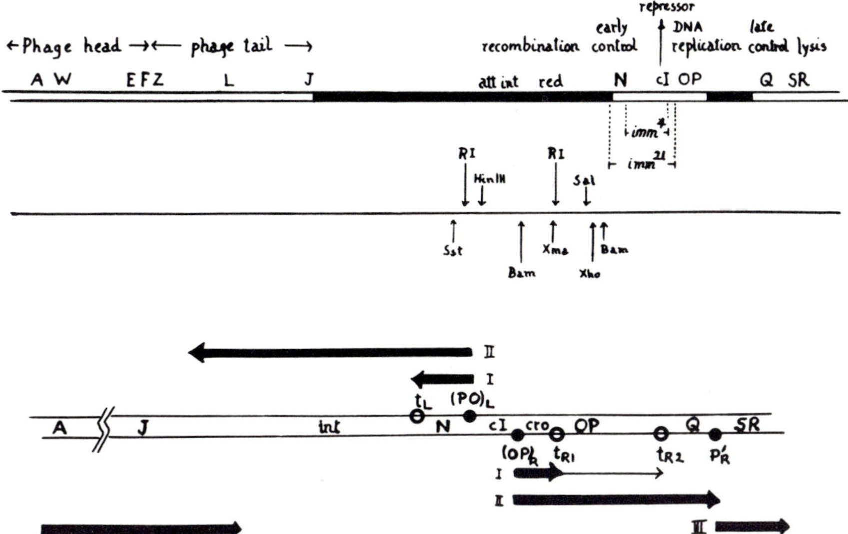

FIGURE 4. The genetic map and transcriptional circuits of bacteriophage λ. The genetic map at the top of the figure shows the more important genetic markers. The map immediately below indicates restriction targets into which DNA fragments may be inserted and transcribed from the λ promoter, P_L. The third map indicates, in more detail, the organization of the phage control region, including the sites at which transcription is initiated and terminated. The heavy arrows represent the major transcripts: (I) the immediate early transcripts initiated, respectively, from P_L or P_R and terminating at t_L or t_{R1} and t_{R2}, in the absence of the phage's N gene product; (II) the early transcripts initiated respectively from P_L or P_R and continuing through t_L or t_{R1} and t_{R2} in the presence of the N protein; and (III) late transcripts, dependent on the Q gene product, continue through genes S, R, A, and J of the circular phage genome.

D. Plasmid Derivatives of λ

Encapsidation of the phage λ genome has both a minimum[33] and maximum size requirement.[9] The obligatory minimal size provides the biological basis for the enrichment of recombinant molecules using a replacement vector; the maximum size places a strict upper limit on the size of the DNA fragment incorporated. Currently, this is 18 to 20kb for a plaque-forming derivative of phage λ.

Mutants of λ defective in gene N, a gene whose product is necessary for maximal expression of most phage genes, or even small derivative genomes (λdvs),[34,35] in which only three phage genes (cro, O, and P) are essential (Figure 4), will propagate autonomously in E. coli as a plasmid. Removing the need for encapsidation relieves the constraint imposed on the maximum size of the phage chromosome. The original λdv is approximately 4kb in length;[36] more recent λdvs vary from 1.5 to 6kb.[37] The λdv plasmids resemble the colE1 plasmid in that there are many copies (50 to 80) per cell.[36] Furthermore, the transformants carrying plasmids are easily selected by superinfection with phage λ.[38] The λdvs include a target for R.EcoRI, and although this target is thought to be within one of the genes essential for plasmid replication (gene O or P), λdv still provides a useful cloning vehicle. This depends on the propagation of λdv as a dimer in which one unit serves as a replicator and the other supplies the site for the insertion of foreign DNA.[34] λdvs have been used for cloning fragments of DNA generated by R.EcoRI, R.BamI, and R.HindIII[37] and should serve for the Xma system.[23]

These plasmids allow the insertion of DNA into regions of phage λ that are well defined with respect to transcriptional circuits. They permit easy selection of transformants and offer scope for the use of genetic features of λ for the further amplification of DNA or gene products.

E. Synthetic Cohesive Ends

In vitro recombination experiments often take advantage of restriction enzymes that facilitate the joining of fragments by endowing them with 5′ extensions which cohere by complementary base-pairing. For some studies, it is disadvantageous to have the fragment content and size defined by the distribution of restriction targets.[39] In these instances, cohesive ends may be attached by means of polynucleotide terminal transferase[40,41] to randomly sheared fragments of chosen lengths and to vector molecules in such a way that the vector molecular will only circularize via the donor fragment.[39] This technique has been elegantly extended to the cloning of complementary DNA (c-DNA) made from a messenger RNA template.[42-44] Small, circular plasmid molecules seem ideally suited to this approach, and all the reported experiments using synthetic cohesive ends have used plasmid vectors. Some laboratories, however, are adapting these techniques for the larger, linear λ chromosome.[45]

III. RECOVERY OF RECOMBINANT DNA MOLECULES

A. Transfection of *E. coli*

The simplest method of recovering phages from λ DNA molecules is a process analogous to bacterial transformation in which bacteria are made competent for uptake of DNA by a starvation treatment in calcium chloride. In the original method, starvation followed growth in a stringent synthetic medium;[46] currently, growth in a rich broth is preferred.[47] The efficiency of transfection is extremely strain-dependent and, even for a given strain, shows considerable fluctuation. Approximately 10^5 to 2.10^6 molecules are recovered as plaques from 1 μg of phage λ DNA; in other words, a maximum of only 10^{-4} of the input molecules are retrieved.

B. Infection of Spheroplasts

Recovery of DNA by infection of spheroplasts offers the advantage that a preparation of spheroplasts can be stored for use on subsequent occasions.[48] The efficiency with which molecules are recovered is, however, no greater than for the transfection method and spheroplasts are more troublesome to prepare.

C. Helper-mediated Transformation

The first procedure developed for the effective uptake of phage λ DNA into *E. coli* was dependent on the preinfection of host cells with phage, so-called helper phage.[49] This method has achieved the most efficient rescue of λ DNA; recoveries have been as high as 10^{-2},[50] although figures between 10^{-3} and 10^{-4} are more common.[51] The disadvantage of this method is the dependence on a helper phage whose recovery is not desired. This system can, however, be used to advantage since λ molecules lacking one cohesive end can be rescued efficiently by recombination with the genome of the helper phage.[51] This has permitted the use of a replacement vector for the BamI system[51] despite the presence of the additional target in the left arm of the genome (Figure 3). The exonucleolytic degradation of an infecting λ DNA molecule lacking a left-hand terminus, and hence unable to circularize, is prevented by the use of bacteria defective in the *recBC* nuclease.[52]

D. In Vitro Encapsidation *or* "Packaging"

Recent methods for packaging λ DNA in vitro recover phage with efficiencies as high as 10^{-3}.[53,54] The required packaging extract takes more time to prepare than a batch of transfection cells, but, once made, it can be stored frozen for some months.[55] A pretested, guaranteed batch of packaging proteins currently offers the most efficient way of recovering recombinant molecules.[55,56] The resulting phage may be plated on any λ-sensitive strain of *E. coli*, even those which are inefficient for transfection. The safety features of in vitro packaging are considered in Section VIII.

Two slightly different packaging procedures have been developed. Both use a mixture of phage proteins resulting from the induction of two differently defective prophages. Each induced strain is lacking one essential head component, but when mixed, the extracts are capable of packaging both endogenous and exogenous DNA. The lysogens used are heat-inducible, and the prophages always lack a function necessary for cell lysis. In the absence of lysis, the phage proteins are readily concentrated.

Hohn and Murray[55] use one prophage with a defect in gene *D* and a second with a defect in gene *E*. Ultraviolet irradiation of the extract inactivates the endogenous DNA and the recovered phage do not then include the genetic markers present in the prophages. In vitro packaging has been used to recover recombinant molecules from ligase reactions with efficiencies that consistently exceed those achieved by transfection. This system shows little, or only a slight, preference for particular sizes of molecules, at least within the range of 78 to 100% the length of λ wild-type.

Alternatively, Sternberg and co-workers[56] use protein extracts resulting from the induction of A^- and E^- prophages. The encapsidation of endogenous DNA is prevented by a genetic trick which keeps the replicated phage DNA in the bacterial chromosome. Recombination between the exogenous DNA and the prophage chromosomes is prevented by genetic lesions in both the host and prophage recombination pathways. Again, this in vitro system offers a more efficient recovery than the usual transfection method. Furthermore, large DNA molecules, which in many instances may be equated with recombinant molecules, are preferentially recovered (see Section IV).

IV. DETECTION OF RECOMBINANT MOLECULES

A. General Approaches that Simplify the Recovery of In Vitro Recombinants

From a ligation reaction with fragments of donor DNA and a cleaved λ vector, many phage molecules will result merely from the restitution of the parental phage chromosome. Various genetic approaches are used to increase the efficiency with which recombinants are both recovered and recognized.

In the first instance, replacement vectors themselves increase the efficiency with which recombinant molecules are recovered. All plaque-forming molecules necessarily include a fragment of DNA inserted between the two arms of the vector; the recovered phages thus include either a donor fragment of DNA or have regained (or retained) the original dispensable fragment. In the presence of excess donor DNA, over half and perhaps as many as 90% of the recovered phage are recombinants. The efficiency depends on *both* targets of the vector molecule being cleaved. Easy ways of distinguishing recombinant from parental phages have been described (Section II.B). A change in phenotype associated with the insertion of donor DNA also facilitates the recognition of recombinants using an insertion vector. The example already cited depends on a change in plaque morphology from turbid to clear (see Section II.A).

Phage λ grows normally in the absence of the capsid protein, pD, provided the phage contains no more than 82% of the wild-type DNA content.[56] The addition of a *Dam* mutation to the vector depicted in Figure 1, phage 1 provides alternative ways of detecting recombinants. It will be recalled that with this vector, in the absence of *Dam*, hybrids could be detected as recombination-deficient phages *red⁻* because insertion of DNA destroys the *red⁺* gene.[5] Since this vector has a DNA content of only 78% of that of wild-type despite the presence of the *Dam* mutation, it grows equally well on *sup°* and *sup⁺* hosts. Hybrid derivatives that have increased their DNA content to 82% of the wild-type genome will grow only on a *sup⁺* host.[56] Recombinants that have acquired a smaller fragment are *red⁻* but retain their ability to grow on a suppressor-free host.[56] For the EcoRI system, additional evidence for the presence of a small fragment can be obtained by an increase in the efficiency of biological restriction, since the extent of restriction by this system depends on the number of restriction targets.[5]

The in vitro packaging system of Sternberg and colleagues[56] shows a very marked preference for larger molecules; λ wild-type DNA is packaged 200-fold more efficiently than a vector with 78% of the wild-type DNA content. Using either an insertion vector or a replacement vector, in which the central replaceable fragment is small, almost all the phages recovered will be recombinants.

It is evident from these discussions, that the size of an acceptable donor fragment may be dictated to advantage by the choice of vector and recovery system.

B. Recognition of Recombinants by Expression of the Cloned Genes

Specialized transducing derivatives of phage λ were first detected by their ability to carry *E. coli gal* genes into a Gal⁻ strain of *E. coli*.[57] Where appropriately defective strains of *E. coli* are available, compensation for a genetic lesion in the host cells remains a good selective system for the isolation of in vitro recombinants. It was used to detect the transfer in vitro of *E. coli trp* genes into both phage[5] and plasmid vectors[58] and has formed the basis for the isolation of recombinants containing various bacterial,[59-62] and even yeast, genes.[63]

The efficiency with which transducing derivatives are recovered as lysogenic transductants is greatly enhanced by the presence of a normal phage integration system. This can be achieved either with an integration-proficient vector or by supplying an integration-deficient transducing phage with a wild-type helper phage when dilysogens will frequently include the transducing phage. Alternatively, extra homology may be provided by using a lysogen.[56] When a transducing phage is integrated into the *E. coli* chromosome at the phage attachment site, rather than by recombination between homologous bacterial sequences, efficient expression of the incorporated genes usually depends on the inclusion of their normal promoter.

Frequently, the expression of inserted genes from either a phage or an incorporated promoter may be detected in the lytic phase. Thus, Bio⁺ phage can be detected through use of a Bio⁻ host on appropriate indicator medium by the Bio⁺ cells in the center of a turbid plaque. Trp⁺ phages are readily selected as plaques, "Trp⁺ plaques", on a Trp⁻ host in the absence of exogenous tryptophan.[65] In the latter case, the bacterial genes within the phage compensate for the host defect by providing the necessary enzymes for bacterial growth and phage production. This selection works efficiently for genes divorced from their normal promoters and negates the need for stable lysogens. The establishment of repression, however, facilitates the ease of detection by producing a plaque, more appropriately called a "galaxy," which is conspicuous by the growth of prototrophic, immune cells. Galaxies are easily distinguished from revertant colonies. This powerful selection system has been used extensively for *E. coli* genes[61] and may be used for genes from other bacteria. Transducing phages, including genes from Gram-positive bacteria, have been selected.[66] An alternative selection relies on the visualization of plaques in the absence of a visible lawn by staining with ethidium bromide. The selective propagation of transducing phage within defective bacteria is detected by fluorescence in ultraviolet light.[67] This may increase the sensitivity of the screen and is the preferred method when clear-plaque mutants are used.

Tests relying on the functional expression of the incorporated genes are limited by the availability of suitably defective strains of *E. coli*, although in some cases the potential of a wild-type bacterium may be extended, e.g., in the acquisition of resistance to certain drugs.

C. Detection of Recombinants by Immunoassay

Assays have recently been developed for detecting the translation of cloned genes independent of expression of protein function. These methods are based on specific cross-reactions with antibodies and are applicable within a phage plaque or surrounding a lysed bacterial colony.[68,69] Such interactions may even detect parts of a polypeptide chain.[68] A screen of this sort has great potential in detecting genetic changes that may be necessary to permit the translation of a cloned gene.

D. Screening for Hybrids by In Situ Nucleic Acid hybridization

The *in situ* hybridization methods used so successfully in cytological analyses[70,71] are applicable to individual plaques[72] or bacterial colonies.[73] This technique is extremely sensitive (sequences of less than 1kb are detectable) and limited only by the availability of a labeled RNA or DNA probe. Jones and Murray[72] used the conventional hybridization technique on single plaques cut from a bacterial lawn and either autoradiography or scintillation counting to detect the retention of labeled probe. A recent modification of the method described for bacterial colonies[73] has been reported in which phage are propagated on nitrocellulose filters, denatured, and fixed for *in situ* hybridization.[74] Hybrid phages retain the labeled probe and are easily detected by autoradiography. These methods enable the rapid and sensitive screening of many plaques. The nitrocellulose filter method is a little quicker in terms of the rapidity with which plaques or colonies may be transferred to the filter but takes longer to process than the earlier procedure.[72] Both techniques are slightly more laborious for plaques than colonies since duplicate transfers are made for the former, rather than a replica plating from a master grid. However, the simplest and quickest method of screening by *in situ* hybridization follows the direct transfer of phage DNA from plaques to filters merely by contact. The surface of an agar plate covered with 10^4 plaques may be screened in this way.[74a] *In situ* hybridization techniques have proved invaluable in the isolation and study of eukaryotic sequences of DNA.

V. AMPLIFICATION AND ANALYSIS OF λ HYBRIDS

A. Amplification of DNA

Small volumes of high-titer lysates of phage λ are traditionally prepared as plate lysates. If DNA is required, this lysate is amplified by the infection of sensitive cells in liquid culture. Current procedures obviate the need for plate lysates in some instances, since bacterial cells infected with the phage from a single plaque are diluted into as much as a liter of broth and incubated for 4 to 12 hr.[18,56] A liter of lysate containing between 10^{13} to 10^{14} phages is obtained. The phage may be concentrated directly by high-speed centrifugation[49] or by low-speed centrifugation following precipitation with polyethylene glycol.[75]

In some cases, the introduction of a suppressible mutation in the phage gene, *S*, whose product is necessary for cell lysis, leads to even higher yields of phage.[75a] Plate lysates are first prepared in a host permitting suppression of the lysis defect and provide inocula for infecting, in liquid culture, bacteria lacking the ability to suppress the mutation in gene *S*. The phage are concentrated simply by harvesting the cells prior to artificial lysis with chloroform. Increased yields of phage are contained within a much smaller volume with this procedure.

The phage suspension is subsequently treated with DNAase and RNAase and spun to equilibrium in CsCl[49] to gain further concentration and purification. The DNA is extracted with phenol and should be free of contaminating bacterial nucleic acids. A few milligrams of DNA are obtained from a liter culture (i.e., prior to concentration of the cells when a lysis block is used). This total yield exceeds that normally attained with *colE*1 hybrids; however, it must be remembered that the ratio of vector to cloned DNA differs in the two cases. For small fragments of cloned DNA, the relaxed *colE*1 plasmid provides the better amplification of the heterologous DNA; however λ gives better amplification than the large plasmids.

B. Analysis of DNA

The incorporated DNA may be separated from the vector by electrophoresis in agarose gels[76] after digestion with the appropriate restriction enzyme. This technique also gives a size estimate of the included fragment. Separation in a sucrose gradient may be preferred when the fragment is considerably smaller than the arms of the vector, as a reasonable quantity of the fragment can be recovered very efficiently.

The linear duplex molecules of phage λ are excellent substrates for heteroduplex analysis.[77,78] Preparations of phage are mixed and treated with alkali so that the DNA is simultaneously released and denatured.[79] Heteroduplexes are formed during neutralization, and in the presence of formamide, the lengths of both unpaired and paired strands may be measured.[78] The well characterized deletion and substitution derivatives of phage λ provide excellent markers.[13]

A pronounced AT-rich region within the right arm of the λ chromosome provides a strikingly convenient marker for partial denaturation mapping.[80] This greatly facilitated the construction of an elegant denaturation map of the histone genes of *Psammechinus miliaris* by defining the orientation of the inserted genes.[81] Early studies of λ DNA provide another useful analytical feature. The two strands of phage λ DNA are readily separated in caesium chloride in the presence of polyrUG,[82] due to the asymmetric distribution of polyrUG binding sites. Since the polarity of each strand has been defined, hybridization of mRNA to the separated strands of the λ hybrid identifies both the coding strand and the polarity of a cloned gene. This simple procedure was used to identify the coding strand for the five histone genes of *Echinus esculentus* cloned within a λ vector.[83]

VI. GENETIC ANALYSIS AND EXPRESSION OF PHAGE λ

A. Essential Features of Phage λ

The genome of phage λ (see Figure 4) is divided into three regions for descriptive purposes. The left-hand region includes all the genes (from *A* through *J*) necessary to convert phage DNA into a mature, encapsidated virion. A central region between genes *J* and *N* is nonessential in the sense that no function coded within this region is necessary to maintain plaque-forming ability. This region does, however, include genes concerned in general recombination (e.g., *redα* and *redβ*) and those (i.e., *int* and *xis*) necessary for the integration of the phage into the *E. coli* chromosome and the excision of prophage from the chromosome. The latter recombination events are confined to specific sites, i.e., the attachment regions (*att*) of the phage and bacterial chromosomes. In the vectors described in Section II, most of the central region of the phage genome, including the integration system, has been sacrificed to provide space for heterologous DNA. Receptor chromosomes retaining *att* are available for fragments of DNA generated by R.EcoRI, R.HindIII,[14] and R.SstI.[19] It is advantageous to use an integration proficient phage for some genetic analyses and the amplification of gene products, (see Sections VI.B, VII.A, and VII.B). The portion of the genome from gene *N* rightwards including all the major control elements, the genes necessary for phage replication (*O* and *P*) and for cell lysis (*S* and *R*). The basic features[84,85] of these control circuits may be summarized as follows.

The essential genes of phage λ are transcribed from three major promoters: P_L, P_R, and P'_R (see Figure 2.4). In the prophage state or on infection of an immune host, the λ repressor (the product of the *cI* gene) binds to the operators O_L and O_R preventing transcription from P_L and P_R. Following the infection of a sensitive host, transcription proceeds leftwards from P_L through gene N and rightwards from P_R through gene *cro*. In the absence of the product of gene N, most transcripts terminate just beyond these genes at the sites t_L and t_{RI}. A second termination site, t_{R2}, impedes those transcripts from P_R that escape the first termination signal, t_{RI}.[86] Gene N protein exerts its positive regulatory role by interacting with RNA polymerase to permit transcription to ignore these "stop" signals.[87-89] When available in a sufficient concentration, the *cro* gene product depresses transcription from P_L[65,90,91] and P_R.[92] In the presence of N protein, transcription continues leftwards beyond t_L into the recombination genes, and rightwards through genes O and P, to permit efficient replication of phage DNA, and subsequently through t_{R2} and gene Q. The product of the latter gene is essential for the transcription of the "late genes" of λ.[93,94] Since the linear chromosome of the phage circularizes on infection, Q-dependent transcription of P'_R (located between genes Q and S) continues through genes S and R, into A, and beyond gene J. Genes N and Q play essential positive regulatory roles. *nin5*, a small deletion present in many λ vectors, removes t_{R2}, and an N^- *nin5* phage can rely on leakage of transcripts through t_{R1} to provide $O, P,$ and Q functions.[95]

If the temperate lysogenic state is to be achieved, repression of the phage by binding of the repressor protein to O_L and O_R must be established before the phage is committed to the lytic pathway (see, e.g., the review by Herskowitz[85]).

B. Genetic Analysis

Recombination between phages permits high resolution in genetic analyses, and it is easy to overlook the expediency that results from the ready transfer of a phage to any sensitive *E. coli* strain. Genetic analysis of heterologous genes within λ vectors is most incisive when the functional expression of these genes supplies a basis for the selection of recombinants. This is illustrated by the detection of $\lambda trpB^+$ recombinants, which arose with frequencies as low as 10^{-6}, from crosses between two $trpB^-$ transducing phages ($\lambda trpA^+B^-C^-$ and $\lambda trpA^-B^-C^+$) as TrpB$^+$ plaques on $trpB^-$ indicator cells.[96]

In the absence of direct selection procedures, flanking markers can be used to select exchanges in an intervening interval that includes heterologous DNA. The most stringent selective system results when the two vectors have no homology within the selected interval. This principle was used to orient fragments of the *trp* operon of *E. coli* cloned within λ vectors (Figure 5). Crosses to a well-characterized $\phi80trp$ phage divided the λtrp phages into two classes, since the only homology was provided by the incorporated bacterial DNA. The class that permitted recovery of $h^\lambda imm^{\bullet80}$ recombinants contained its *trp* genes in the same orientation as the biologically derived $\phi80trp$.[96]

Suitable pairs of λ vectors lacking homology in the region between host range and immunity would have general application by providing a genetic test for the identification of overlapping fragments.[30]

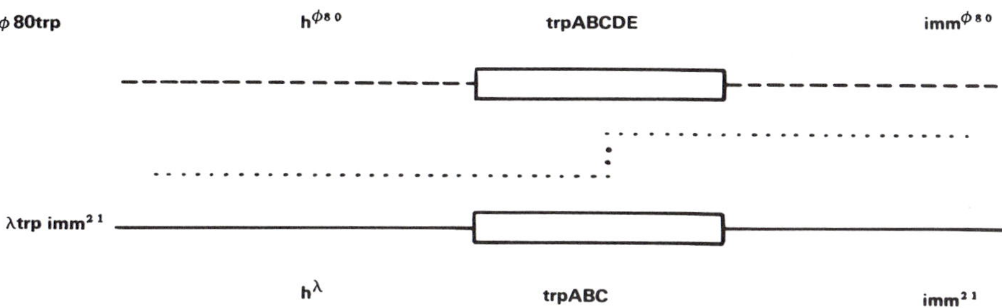

FIGURE 5. Phage cross to determine the orientation of the *trp* genes within a λtrp phage. The homology is confined to the bacterial DNA, and $h^\lambda imm^{\bullet80}$ recombinants require the *trp* genes to be in the same orientation.

Mutant derivatives of λ hybrids can be isolated following either in vitro or in vivo mutagenesis. Those arising during in vivo propagation are readily isolated, since dilution of a phage lysate achieves immediate recloning. Relaxed plasmids are not ideal substrates for in vivo mutagenesis. The propagation of phage λ as a prophage, rather than a lytic virus, offers considerable advantage in the maintenance of mutations or hybrids that confer a selective disadvantage on the phage and in the controlled amplification of gene products (Sections VII.A and B). Dilysogens can be a useful alternative when a transducing phage lacks its own integration system.

Phage λ and its hybrids are, perhaps, uniquely suited for deletion analysis. Deletion derivatives of λ hybrids may be analysed by genetic recombination, gene expression, and the physical techniques of heteroduplex analyses. The suitability of phage λ derives from the finding that deletion derivatives of λ are much less sensitive to chelating agents than is wild-type λ.[97] When added to agar medium, a chelating agent such as sodium pyrophosphate permits direct selection of deletion mutants as plaques. Within a certain size range, the concentration of chelating agents tolerated increases with the size of the deletion. Bacterial strains that are nonpermissive for extensively deleted phages[98] provide selective systems for the isolation of recombinants with increased DNA content.

Deletion analyses are being applied to incorporated fragments.[62] Deletions that extend from the fragments into the phage orient the incorporated genes with respect to the vector chromosome and are readily isolated by coselecting for a deletion and loss of λ genes. Loss of the adjacent phage genes, *red* and *gam*, permits phage λ to grow on a strain of *E. coli* that is lysogenic for phage P2.[15] Deletions extending from the fragment through to the phage gene *gam* (Figure 4) are potentially useful for fusing the inserted genes to the phage promoter P_L.

Currently, there is much interest in the expression of prokaryotic genes in different bacterial environments. The P group plasmids transfer between many Gram-negative species. A derivative of the P group plasmid, RP4, containing the attachment region of phage λ has provided the means of transferring λ-*E. coli* hybrids into other bacteria.[99]

VII. EXPRESSION OF BACTERIAL GENES IN TRANSDUCING PHAGES

Bacterial genes within phage λ can be expressed either from their own promoter or a phage promoter.[65,100,101]

A. Amplification of Gene Expression from an Included Promoter

To maximize the yield of products from a bacterial promoter included in a transducing phage, it is necessary to delay cell lysis but permit DNA replication so that the number of gene copies is greatly enhanced. This end was originally achieved by the introduction of a mutation in the phage gene *S*.[102] A defect in gene *S* prevents cell lysis but permits DNA replication and protein synthesis to continue for some hours.[103] Moir and Brammar[104] recently obtained a significantly better amplification of gene products by using a mutation in gene *Q*. In this case, all late functions are blocked, and while lysis is prevented, so too are other late functions and the consequent packaging of the replicated DNA. Using either biologically derived *trp* phages[104] or biochemically produced, integration proficient λ*trp* phages,[96] the products of the *trpE* and *D* genes comprised more than 25% of the total soluble protein of infected cells. Amplification of the *trp* genes using the *colE*1 plasmid has produced similar yields of *trp* enzymes.[58]

The general use of this approach is limited by the efficiency of the incorporated promoter and other features of the normal control system. The following observations give some insight into problems consequent upon the use of an included promoter. First, the maintenance of the *colE1-trp* plasmid in the *E. coli* cell is dependent on the presence of the *trp* repressor protein.[58] Presumably, the uncontrolled expression of 20 copies of the *trp* operon is not advantageous to *E. coli*. The overproduction of a normally nontoxic protein could be the lethal outcome of amplifying a negatively controlled gene from bacteria other than *E. coli*. Secondly, good yields of the *trp* enzymes are dependent on the addition of a derepressing agent, a condition readily fulfilled for the *trp* operon. In both these respects, λ has some advantages. The phage may be propagated in the prophage state, one copy per bacterial chromosome, and induced to achieve amplification of gene products. In some cases, notably for *gal*[105] and *trp*[104] the amplification of the phage genome has been sufficient to override repression by titration of the repressor protein. More difficult to overcome in either system is autogenous control or the requirement for some missing component of a positive control system. All these considerations indicate the desirability of controlling expression of the required enzyme by strategically chosen operator and promoter regions. One such control system is the leftward operator and promoter region of phage λ.

B. Expression of Bacterial Genes from P_L of λ

The most extensive information concerning the transcription of bacterial genes from the promoters of phage λ concerns expression of the *trp* genes of *E. coli* from P_L.[65,88,96,104,106] These data illustrate the useful features that P_L and the *N* operon of λ offer for the expression of genes inserted into the central region of the λ genome (see Figure 4).

Experiments with λ*trp* phages[88,89] and others[87] have shown that a transcript initiated at P_L in the presence of the *N* protein will read through sequences that would otherwise terminate transcription. This role of *N* protein as an antiterminator may enable the transcription of genes separated from the phage promoter by sequences that would otherwise impede RNA polymerase. Transcription from p_L mediated in the presence of *N* protein of λ is undeterred either by repression of the *trp* operon[65] or the *trp* attenuator sequence.[107] Transcription from P_L proceeds unabated through the phage attachment site to the R.*Hind*III target *shn*λ3 (see Figures 2 or 4), into which the *trp* genes of *E. coli* have been inserted.[96] Amplification from the phage promoter should, therefore, be possible following induction of a prophage.

In addition to providing the peculiarly useful *N* protein, P_L promotes a very high rate of expression. The maximum expression rate may be an order of magnitude higher than that from an efficient, derepressed bacterial promoter such as the *trp* promoter.[106,107] The achievement of this expression does, however, depend on both the absence of the λ repressor and the loss of the additional negative control normally provided by the *cro* gene product.[88,106]

λ*trp* phages lacking the *trp* promoter have been used to investigate ways of optimizing gene expression initiated at P_L.[104] Derepression of transcription from P_L was obtained by the use of a mutation in gene *cro*, and cell lysis was delayed by blocks in genes *Q* and *S*. Cells infected with λ*trp cro⁻ Q⁻ S⁻* phages, in which the *trp* genes were expressed exclusively from P_L, can contain about 10% of their soluble protein as anthranilate synthetase (the products of the *trpD* and *E* genes), representing nearly a tenfold increase above that achieved in derepressed cells.

This is a lower yield than might be anticipated; nevertheless, such levels of specific proteins would provide an advantageous start in the purification of a protein, particularly one that is normally present in small amounts. The success in the use of *cro⁻* derivatives of λ is tempered by the finding that λ*cro⁻ Q⁻ S⁻* phages are not always easy to construct or propagate. These problems may be overcome by using the temperature-sensitive *cro* allele recently described by Matsubara.[108] An integration-proficient vector defective in cell lysis but carrying temperature sensitive alleles in the *cI* and *cro* genes may prove easier to handle. The use of any of the restriction enzymes listed in Figure 3 enable the insertion of fragments of DNA so that they can be transcribed from P_L of λ.

An alternative to a *cro⁻* mutation is a phage with a "hybrid immunity" region.[109] Since the *cro* gene is immunity specific,[110] it follows that a phage having P_L from λ, but P_R and *cro* from phage 434, should be phenotypically cro⁻ for only leftward transcription. Such a recombinant has been isolated from an elegantly designed phage cross; however, this phage included a defect in the leftward promoter to reduce transcription from P_L.[111] With the objective of both controlling and maximizing expression from P_L, a hybrid immunity phage with a suppressible defect in gene *N* was made.[109] An *N⁻* phage having a hybrid immunity region and a deletion (*nin*) removing t_{R2} will propagate in a nonsuppressing host; on infection of an *N*-suppressing host, uncontrolled expression of *N*-dependent gene products is achieved. When cell lysis was prevented by a block in the phage gene *S*, the amplification obtained from this system was two- to threefold more than that from *cro⁻* phages. The dependence of transcription, but not replication, on suppression of gene *N* allows the propagation of the *N⁻ nin* phage in the absence of excessive expression from P_L.

Helinski and co-workers are seeking to use the lambda promoter (P_L) to moderate and amplify the expression of genes cloned within the relaxed plasmid, *colE*1.[112] Bacteria containing about 20 copies of the plasmid may be grown in the absence of expression from P_L. Elevation of the temperature inactivates the heat-labile λ repressor and leads to transcription from P_L, which in the absence of *cro* gene product is derepressed. In the near future, it will be possible to clone directly into a *colE*1-λ hybrid to take advantage of both the relaxed replication of *colE*1 and high level of controlled expression from the λ promoter.

Alternatively, deletion derivatives of λ transducing phages may be isolated to remove unwanted DNA or restriction targets between P_L and the gene of interest. Such deletions may be necessary to remove elements of the normal control system. Bacterial genes fused to the regulatory elements of phage λ may then be transposed in vitro within a single fragment of DNA to the *colE*1 vector. However, some caution is provided by a well-documented case where good transcription of the *lacZ* gene of *E. coli* was achieved from the phage promoter, P_L, but translation of the mRNA to give active enzyme failed.[113]

C. Use of Other Promoters

Other efficient promoter regions whose operators can be experimentally controlled may be used. A good example is the *lac* control region, which is already present within a λ*lac* transducing phage,[2,9] and has recently been transferred to a relaxed plasmid.[114,115]

D. Analytical Screening of Proteins Synthesized by Incorporated Genes

It is frequently of interest to chart or identify the polypeptides coded by a fragment of DNA. The proteins are usually labeled with a radioactive isotope and detected by autoradiography following separation on a polyacrylamide gel. When a plasmid is the vector, the background of bacterial polypeptides is removed by following protein synthesis in minicells.[116,117] Proteins synthesised by λ or λ hybrids are detected in a similar way, but following the infection of a UV-irradiated host.[118-120] Transcription from phage promoters can be avoided by infecting a λ lysogen.[120]

VIII. BIOHAZARDS AND SAFER VECTORS

Phage λ can replicate either lytically or as a permanent resident within a bacterium,

but outside the bacterial cell, it has no capacity to propagate and, indeed, is poorly suited to survive. Since the infectivity of phage is extremely sensitive to low pH and proteases, it is unlikely that infectious particles will traverse the stomach. Phage λ is sensitive to desiccation and survives poorly in water or even sewage.[18] It has been calculated that a phage entering a sewer is most likely to lose infectivity before encountering an *E. coli* cell,[18] and most of the *E. coli* cells will be resistant to infection by laboratory strains of phage λ.[121] In summary, phage λ probably has limited capabilities for lytic growth in a natural environment; however, once established in a bacterium as a plasmid or lysogen, phage λ and the cloned segment could be perpetuated at least as well as the bacterial host.

The principal strategy, therefore, has been to develop vectors which will propagate efficiently via the lytic mode in *chosen* laboratory strains but fail to make lysogens or plasmids in any bacterial cell. To reduce the possibility of a λ hybrid being established as a prophage, vectors are defective in the gene (*cI*) that codes for the phage repressor; preferably all or part of the *cI* gene is deleted.[16,18] The lysogenic state cannot be maintained in the absence of repression. Most vectors are also missing both structural and functional components of the phage integration system and are devoid of bacteria DNA that would provide homology with the host chromosome.[16]

The possibility of plasmid formation is reduced by ensuring the expression of the lysis gene *R*. This end is achieved even in the absence of gene *N* if the termination site, t_{R2}, (see Figure 4) is removed by the *nin5* deletion.[18]

The host range of the vectors in the lytic phase is reduced by the inclusion in the vector of one or more amber mutations.[18,122,123] A suitable bacterial host must then carry appropriate suppressor genes to encapsidate the phage DNA. Propagation of the phage on nonmodifying bacteria provides an additional barrier to lytic propagation on bacterial strains containing a restriction system.

In the light of the above considerations and precautions, dissemination of escaped phages seems unlikely. The most vulnerable stage is the lytic propagation of the phage. In practice, rare lysogens, like all the bacteria, should be killed by the addition of chloroform. It has, however, been suggested that the host bacteria should also be crippled to further reduce the possible spread of infected bacteria, should the experimenter fail to kill all the bacteria with chloroform. Regrettably, crippled bacteria are generally less efficient hosts for the recovery of recombinant molecules by transfection. In vitro packaging offers a considerable advantage in this respect, since the recombinant molecules are packaged in the absence of bacteria and may be recovered with reproducible efficiency on any λ-sensitive host.[55,56]

The subject of safer vectors is discussed in more detail in References 16, 18, 121, 122, and 123.

IX. DISCUSSION

In experiments where it is necessary to attach cohesive ends synthetically to fragments of DNA and vector molecules, the small circular plasmids seem the obvious choice of vector. In others, where it may be advantageous to recover the cloned fragment of DNA by treating with the restriction endonuclease, phage vectors provide simple and very efficient ways of recovering and detecting recombinant DNA molecules. A phage chromosome is recovered as a plaque, and hybrid molecules can be detected merely by a change in plaque morphology. The vector and recovery system may be so chosen that almost all the recovered molecules will be hybrids, even hybrids incorporating a fragment within a particular size range. However, the phage system imposes an upper limit on the size of the fragment that can be cloned. This upper limit of 18 to 20 kb is of decreasing concern for many purposes, as restriction endonucleases other than R.EcoRI become available. Indeed, large fragments (20 kb or more) can be disadvantageous; the propagation of only the coveted, identified segment of DNA may be preferred.

Currently, many λ vectors are designed for use with only one, or at the most two or three, enzyme systems. The number of useful restriction endonucleases is increasing rapidly, and vectors will be developed accordingly to provide a more multipurpose service. The recent use of in vitro packaging procedures has reliably increased the efficiency with which recombinant molecules are recovered. It is particularly important that this high efficiency is achieved when multiply defective hosts are used to reduce the risk of disseminating novel recombinant molecules.

Phage vectors are at their best in those studies that benefit from genetic analyses of the incorporated DNA. This includes experiments based on genetic recombination, deletion analyses, and, in some instances, expression of the incorporated DNA. Phage systems provide opportunity for both the moderation and amplification of gene expression. Of course, the general applicability of these approaches remains to be proven.

Acknowledgements

I am grateful to those who made unpublished information available; to the Department of Biochemistry, the University of Adelaide, South Australia, for their hospitality during the preparation of this review; and to many colleagues in Edinburgh, students and staff, who have generously given their time to provide me with useful criticisms.

REFERENCES

1. **Hershey, A. D., Ed.,** *The Bacteriophage Lambda,* Cold Spring Harbor Laboratory, New York, 1971.
2. **Hedgpeth, J., Goodman, H. M., and Boyer, H. W.,** DNA nucleotide sequences restricted by the RI endonuclease, *Proc. Natl. Acad. Sci. U.S.A.,* 69, 3448, 1972.
3. **Bigger, C. H., Murray, K., and Murray, N. E.,** Recognition sequence of a restriction enzyme, *Nature New Biol.,* 244, 7, 1973.
4. **Old, R., Murray, K., and Roizes, G.,** Recognition sequence of restriction endonuclease III from *Haemophilus influenzae, J. Mol. Biol.,* 92, 331, 1975.
5. **Murray, N. E. and Murray, K.,** Manipulation of restriction targets in phage λ to form receptor chromosomes for DNA fragments, *Nature,* 251, 476, 1974.
6. **Thomas, M., Cameron, J. R., and Davis, R. W.,** Viable molecular hybrids of bacteriophage λ and eukaryotic DNA, *Proc. Natl. Acad. Sci. U.S.A.,* 71, 4579, 1974.
7. **Rambach, A. and Tiollais, P.,** Bacteriophage λ having *EcoRI* endonuclease sites only in the non-essential region of the genome, *Proc. Natl. Acad. Sci. U.S.A.,* 71, 3927, 1974.
8. **Weil, J., Cunningham, R., Martin, R. III, Mitchell, E., and Bolling, B.,** Characteristics of λp4, a λ derivative containing 9% excess DNA, *Virology,* 50, 373, 1973.
9. **Smith, H. O. and Nathans, D.,** Nomenclature for restriction enzymes, *J. Mol. Biol.,* 81, 419, 1973.
10. **Mertz, J. E. and Davis, R. W.,** Cleavage of DNA by RI restriction endonuclease generates cohesive ends, *Proc. Natl. Acad. Sci. U.S.A.,* 69, 3370, 1972.
11. **Arber, W. and Kühnlein, U.,** Mutationeller Verlust B-Spezifischer Restriction des Bakteriophagen fd., *Pathol. Microbiol.,* 30, 946, 1967.
12. **Murray, N. E., Manduca de Ritis, P., and Foster, L. A.,** DNA targets for the *Escherichia coli K* restriction system analysed genetically in recombinants between phages phi80 and lambda, *Mol. Gen. Genet.,* 120, 261, 1973.
13. **Fiandt, M., Hradecna, Z., Lozeron, H. A., and Szybalski, W.,** Electron micrographic mapping of deletions, insertions, inversions and homologies in the DNAs of coliphages lambda and phi80, in *The Bacteriophage Lambda,* Hershey, A. D., Ed., Cold Spring Harbor Laboratory, New York, 1971, 329.

14. **Murray, K. and Murray, N. E.,** Phage lambda receptor chromosomes for DNA fragments made with restriction endonuclease III of *Haemophilus influenzae* and restriction endonuclease I of *Escherichia coli, J. Mol. Biol.,* 98, 551, 1975.

15. **Zissler, J., Signer, E. R., and Schaefer, F.,** The role of recombination in growth of bacteriophage lambda, in *The Bacteriophage Lambda,* Hershey, A. D., Ed., Cold Spring Harbor Laboratory, New York, 1971, 455.

16. **Murray, N. E., Brammar, W. J., and Murray, K.,** Lambdoid phages that simplify the recovery of *in vitro* recombinants, *Mol. Gen. Genet.,* 150, 53, 1977.

17. **Beggs, J. D., Guerineau, M., and Atkins, J. F.,** A map of the restriction targets in yeast 2 micron plasmid DNA cloned on bacteriophage lambda, *Mol. Gen. Genet.,* 148, 287, 1976.

18. **Williams, B. G., Moore, D. D., Schumm, J. W., Grunwald, D. J., Blechl, A. E., and Blattner, F. R.,** Construction and Testing of Safer Phage Vectors for DNA cloning, in *Proceedings of the Tenth Miles International Symposium,* Raven Press, New York, in press.

19. **Blattner, F. R., Williams, B. G., Blechl, A. E., Denniston-Thompson, K., Faber, H. E., Furlong, L., Grunwald, D. J., Kiefer, D. O., Moore, D. D., Schumm, T. W., Sheldon, E. L., and Smithies, O.,** Charon phages: safer derivatives of bacteriophage lambda for DNA cloning, *Science,* 196, 161, 1977.

20. **Smith, J. D., Barnett, L., Brenner, S., and Russell, R. L.,** More mutant tyrosine transfer RNAs, *J. Mol. Biol.,* 54, 1, 1970.

21. **Philippsen, P. and Davis, R. W.,** personal communication, 1976.

22. **Hughes, S. G., Roberts, R. J., and Sanger, F.,** personal communication, 1976.

23. **McParland, R. H., Brown, L. R., and Pearson, G. D.,** Cleavage of lambda DNA by a site-specific endonuclease from *Serratia marcescens, J. Virol.,* 19, 1006, 1976; Kopecka, H., Yot, P., and Collins, J., personal communication, 1976.

24. **Smith, D. I., Blattner, F. R., and Davies, J.,** The isolation and partial characterisation of a new restriction endonuclease from *Providencia stuartii, Nucl. Acids Res.,* 3, 343, 1976.

25. **Boyer, H. W., Betlach, M., Bolivar, F., Rodriguez, R. L., Heyneker, H. L., Shine, T., and Goodman, H. W.,** The construction of molecular cloning vehicles in recombinant molecules; impact on science and society, in *Tenth Miles International Symposium 1977,* Beers, R. F., Jr. and Bassett, E. G., Eds., Raven Press, New York, 1977, 9.

26. **Murray, K.,** Restriction enzymes and their uses in molecular cloning, this volume, chapter 5.

27. **Haggerty, D. M. and Schleif, R. F.,** Location in bacteriophage lambda DNA of cleavage sites of the site-specific endonuclease from *Bacillus amyloliquefaciens H., J. Virol.,* 18, 659, 1976.

28. **Roberts, R. J.,** Restriction endonucleases, *Crit. Rev. in Biochem.,* 4, 123, 1976.

29. **Tiollais, P.,** personal communication, 1976.

30. **Brenner, S.,** personal communication, 1975.

31. **Sgaramella, V.,** Enzymatic oligomerisation of Bacteriophage P22 DNA and of Linear Simian Virus 40 DNA, *Proc. Natl. Acad. Sci. U.S.A.,* 69, 3389, 1972.

32. **Polisky, B., Greene, P., Garfin, D. E., McCarthy, B. J., Goodman, H. M., and Boyer, H. W.,** Specificity of substrate recognition by the *Eco*RI restriction endonuclease, *Proc. Natl. Acad. Sci. U.S.A.,* 72, 3310, 1975.

33. **Bellett, A. J. D., Busse, H. G., and Baldwin, R. L.,** Tandem Genetic Duplications in a derivative of phage lambda, in *The Bacteriophage Lambda,* Hershey, A. D., Ed., Cold Spring Harbor Laboratory, New York, 1971, 501.

34. **Matsubara, K. and Kaiser, A. D.,** λdv an autonomously replicating DNA fragment, *Cold Spring Harbor Symp. Quant. Biol.,* 33, 769, 1968.

35. **Berg, D.,** Genes of phage λ essential for λdv plasmids, *Virology,* 62, 224, 1974.

36. **Mukai, T., Matsubara, K., and Takagi, Y.,** Cloning bacterial genes with plasmid λdv, *Mol. Gen. Genet.,* 146, 269, 1976.

37. **Streeck, R. E. and Hobom, G.,** Mapping of cleavage sites for restriction endonucleases in λdv plasmids, *Eur. J. Biochem.,* 57, 595, 1975, Philippsen, P., personal communication, 1976.

38. **Hashimoto, T. and Matsubara, K.,** Transformation of Ca^{++} treated *recA* derivatives of *Escherichia coli* K12 by λdv DNA, *Jpn. J. Genet.,* 49, 97, 1976.

39. **Wensink, P. C., Finnegan, D. J., Donelson, J. E., and Hogness, D. S.,** A system for mapping DNA sequences in the chromosomes of *Drosophila melanogaster Cell,* 3, 315, 1974.

40. **Lobban, P. E. and Kaiser, A. D.,** Enzymatic end-to-end joining of DNA molecules, *J. Mol. Biol.,* 78, 453, 1973.

41. **Jackson, D. A., Symons, R. H., and Berg, P.,** Biochemical method for inserting new genetic information into DNA of simian virus 40, *Proc. Natl. Acad. Sci. U.S.A.,* 69, 3904, 1972.

42. **Rougeon, F., Kourilsky, P., and Mach, B.,** Insertion of a rabbit β-globin gene sequence into an *E. coli* plasmid, *Nucl. Acids Res.,* 2, 2365, 1975.

43. **Rabbitts, T. H.,** Bacterial cloning of plasmids carrying copies of rabbit globin messenger RNA, *Nature,* 260, 221, 1976.

44. **Maniatis, T., Kee, S. G., Efstratiadis, A., and Kafatos, F. C.,** Amplification and characterisation of a β-globin gene synthesised *in vitro, Cell,* 8, 163, 1976.

45. **Blattner, F. R. and Smithies, O.,** personal communication, 1977.

46. **Mandel, M. and Higa, A.,** Calcium-dependent bacteriophage DNA infection, *J. Mol. Biol.,* 53, 159, 1970.

47. **Lederberg, E. M. and Cohen, S. N.,** Transformation of *Salmonella typhimurium* by plasmid deoxyribonucleic acid., *J. Bacteriol.,* 119, 1072, 1974.

48. **Benzinger, R., Enquist, L. W., and Skalka, A.,** Transfection of *Escherichia coli* spheroplasts. V. Activity of *recBC* nuclease in *red⁺* and *red⁻* spheroplasts measured with different forms of bacteriophage DNA, *J. Virol.,* 15, 861, 1975.

49. **Kaiser. A. D. and Hogness, D. S.,** The transformation of *Escherichia coli* with deoxyribonucleic acid isolated from bacteriophage λdg, *J. Mol. Biol.,* 2, 392, 1960.

50. **Egan, J. B.,** personal communication, 1976.

51. **Ward, D.,** personal communication, 1976.

52. **Pilarski, L. M. and Egan, J. B.,** Role of DNA topology in transcription of coliphage λ *in vivo.* II. DNA topology protects the template from exonuclease attack, *J. Mol. Biol.,* 76, 257, 1973.

53. **Hohn, B. and Hohn, T.,** Activity of empty, headlike particles for packaging of DNA of bacteriophage λ *in vitro, Proc. Natl. Acad. Sci. U.S.A.,* 71, 2372, 1974.

54. **Becker, A. and Gold, M.,** Isolation of the bacteriophage lambda A-gene protein, *Proc. Natl. Acad. Sci. U.S.A.,* 72, 581, 1975.

55. **Hohn, B. and Murray, K.,** Packaging recombinant DNA molecules into bacteriophage particles *in vitro, Proc. Natl. Acad. Sci. U.S.A.,* 74, 3259, 1977.

56. **Sternberg, N., Tiemeier, D., and Enquist, L.,** *In vitro* packaging of a λDam vector containing *EcoRI* DNA fragments of *E. coli* and phage Pl, *Gene,* 1, 255, 1977.

57. **Morse, M., Lederberg, E., and Lederberg, J.,** Transduction in *Escherichia coli* K-12, *Genetics,* 41, 121, 1956.

58. **Hershfield, V., Boyer, H. W., Yanofsky, C., Lovett, M. A., and Helinski, D. R.,** Plasmid *colE1* as a molecular vehicle for cloning and amplification of DNA, *Proc. Natl. Acad. Sci. U.S.A.,* 71, 3455, 1974.

59. **Cameron, J. R., Panasenko, S. M., Lehman, I. R., and Davis, R. W.,** *In vitro* construction of bacteriophage λ carrying segments of the *Escherichia coli* chromosome; selection of hybrids containing the gene for DNA ligase, *Proc. Natl. Acad. Sci. U.S.A.,* 72, 3416, 1975.

60. **Clark, L. and Carbon, J.,** Biochemical construction and selection of hybrid plasmids containing specific segments of the *Escherichia coli* genome, *Proc. Natl. Acad. Sci. U.S.A.,* 72, 4361, 1975.

61. **Borck, K., Beggs, J. D., Brammar, W. J., Hopkins, A. S., and Murray, N. E.,** The construction *in vitro* of transducing derivatives of phage lambda, *Mol. Gen. Genet.,* 146, 199, 1976.

62. **Silverman, M., Matsumara, P., Draper, R., Edwards, S., and Simon, M.,** Expression of flagellar genes carried by bacteriophage lambda, *Nature (London),* 261, 248, 1976.

63. **Struhl, K., Cameron, J. R., and Davis, R. W.,** Functional genetic expression of eukaryotic DNA in *E. coli, Proc. Natl. Acad. Sci. U.S.A.,* 73, 1471, 1976.

64. **Kayajanian, G.,** Studies on the genetics of biotin-transducing, defective variants of bacteriophage λ, *Virology,* 36, 30, 1968.

65. **Franklin, N. C.,** The *N* operon of lambda; extent and regulation as observed in fusions to the tryptophan operon of *E. coli,* in *The Bacteriophage Lambda,* Hershey, A. D., Ed., Cold Spring Harbor Laboratory, New York, 1971, 621.

66. **Brammar, W. J. and Muir, S.,** personal communication, 1976; Windass, J. D., personal communication, 1976.

67. **Struhl, K. and Davis, R. W.,** Genetic selections and the cloning of prokaryotic and eukaryotic genes, *Molecular Mechanisms in the Control of Gene expression,* Academic Press, New York, 1976, 495.

68. **Sanzey, B., Mercereau, O., Ternynzk, T., and Kourilsky, P.,** Methods for identification of recombinants of phage λ, *Proc. Nat. Acad. Sci. U.S.A.,* 73, 3394, 1976.

69. **Skalka, A. and Shapiro, L.,** *In situ* immunoassays for gene translation products in phage plaques and bacterial colonies, *Gene,* 1, 65, 1977.

70. **John, H., Birnstiel, M. L., and Jones, K. W.,** RNA-DNA hybrids at the cytological level, *Nature,* 223, 582, 1969.

71. **Gall, J. G. and Pardue, M. L.,** Formation and detection of RNA-DNA hybrid molecules in cytological preparation, *Proc. Natl. Acad. Sci. U.S.A.,* 69, 378, 1969.

72. **Jones, K. W. and Murray, K.,** A procedure for detection of heterologous DNA sequences in lambdoid phage by *in situ* hybridisation, *J. Mol. Biol.,* 96, 455, 1975.

73. **Grunstein, M. and Hogness, D.,** Colony hybridisation: a method for the isolation of cloned DNAs that contain a specific gene, *Proc. Natl. Acad. Sci. U.S.A.,* 72, 3961, 1975.

74. **Kramer, R. A., Cameron, J. R., and Davis, R. W.,** Isolation of bacteriophage λ containing yeast ribosomal RNA genes: screening by *in situ* RNA hybridisation to plaques, *Cell,* 8, 227, 1976.

74a. **Benton, W. D. and Davis, R. W.,** Screening λgt recombinant clones by hybridization to single plaques *in situ, Science,* 196, 180, 1977.

75. **Yamamato, K. R., Alberts, B. M., Benzinger, R., Lawholne, L., and Treiber, G.,** Rapid sedimentation in the presence of polyethylene glycol and its application to large scale virion purification, *Virology,* 40, 734, 1970.

75a. **Goldberg, A. R. and Howe, M.,** New mutations in the *S* cistron of bacteriophage λ affecting host cell lysis, *Virology,* 38, 200, 1969.

76. **Hayward, G. S. and Smith, M. G.,** The chromosome of bacteriophage T5. I. Analysis of single-stranded DNA fragments by agarose gel electrophoresis, *J. Mol. Biol.,* 63, 383, 1972.

77. **Davis, R. S. and Davidson, N.,** Electron-microscopic visualisation of deletion mutations, *Proc. Natl. Acad. Sci. U.S.A.,* 60, 243, 1968.

78. **Westmoreland, B. C., Szybalski, W., and Ris, H.,** Mapping of deletions and substitutions in heteroduplex DNA molecules of bacteriophage lambda by electron microscopy, *Science,* 163, 1343, 1969.

79. **Davis, R. W., Simon, M., and Davidson, N.,** Electron microscope heteroduplex methods for mapping regions of base sequence homology in nucleic acids, in *Methods in Enzymology, Part D,* Grossman, L. and Moldave, K., Ed., 21, Academic Press, New York, 1971, 413.

80. **Schnös, M. and Inman, R.,** Position of branch points in replicating λ DNA, *J. Mol. Biol.,* 51, 61, 1970.

81. **Portmann, R., Schaffner, W., and Birnstiel, M.,** Molecular analysis of the histone gene cluster of *Psammechinus miliaris.* IV. Partial denaturation mapping of cloned histone DNA, *Nature,* 264, 31, 1976.

82. **Hradecna, Z. and Szybalski, W.,** Fractionation of the complementary strands of coliphage λ DNA based on the asymmetric distribution at the poly 1, G-binding sites, *Virology,* 32, 633, 1967.

83. **Mounts, P., Southern, E., and Murray, K.,** personal communication, 1975.

84. **Szybalski, W.,** Controls of transcription and replication in coliphage λ, in *Karl-August-Forster Lectures,* Academie der Wissenshaften und der Literatur, Mainz, W. Germany, 1971, 11.

85. **Herskowitz, I.,** Control of gene expression in bacteriophage lambda, *Annu. Rev. Genet.,* 7, 289, 1973.

86. **Roberts, J. W.** The rho factor: termination and antitermination in lambda, *Cold Spring Harbor Symp. Quant. Biol.,* 35, 121, 1971.

87. **Adhya, S., Gottesman, M., and de Crombrugghe, B.,** Release of polarity in *Escherichia coli* by gene *N* of phage λ: termination and antitermination of transcription, *Proc. Natl. Acad. Sci. U.S.A.,* 71, 2534, 1974.

88. **Franklin, N. C.,** Altered reading of genetic signals fused to the *N* operon of bacteriophage λ: genetic evidence for modification of polymerase by the protein product of the *N* gene, *J. Mol. Biol.,* 89, 33, 1974.

89. **Segawa, T. and Imamoto, F.,** Diversity of regulation of genetic transcription. II. Specific relaxation of polarity in read-through transcription of the translocated *trp* operon in bacteriophage lambda *trp, J. Mol. Biol.,* 87, 741, 1974.

90. **Pero, J.,** Deletion mapping of the site of action of the *tof* gene product, in *The Bacteriophage Lambda* Hershey, A. D., Ed., Cold Spring Harbor Laboratory, New York, 1971, 599.

91. **Radding, C. M. and Shreffler, D. C.,** Regulation of λ exonuclease. II. Joint regulation of exonuclease and a new λ antigen, *J. Mol. Biol.,* 18, 251, 1966.

92. **Echols, H., Green, L., Oppenheim, A. B., Oppenheim, A., and Honigman, A.,** The role of *cro* gene in bacteriophage λ development, *J. Mol. Biol.,* 80, 203, 1973.

93. **Herskowitz, I. and Signer, E. R.,** A site essential for expression of all late genes in bacteriophage λ, *J. Mol. Biol.,* 47, 545, 1970.

94. **Thomas, R.,** Control of development in temperate bacteriophages. III. Which prophage genes are, and which are not *trans*activable in the presence of immunity?, *J. Mol. Biol.,* 49, 393, 1970.

95. **Court, D. and Sato, K.,** Studies of novel transducing variants of lambda dispensability of genes N and Q, *Virology,* 39, 348, 1969.

96. **Hopkins, A. S., Murray, N. E., and Brammar, W. J.,** Characterisation of λtrp-transducing bacteriophage made *in vitro, J. Mol. Biol.,* 107, 549, 1976.

97. **Parkinson, J. S. and Huskey, R. J.,** Deletion mutants of bacteriophage lambda. I. Isolation and initial characterisation, *J. Mol. Biol.,* 56, 369, 1971.

98. **Emmons, S. W., MacCosham, V., and Baldwin, R. L.,** Tandem genetic duplications in phage lambda. III. The frequency of duplication mutants in two derivatives of phage lambda is independent of known recombination systems, *J. Mol. Biol.,* 91, 133, 1975.

99. **Watson, M. D. and Scaife, J. G.,** Chromosomal transfer by R factor RP4 in *E. coli* K12, *Proc. Soc. Gen. Microbiol.,* 3, 184, 1976; Pastrana, R. and Brammar, W. J. personal communication, 1976.

100. **Sato, K. and Matsushiro, A.,** The tryptophan operon regulated by phage immunity, *J. Mol. Biol.,* 14, 608, 1965.

101. **Schleif, R., Greenblatt, J., and Davis, R. W.,** Dual control of arabinose genes on transducing phage λdara, *J. Mol. Biol.,* 59, 127, 1971.

102. **Muller-Hill, B., Crapo, L., and Gilbert, W.,** Mutants that make more *lac* repressor, *Proc. Natl. Acad. Sci. U.S.A.,* 59, 1259, 1968.

103. **Adhya, S., Sen. A., and Mitra, S.,** The role of gene *S,* in *The Bacteriophage Lambda,* Hershey, A. D., Ed., Cold Spring Harbor Laboratory, New York, 1971, 743.

104. **Moir, A. and Brammar, W. J.,** The use of specialised transducing phages in the amplification of enzyme production, *Mol. Gen. Genet.,* 149, 87, 1976.

105. **Willard, M. and Echols, H.,** Role of bacteriophage DNA replication in λ*dg* escape synthesis, *J. Mol. Biol.,* 32, 37, 1968.

106. **Davison, J., Brammar, W. J., and Brunel, F.,** Quantitative aspects of gene expression in a λ-*trp* fusion operon, *Mol. Gen. Genet.,* 130, 9, 1974.

107. **Franklin, N. C. and Yanofsky, C.,** The *N* protein of λ: evidence bearing on transcription termination, polarity and the alteration of *E. coli* RNA polymerase, in *RNA Polymerase,* Cold Spring Harbor Laboratory, New York, 1971, 693.

108. **Matsubara, K.,** in Curtiss, III, R., Szybalski, W., Helinski, D. R., and Falkow, S., Workshop on design and testing of safer prokaryotic vehicles and bacterial hosts for research on recombinant DNA molecules, *Am. Soc. Microbiol. News.,* 42, 134, 1976.

109. **Brammar, W. J. and Murray, N. E.,** unpublished, 1976.

110. **Pero, J.,** Location of the phage λ gene responsible for turning off λ-exonuclease synthesis, *Virology,* 40, 65, 1970.

111. **Wilgus, G. S., Mural, R. J., Friedman, D. I., Fiandt, M., and Szybalski, W.,** λ*imm*$^{\lambda434}$ A phage with a hybrid immunity region *Virology,* 56, 46, 1973.

112. **Helinski, D. R., Hershfield, V., Figurski, D., and Meyer, R. J.,** Construction and properties of plasmid cloning vehicles, in *Recombinant Molecules; Impact on Science and Society, the Tenth Miles International Symposium 1977,* Beers, R. F. Jr. and Bassett, E. G., Eds., Raven Press, New York, 1977, 151.

113. **Mercereau-Puijalon, O. and Kourilsky, P.,** Escape synthesis of β-galactosidase under the control of bacteriophage lambda, *J. Mol. Biol.,* 108, 733, 1976.

114. **Backman, K., Ptashne, M., and Gilbert, W.,** Construction of plasmids carrying the c*I* gene of bacteriophage λ, *Proc. Natl. Acad. Sci. U.S.A.,* 73, 4174, 1976.

115. **Polisky, B., Bishop, R. J., and Gelfand, D. H.,** A plasmid cloning vehicle allowing regulated expression of eukaryotic DNA in bacteria, *Proc. Natl. Acad. Sci. U.S.A.,* 73, 3900, 1976.

116. **Adler, H. I., Fisher, W. D., Cohen, A., and Hardigree, A. A.,** Miniature *Escherichia coli* cells deficient in DNA, *Proc. Natl. Acad. Sci. U.S.A.,* 57, 321, 1966.

117. **Chang, A. C. Y., Lansman, R. A., Clayton, D. A. and Cohen, S. N.,** Studies of mouse mitochondrial DNA in *Escherichia coli*: structure and function of the eukaryotic-prokaryotic chimeric plasmids, *Cell,* 6, 231, 1975.

118. **Ptashne, M.,** Isolation of the λ phage repressor, *Proc. Natl. Acad. Sci. U.S.A.* 57, 306, 1967.

119. **Hendrix, R. W.,** Identification of proteins coded in phage lambda, in *The Bacteriophage Lambda,* Hershey, A. D., Ed., Cold Spring Harbor Laboratory, New York, 1971, 355.

120. **Jaskunas, S. R., Lindahl, L., Nomura, M., and Burgess, R. R.,** Identification of two copies of the gene for the elongation factor EF-Tu in *E. coli, Nature,* 257, 458, 1975.

121. **Szybalski, W.,** Safety of coliphage lambda vectors carrying foreign genes, in *Recombinant Molecules; Impact on Science and Society, Proceedings of the 10th Miles International Symposium,* Beers, R. F. Jr. and Bassett, E. G., Eds., Raven Press, New York, 1977, 137.

122. **Enquist, L., Tiemeier, D., Leder, P., Weisberg, R., and Sternberg, N.,** Safer derivatives of bacteriophage λgt.λC for use in cloning of recombinant DNA molecules, *Nature,* 259, 596, 1976.

123. **Tiemeier, D., Enquist, L., and Leder, P.,** Improved derivatives of phage λ EK2 vector for cloning recombinant DNA, *Nature,* 263, 526, 1976.

Chapter 3

CLONING cDNA SEQUENCES: A GENERAL TECHNIQUE FOR PROPAGATING EUKARYOTIC GENE SEQUENCES IN BACTERIAL CELLS

W. Salser

TABLE OF CONTENTS

I. INTRODUCTION

In 1972, three laboratories[1-3] independently demonstrated that the reverse transcriptase from avian myeloblastosis virus (AMV) could be used to synthesize a complementary DNA (cDNA) copy of globin mRNAs. Since that time, this reaction has been utilized to synthesize more or less complete cDNA copies of a great variety of polyadenylated mRNAs. My laboratory originally became interested in such cDNAs since they provided convenient templates for the in vitro synthesis of highly labeled products used for a variety of nucleotide sequencing approaches. The in vitro products which proved useful in our initial nucleotide sequence analyses include RNA[4,5] as well as unconventional products such as ribosubstituted DNA[6] and deoxysubstituted RNA.[7,8] These methods enabled progress in the nucleotide sequence analysis of selected mammalian mRNAs to proceed several times more rapidly than had previously been possible. They are currently eclipsed by still more rapid sequencing techniques[9,10] which can be applied once the mRNA sequence is cloned as part of a bacterial plasmid. This chapter will discuss the techniques that were developed to permit the construction of plasmids containing gene copies made from cDNA, with primary emphasis on the results obtained by myself and my collaborators.

Our procedure for cloning gene copies made from cDNAs resulted from a chance observation made in my laboratory during the investigation of a new approach to DNA sequence analysis. This approach required the digestion of ribosubstantiated DNA[6] by bacteriophage T4 endonuclease IV, which will cleave single-stranded DNA at specific locations. This investigation was made much more difficult by the discovery that the ribosubstituted DNA synthesized from a cDNA template was invariably a rapidly renaturing hairpin structure. The experiments suggested that the cDNA molecules made with reverse transcriptase contained a short, double-stranded hairpin region or "hook" which provided a very efficient primer for second-strand synthesis by DNA polymerase I.[11] Since the existence of a double-stranded hook had not been reported in the literature, a variety of control experiments were carried out to further characterize this phenomenon. These experiments showed that the self-priming observed was fundamentally different from that observed when other single-stranded DNAs folded back upon themselves in a random fashion, even though either can prime synthesis by DNA polymerase I. It was concluded[11] that reverse transcriptase must leave a short double-stranded hook at the point where it terminates synthesis of the cDNA molecule and that this is so even when actinomycin D is used in attempts to block second-strand synthesis. This and similar later observations[12,13] prompted several laboratories to use these self-priming characteristics to convert globin cDNAs to duplex gene copies which could be inserted into bacterial plasmids.[13-19]

This approach has already provided advantages for several research areas. As mentioned above, it permits the use of extremely rapid DNA sequencing techniques for the sequence analysis of mammalian mRNAs. Also, since each clone carries only a single mRNA sequence, this approach is more powerful than standard biochemical techniques in resolving complex mixtures of different mRNA sequences. Finally, the cloning of cDNA sequences provides pure probes for the isolation and cloning of the larger DNA sequences surrounding the corresponding structural genes in nuclear DNA.

The basic approach which we have used[14-16] is outlined in Figure 1, using the cDNA cloning of the rabbit globin genes as an example. Globin cDNA is prepared from globin mRNA using AMV reverse transcriptase. After purification, this cDNA is used as the template for second-strand synthesis by *Escherichia coli* DNA polymerase I. Since AMV reverse transcriptase leaves a hook on the 3′-OH end of its product, this template is self-priming. A potentially full-size duplex gene copy results; however, it is a hairpin structure with only one open end. Treatment with S_1 nuclease is used to open the hairpin loop so that the gene copy can be inserted into bacterial plasmids. The gene copy is subsequently joined to a plasmid vector which has been converted to linear form by EcoRl digestion. The joining of the S_1-treated globin gene copy to the plasmid DNA is accomplished by a modification of the poly dA-poly dT tailing method first used by Jackson et al.[20] and Lobban and Kaiser[21] and refined by Wensink et al.[22] In this approach, polynucleotide terminal transferase is used to add poly dA tails to the S_1 nuclease-treated globin gene copy and poly dT tails to the linearized plasmid vector. The poly dA-tailed globin gene copies are subsequently mixed with the poly dT-tailed plasmids and annealed to give circular complexes which can be repaired and replicated in *E. coli.* These complexes are introduced into the bacteria using a modification of the transformation procedure of Mandel and Higa.[23]

The poly dT-poly dA tailing approach has hitherto been used mainly for the construction of plasmids from naturally occurring duplex DNA. The efficiencies of transformation observed are frequently two orders of magnitude less than that obtained for intact plasmid-twisted circles, an effect which is possibly due to difficulties in sealing the gaps in the resulting plasmid-insert complexes. Cloning cDNA sequences involves an additional series of steps which we feared might cause substantial further reductions in the cloning efficiency. The data in Table 1 show that this is not a serious problem. In two different experiments,[15] the complexes containing globin gene copies showed transformation efficiencies 0.6 and 1.8% as great as covalently closed plasmid DNA, similar to results obtained when DNA fragments from other sources are inserted by this method. It should be noted that the overall efficiency of transformation is strongly influenced by the bacterial host used, the rec A⁻ host *E. coli* HB101 consistently showing transformation efficiencies 10- to 40-fold lower than *E. coli* C600. This difference between the two strains is noted with either intact plasmids (covalently closed twisted circles) or complexes containing gene copies made from cDNA.

II. DETAILS OF THE BASIC PROCEDURE

A. Enzymes

Purified RNA-dependent DNA polymerase (reverse transcriptase) from AMV[26] was provided by the Office of Program Resources and Logistics.* This enzyme has been purified through a variety of steps, including chromatography on DEAE cellulose and phospho-cellulose and sedimentation on glycerol gradients, which insure its freedom from contaminating viral RNA. Reverse transcriptase is also commercially available

* Viral Cancer Program, Viral Oncology, Division of Cancer Cause and Prevention, National Cancer Institute, Bethesda, Maryland, 20014.

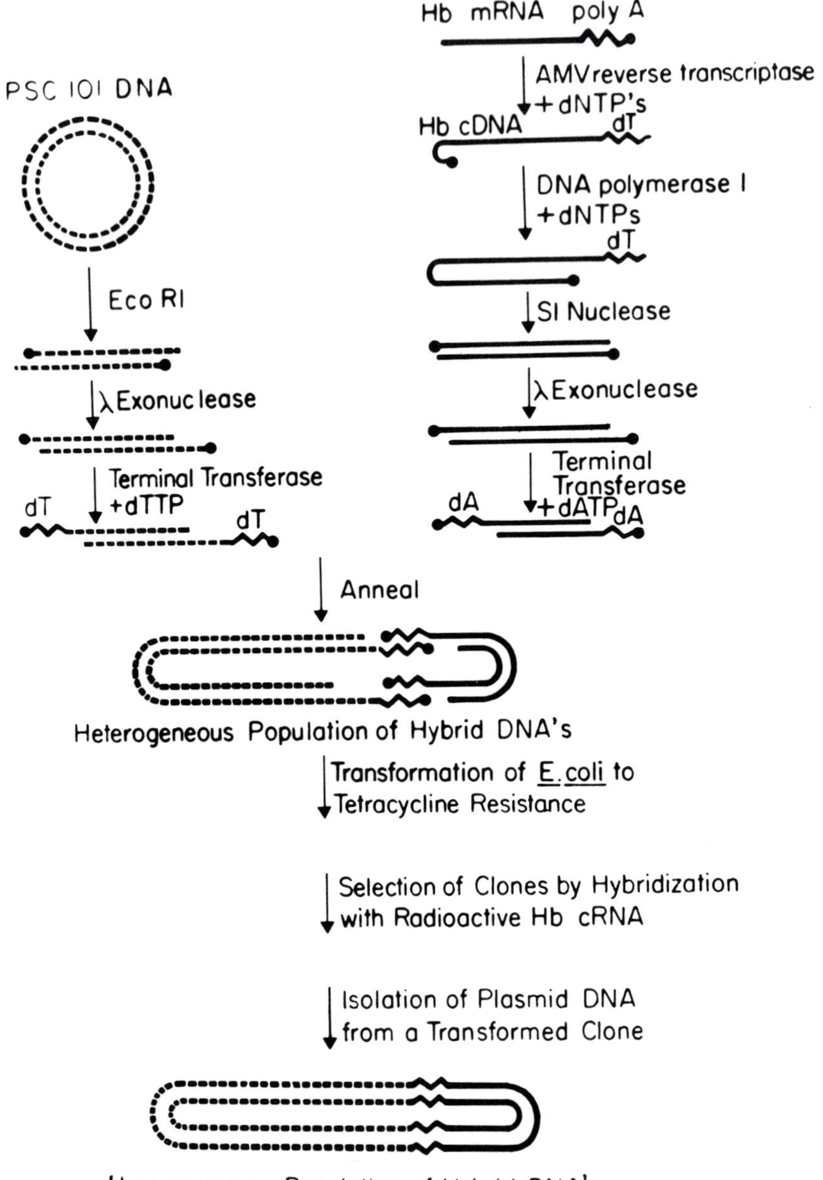

FIGURE 1.   cDNA Cloning approach utilizing the fold-back "hook" to prime second-strand synthesis.

from Boehringer-Mannheim in Germany. We have had no experience with the commercial preparation. It is absolutely essential that only enzyme preparations which have been rigorously purified and are free from viral RNA be utilized for these procedures.

DNA polymerase I, fraction 7, was prepared by K. Fry according to the procedure of Richardson et al.[27] In our experiments, the *Aspergillis* S1 nuclease used was purified by the methods of Vogt.[28] Other laboratories have successfully used various commercial preparations of this enzyme. Bacteriophage λ exonuclease preparations[29] used in some early experiments were given to us by R. Firtel and by T. Maniatis. However,

TABLE 1

Transformation of the *rec* A⁻ Host *Escherichia coli* HB101 by Recombinant Globin DNA/pSC101 Hybrids

DNA used in transformation	No. of tetracycline resistant colonies (per picamole of DNA complexes[a])
pSC101 intact circles	46,000
poly dT-tailed pSC101	0
poly dT-tailed pSC101 annealed with poly dA-tailed double-stranded globin DNA	270

Transformation of *E. coli* C600 by Recombinant Globin DNA/pMB9 Hybrids

DNA used in transformation	No. of tetracycline resistant colonies (per picamole of DNA complexes)
pMB9 intact circles	610,000
pMB9 EcoR1 cleaved	2,000
pMB9 poly dT-tailed	1,300
pMB9 poly dT-tailed annealed with poly dA-tailed double-stranded globin DNA	11,000

Note: These experiments were carried out in P3 physical containment under Asilomar guidelines[24] before bacterial hosts suitable for EK2 biological containment had been created. More recent experiments with mammalian cDNA gene copies have been carried out in P3 physical containment using vector-host combinations approved by the Director of National Institutes of Health, Bethesda, Maryland as meeting EK2 biological containment requirements.[25]

[a] Assumed pSC101 = 9000 nucleotides.

Adapted from Higuchi, R., Paddock, G. V., Wall, R., and Salser, W., *Proc. Natl. Acad. Sci. U.S.A.*, 73, 3146, 1976.

the λ exonuclease was discontinued when we discovered[16] that terminal transferase does not require protruding 3′-OH ends when used under the original assay conditions of Chang and Bollum,[31] which involve a cobalt buffer.

Using a modification of the procedure of Chang and Bollum,[31] terminal transferase was prepared in the laboratory of R. L. Ratliff by J. Isaacson, C. Manske, and W. Salser. Personal communication from several other laboratories suggest that some of the commercial preparations of this enzyme may contain quantities of endonuclease which make them unsuitable for recombinant DNA research (e.g., see Reference 32). This is perhaps not surprising since the most frequent research use of this enzyme prior to the advent of recombinant DNA studies was to prepare artificial copolymers of random sequence. An enzyme preparation free from nicking activity is not essential for such purposes. After preparing a large supply of the enzyme, we became aware that three other groups with considerable experience in enzyme isolations had failed to obtain acceptable preparations of terminal transferase. In some cases, the problem was a loss of enzyme activity, while in others, it was a failure to remove endonuclease activity. Therefore, we have made our enzyme preparation widely available to other investigators, albeit on a basis which tries to avoid intolerable drains on our own re-sources. As far as I am aware, all of the successful poly dA-poly dT tailing experiments carried out in the U.S., except for that of Wood and Lee,[53] have used either the original Ratliff preparation, which is now exhausted, or the enzyme subsequently prepared by Isaacson, Manske, Salser, and Ratliff.

The endonuclease activity of the Ratliff preparation has been compared with that which we prepared subsequently. Both preparations showed the save low level of nicking (less than one nick per 10 kilobases (kb)/2 hr of incubation) when assayed by following the nicking of twisted circles of the plasmid pCRI under the conditions specified by Lobban and Kaiser[21] for tailing by terminal transferase.

B. Bacterial Strains and Plasmids

E. coli strains HB101 and C600 are standard strains described elsewhere.[33,34] The biocontainment strains Xl849 and X1776 were constructed by Roy Curtiss III and colleagues and are designed to self-destruct, along with any plasmid they may be carrying, should they escape to a natural environment.[35] Strain Xl776 was certified by the Director of the National Institutes of Health (NIH) in December 1976 for use with the plasmids pSC101[36] and pCRI[37] as EK2 vector host systems (see Reference 25 for a definition of EK2). Plasmids pSC101, pMB9, and pCRI have been described elsewhere.[37,38] Plasmid pMB9 and two new plasmids, pBR313[39] and pBR322[39a] were certified in 1977 for use with *E. coli* X1776 as EK2 systems.

C. Purification of mRNA

Reticulocytes from phenylhydrazine-treated rabbits[40] were washed and broken by cavitation in a nitrogen disruption bomb.[41] Polysomes were magnesium precipitated,[42] and the 10S globin mRNA was purified by repeated phenol/$CHCl_3$ extractions, poly U-sepharose selection of poly A-containing mRNA, and two sodium dodecyl sulfate/sucrose gradient velocity sedimentations.[15]

D. Synthesis of cDNA

When cDNA of maximal length was desired, the reaction was carried out with 1000 μM concentrations of each deoxyribonucleoside triphosphate.[15] In our original cloning experiments, reactions were carried out for 2 hr at 37°C in reaction volumes of 0.05 to 1.00 mℓ. The reaction mixtures contained the following components: dT_{12-18} primer at a final concentration of 20 μg/mℓ; 60 μg/mℓ globin mRNA; 120 units reverse transcriptase per milliliter; 50 mM Tris-HCl, pH 8.3; 10 mM $MgCl_2$; 20 mM 2-mercaptoethanol; and 1 mM concentrations of each of the four deoxyribonucleoside triphosphates. Usually 250 μCi of ^{3}H label per milliliter is added in one of the dNTPs so that the cDNA can be followed in preparation for the next step. This low specific activity permits the use of a ^{3}H incorporation (at 20-fold higher specific activity) to monitor second-strand synthesis by DNA polymerase I and ^{32}P incorporation to monitor the addition of homopolymer tails. With normal efficiencies of synthesis, the cDNA yield as measured by acid precipitable radioactivity is roughly 50% of theoretical and corresponds to the incorporation of approximately 2% of the radioactivity into cDNA. We have cloned cDNAs made either in the presence or absence of actinomycin D. The effect of added actinomycin D (100 μg/mℓ) was tested by fingerprint analysis of RNA transcribed in vitro from cDNA templates synthesized in the presence or absence of 100 μg/mℓ actinomycin. The RNA fingerprint patterns obtained were the same in either case;[15] however, the yields of cDNA were about twofold higher when actinomycin D was omitted. It is suspected, however, that actinomycin D may sometimes encourage a greater fractional yield of full-length cDNA product by inhibiting termination caused by premature synthesis of the double-stranded hook.

Accordingly, my laboratory has recently started using what we term "high temperature-low actinomycin incubation conditions," which give the same high yield obtained when actinomycin is omitted.[87] This reaction mixture contains 50 mM Tris-HCl, pH 8.3; 20 mM 2-mercaptoethanol; 10 mM $MgCl_2$; and 1 mM concentration of each of the four deoxynucleotide triphosphates. In addition, it contains (per milliliter of final volume) 30 μg of actinomycin D and 120 units of reverse transcriptase. The enzyme is added to the reaction mixture at 4°C, mixed gently, and immediately incubated at 46°C for 15 min. We believe that it may be important to maintain the relatively high concentrations of mRNA and reverse transcripterase specified above.

After completion of cDNA synthesis, the reaction mixtures are treated with 0.3 M NaOH at 90°C for 20 min, neutralized with HCl, phenol extracted, and passed through a Sephadex® G-100 column (1 × 50 cm for the 1.00 mℓ preparations) to remove salt and unincorporated dNTPs. Column fractions are monitored for radioactivity to locate the cDNA peak. The pooled peak fractions are subsequently ethanol precipitated. The ethanol precipitation procedure, which is repeated several times in the following steps, involves the addition of 0.1 volume of $2M$ NH$_4$Ac and carrier $E.$ $coli$ RNA to give a final nucleic acid concentration of 40 μg/mℓ unless otherwise specified. Two volumes of ethanol are subsequently added, and the mixture is chilled for 10 min in a dry-ice ethanol bath and centrifuged at top speed for 10 min in the cold in an Eppendorf® microfuge for 30 min at 7000 r/min in a Corex® tube in a Sorval® SS 34 motor to pellet the nucleic acid.

E. Synthesis of Double-stranded Globin DNA

The reaction is carried out in a 120 $\mu\ell$ volume for 15 min at 37°C in 66 mM KH$_2$PO$_4$, pH 7, 6.6 mM MgCl$_2$, 1 mM 2-mercaptoethanol, 1 μg of globin cDNA, and 5 μg of DNA polymerase I (7.6 units). Deoxyribonucleoside triphosphates were present, 4 nmol each, with 20 μCi of ^{3}H label in one of the dNTPs so that synthesis could be monitored by measuring the acid-precipitable radioactivity and subtracting the low level of acid-precipitable counts introduced in the preceding step. Approximately 10% of the total radioactivity should be incorporated, corresponding to a double-stranded DNA yield 50% of the theoretical value.* Ten $\mu\ell$ of 1 M EDTA is added to stop the reaction. The reaction mixture is then phenol extracted and passed over a 0.6 × 15 cm Bio-Gel® P60 column, and the excluded peak is ethanol precipitated.

F. Nuclease S1 Digestion

The conditions of Shenk et al.,[44] which are designed to minimize the nicking of duplex DNA, have been used in this procedure. In a typical reaction, 1 μg of duplex DNA is digested with 135 units of S$_1$ nuclease for 1 hr at room temperature in 200 to 400 $\mu\ell$ of a reaction mixture containing 0.28 M NaCl, 0.0045 M ZnSO$_4$, 0.03 M NaOAc at pH 4.6.** The reaction is terminated by adding 0.1 volume of 1 M Tris Base. Carrier $E.$ $coli$ RNA is added to a concentration of 40 μg/mℓ to replace that digested by the nuclease, the mixture is extracted with phenol, and the aqueous phase is ethanol precipitated.

While we have always computed the amount of S$_1$ nuclease on the basis of the micrograms of double-stranded DNA digested, it may also be important to keep the ratio of nuclease S$_1$ to total nucleic acid relatively constant. The greater portion of the total nucleic acid present at the time of the S$_1$ digestion is the $E.$ $coli$ carrier RNA, which comes from additions prior to the ethanol precipitation steps that follow cDNA synthesis and second-strand synthesis. In a number of successful experiments in which the amount of S$_1$ nuclease was held at 135 units/μg of double-stranded DNA, the amount of total nucleic acid per microgram of double-stranded DNA ranged from 18 to 43 μg.

---

* Our yields have typically ranged from 40 to 80% of theoretical. Yields lower than theoretically expected may indicate that DNA polymerase I occasionally terminates second-strand synthesis prematurely (as is also suggested by other data discussed above) or that some of the cDNA molecules are not self-priming.

** Note that a typographical error in Reference 15 caused the Shenk et al. buffer to be incorrectly quoted as containing MgSO$_4$ rather than ZnSO$_4$.

It is important to achieve the proper amount of nuclease S_1 digestion. Failure to cleave all of the hairpins may result in a decreased number of transformants containing the desired gene copy, while overdigestion may nick the gene copy or cause a progressive attack on the ends of the duplex molecules. Since the overall cloning efficiency is high, we prefer to use conditions which tend toward underdigestion, which perhaps sacrifice some efficiency but increase the chances of cloning full-size gene copies.

The progress of the nuclease reaction may be followed by heating samples, rapidly chilling them, and digesting again with S_1. Those hairpins opened in the first digestion, and therefore no longer rapidly renaturing, will be completely degraded in the second S_1 digestion. Alternatively, the proper amount of S_1 nuclease digestion can be determined by electrophoresing the S_1-treated hairpin structures on polyacrylamide 98% formamide denaturing gels.[45] It is convenient to include undigested hairpin structures and single-stranded cDNA markers to indicate the positions of double- and single-length molecules, respectively. For obvious reasons, it is important that such trial digestions be carried out with the same ratio of carrier RNA to duplex gene copy as in the real experiment. The positions of radioactive bands may be followed by radioflourography if the material is ^{3}H labeled. We find it more convenient to use ^{32}P labeling detected by radioautography at $-70°C$ with Dupont Cronex Hi-Plus® brand phosphotungstate image intensification screens. Relatively low ^{32}P specific activities are used to avoid interfering with the measurement of the amount of tailing by terminal transferase in the subsequent step.

G. Attachment of Homopolymer Tails to Double-stranded Globin Gene Copies Treated with λ Exonuclease

Originally, we used the conditions described by Lobban and Kaiser[21] in which bacteriophage λ exonuclease is employed to remove approximately 30 nucleotides from the 5′-OH ends of the molecules.[14-16] Alternative conditions will be discussed in Section IV.D. When λ exonuclease is to be used, the λ exonuclease digestion is carried out as described by Lobban and Kaiser and terminated by phenol extraction and ethanol precipitation. The nucleic acid is subsequently stored as an ethanol precipitate until ready to carry out the terminal transferase step in freshly made transferase buffer. The addition of poly dA to exonuclease-treated globin DNA with terminal transferase was carried out in 150 $\mu\ell$ of 0.05 M KH$_2$PO$_4$, pH adjusted to 6.9 with potassium hydroxide, 8 mM MgCl$_2$, and 1 mM 2-mercaptoethanol with globin DNA at 3.3 μg/mℓ. The average size as measured on 98% formamide-gel electrophoresis was 400 bases; therefore, there were about 4 pmol of free 3′-OH ends in the reaction mixture. One μCi of ^{32}P was present as the alpha phosphate in 9 nmol of dATP. The addition of 150 units of terminal transferase initiated the reaction at 37°C. Under these conditions, the incorporation of 0.4 nmol of dATP (5% of the total radioactivity normally present) occurs in 30 to 40 min (somewhat variable) and should give poly dA tails averaging 100 nucleotides in length. The addition of poly dT to λ exonuclease-treated plasmid DNA was carried out similarly except in a solution containing 0.2M cacodylic acid, 1 mM CoCl$_2$, and 2.5 mM 2-mercaptoethanol, pH 7.2. This solution is made fresh prior to each use, and the pH of the cacodylic acid is adjusted to pH 7.2 with KOH prior to the addition of the CoCl$_2$ and 2-mercaptoethanol to minimize the formation of precipitate. The addition of 100 nucleotides should require approximately 5 min, although we have found that the reaction rates are somewhat variable. The reactions were terminated by phenol extraction after adding the desired number of nucleotides. The aqueous layers were passed over a Bio-Gel agarose A-5 M column (a 5- or 10-mℓ column for 1 μg of tailed DNA) in 0.1 M NaCl and 0.01 M Tris-HCl pH 7.5. Additional carrier *E. coli* RNA was added to bring the total nucleic acid concentration to 40 μg/mℓ, and the peak was precipitated with 2 volumes of ethanol.

Experience suggests that the length of poly dA-poly dT tails giving optimal transformation efficiency is about 50 nucleotides, although longer tails can be used. It may be crucial that the poly dA and dT tail lengths used in a particular reaction be approximately equal to avoid the formation of "lariat" structures which will not transform. Since the kinetics of the tailing reaction are somewhat variable, it may be useful to halt the reaction on ice after a conservatively chosen reaction time and measure the acid precipitable radioactivity. If the desired level of incorporation has not been reached, the reaction can be reinitiated at 37°C by adding an additional 150 units of terminal transferase. Results from some experiments suggest that the enzyme originally present may sometimes be inactivated when a reaction is arrested by chilling.

H. Annealing and Transformation

Equimolar quantities of poly dA-tailed gene copy and poly dT-tailed plasmid DNA were resuspended at 0.7 nM each (2.2 μg of tailed pMB9 and 0.22 μg of tailed globin gene copy per milliliter) in a solution containing the following: 10 mM Tris-HCl, pH 8.1; 100 mM CaCl$_2$; and 1 mM EDTA. This reaction mixture was incubated at 51°C for 12 min and subsequently allowed to cool slowly to room temperature over 3 hr.

In the initial experiments[14-16] carried out under Asilomar guidelines[24] and prior to the construction or certification of EK2 hosts, the complexes were transformed into the EK1 hosts C600 or HB101. Transformations were carried out as described by Cohen et al.[46] with modifications as described by Higuchi et al.[15] As bacterial strains with reduced capabilities for growth outside the laboratory become available, we have used them in our work. All present transformations for which an EK2 host is required use *E. coli* X1776.[35] It is sometimes difficult to obtain efficient transformation in X1776 unless certain additional changes are made in the transformation procedure. These have been described in detail by Manske et al.[47] and will be considered in more detail in Section IV.F.

I. Maintaining Physical and Biological Containment

The Federal Guidelines for Recombinant DNA Research[25] detail the precautions which must be taken for the cloning of gene copies made from cDNA. As with other types of experiments, the level of containment specified depends upon the presumed risk of hazard and is primarily assessed on the basis of the source and purity of the DNA cloned. The cloning of cDNA gene copies of nonprimate mammalian mRNAs of less than 99% purity must be carried out under P3 + EK2 containment.[25] While a review of the general precautions specified by the Federal Guidelines is inappropriate, some precautions not explicitly specified in the guidelines, which we have found useful in insuring the integrity of the EK2 biological containment, will be mentioned. The biological containment of any gene insert carried in the certified EK2 host-vector systems involving *E. coli* X1776 (the presently approved EK2 host) has been estimated well over 10^8, a level far in excess of the approximately 10^6 afforded by P3 physical containment. Thus, the combination of P3 + EK2 should theoretically provide a 10^{14} reduction in the escape of the insert. The enormous margin of safety which biological containment provides over and above the P3 physical containment is especially comforting as P3 physical containment alone is deemed sufficient for research with all but worst known pathogens. However, the biological containment only applies once the plasmid has entered the EK2 host. The weakest link in the system appears to be the small possibility that the EK2 host cells to be transformed could become contaminated. In this case, the transformation process could lead to an insertion of the eukaryotic gene segment into the contaminating bacteria rather than the EK2 host.

If the preparation of the culture of EK2 host cells is carried out with standard bacteriological technique and carefully performed, the probability of contamination is extremely small. To insure that no hazard is created even if contamination should occur, we have specified that the purity of the culture be tested at several stages in our experiments. As a source of our innoculum, we use a culture of X1776 which has been divided into 1-mℓ aliquots and stored in a freezing medium. This insures the availability of a large number of identical individual aliquots for many experiments. Extensive tests are run on one aliquot to insure that the lot is free from contamination. In preparing for a transformation experiment, one of these aliquots is added to medium lacking diaminopimelic acid (DAP) and thymidine. This culture is immediately split into three aliquots: one supplemented with both DAP and thymidine, another with DAP alone, and the third with thymidine alone. It is essential that the dilution into the tubes lacking DAP or thymidine be sufficient to prevent the DAP and thymidine introduced from the innoculum from adequately supporting growth. After overnight growth, the experiment is continued only if turbidity is absent from the culture tubes which lack DAP or thymidine. These controls would detect the presence of bacteria other than X1776 in either the innoculum, medium, or the solutions of DAP or thymidine.

If the overnight control cultures show no contamination, the bacteria are diluted to an A_{600} of about 0.05 and incubated to provide an exponentiallly growing culture for the transformation. To minimize the chance of contamination at this stage, it is useful to either place the cells in a side-arm flask so that growth can be monitored without opening the flask or make two identical cultures and subsequently follow the growth of one and use the other as a source of the cells to be transformed.

Care is taken that all of the materials added to the transformation mixture are sterile. Although the added complexes between the vector and gene copy have been sterilized by ethanol precipitation, it is desirable to make a final test of the transformation mixture for contaminants. In the case of a transformation in which plasmid pCR1 is the vector, the transformants will have acquired kanamycin resistance detectable by plating on agar supplemented with kanamycin, DAP, and thymidine. Contaminants are tested by plating on control plates which lack kanamycin and DAP and plates which lack kanamycin and thymidine. All plates are closed with tape and incubated until colonies are visible. If contamination is present on the control plates, the experimental plates are autoclaved without being opened.

These simple stratagems insure that in most cases of contamination the experiment will be terminated before any cells are transformed. Moreover, if contamination should occur at the very last step, the exposure of the experimentor to the transformed cells will be extremely unlikely since these cells will be autoclaved without ever opening the petri plates.

J. Colony Hybridization Screening

It is convenient to use the colony hybridization assay of Grunstein and Hogness[48] to confirm that the transformants obtained contain the desired gene sequences. In this procedure, colonies are grown on a nitrocellulose filter, a reference set of these colonies is prepared by replica plating or other means, and the original colonies are lysed by brief alkali treatment so that their DNA is denatured and fixed to the filter *in situ*. A probe of radioactive nucleic acid is subsequently hybridized to the filters. In the case of the globin cDNA cloning, autoradiographic detection of the hybridized globin mRNA probe provides proof of globin gene insertion, while the intensity of the signal gives a rough estimate of the size of the insert and/or the frequency of the complementary sequences in the RNA probe preparation. Such characterization is also a very useful precaution against possible biohazards. When the homopolymer tailing method is used for cDNA cloning of monocistronic mRNAs believed to make nonhazardous products, it is frequently easy to show that other, possibly hazardous, genes cannot be present. As will be discussed in Section V, the evidence required to rule out the presence of potentially hazardous sequences in such cases is the demonstration that the desired sequence *is* present.

The colony hybridization procedure as we now use it has been modified somewhat from that described by Grunstein and Hogness, as the apparatus which they described for applying lysing solutions to the nitrocellulose filter carrying the colonies is no longer used. Instead, the nitrocellulose filters are either transferred in sequence to a series of blotter papers saturated with the specified solutions or floated on specified solutions placed in shallow trays: 0.5 M NaOH—5 min; 1.0 M Tris-HCl, (pH 7.4)—1 min, repeated twice; 1.5 M NaCl, 0.5 M Tris·HCl, pH 7.4 — 1 to 5 min. The treatment with proteinase K is omitted. The filters are subsequently washed several times in 95% ethanol followed by chloroform. The published procedure calls for the removal of loose cellular debris; however, this may drastically reduce the signal obtained in the autoradiography without a commensurate decrease in the background. This suggests that much of the denatured DNA may be fixed on cell debris rather than on the nitrocellulose filter. Finally, the filters are baked *in vacuo* for 1 to 3 hr at 80°C.

Hybridization is subsequently carried out using a ^{32}P labeled probe such as that made by in vitro transcription of cDNA. The dry filter is moistened with 10 to 15 $\mu\ell/$cm^2 of filter of a 5XSSC, 50% formamide solution containing the labeled RNA. To prevent evaporation during the hybridization, the moistened filter is covered with mineral oil. Following a 16 hr incubation at 37°C, the filter is washed in chloroform three times with gentle shaking (10 min per wash) to remove mineral oil. This is followed by 10-min washes in 6XSSC and subsequently in 2XSSC and 2XSSC containing 20 μg/mℓ of pancreatic ribonuclease. The filter is blotted to remove excess liquid, dried, covered with Saran Wrap®, and exposed to X-ray film for autoradiography. In a typical experiment in which 3×10^5 cpm of a globin mRNA probe is applied to the filter (3.25-in. diameter) a 1- or 2-day exposure is required. The sensitivity may be increased approximately tenfold in this and other autoradiographic steps by carrying out the autoradiography at −70°C with Dupont Hi-plus X-ray intensifying screens.*

The results which our own and other laboratories have obtained suggest that the colony hybridization procedure gives conditions of great DNA excess. Thus, for inserts of similar size, the radiographic intensity in the colony hybridization should be proportional to the frequency of the complementary sequences in the probe. This would be quite useful in the case of a differentiated cell growing in tissue culture if, for instance, 50% of its mRNA was a single sequence to be cloned while the remainder was split between hundreds of different mRNAs, all present in roughly similar quantities. In this case, approximately half of the clones obtained should be of the many uninteresting sequences present at a low level; however, if the colony hybridization is carried out under conditions of DNA excess, these would give a signal hundreds of times lower than that obtained for the desired clones.

K. Purification and Storage of Clones

Colonies chosen for further study must be streaked out on selective medium containing the antibiotic appropriate for the plasmid used. However, this purification step can be deferred in some cases where it is desired to save a large number of colonies for future reference. Instead, the colonies are innoculated into 1 mℓ aliquots of liquid medium containing the appropriate antibiotic, grown to saturation, and frozen after the addition of an equal volume of 2× freezing medium. The 2× freezing medium is a

---

* Hi-plus has been found to give much better results than Dupont Lighting-plus,® Dupont Lo-dose,® or Kodak X-Omatic® Regular X-ray intensification screens. An additional increase in sensitivity may be obtained by using flash-activated film.

modification[67] of an earlier recipe[80] and contains the following: 6.3 gm K_2HPO_4, 0.45 gm Na Citrate, 0.09 gm $MgSO_4 \cdot 7\ H_2O$, 0.9 gm $(NH_4)_2SO_4$, 1.8 gm KH_2PO_4, 44.0 gm glycerol, and H_2O, giving a final volume of 500 mℓ. For routine storage after the clones have been purified and single colonies isolated, we originally used slants and stab cultures. However, this led to several cases in which plasmids were lost carrying eukaryotic gene sequences were lost and a few cases in which deletions may have occurred in the inserted sequence.[81] Consequently, clones are presently stored at −20 or −70°C in freezing medium or as purified plasmid DNA.

L. Fidelity of the cDNA Cloning Procedure

Theoretically, mistakes which might lead to incorrect sequences could occur at any step of the cDNA cloning procedure: the synthesis of the cDNA, conversion of cDNA to a duplex gene copy, the joining and transformation steps, or the replication of the recombinant DNA within the host bacterium. The fidelity of cDNA synthesis by reverse transcriptase has been investigated by Battula and Loeb.[49] They found an error frequency of 1 in 600 when homopolymer templates were copied and 1 in 6000 when alternating copolymer templates were copied. The fidelity of second-strand synthesis with DNA polymerase I should be greater due to the exonucleolytic editing functions of this enzyme. The main uncertainties in the procedure may be (1) the fidelity with which gene copies joined to plasmid vectors by the homopolymer tailing reaction can be converted to covalently closed self-replicating molecules and (2) the fidelity with which these molecules can be propagated in a particular host over a large number of generations.

Our work provides a convenient basis for assessing the fidelity as we have dealt with genes coding for proteins of known amino acid sequence and amassed a large body of nucleotide sequence data by conventional techniques.[16,50,51] It is of special interest to measure the degree of fidelity with which an unselected nucleotide sequence can be maintained over many generations in an *E. coli* host since this is of importance for all recombinant DNA studies. Preliminary indications are that the fidelity at all stages of the process is relatively high; only one "error" has been observed in several hundred nucleotides of sequence.[16,31,50,51] This error was the elimination of a single nucleotide coding for amino acid residue 142 ala in plasmid pHb 13.[50] Of the nucleotide sequences studied, approximately equal numbers have been propagated in *rec* A$^-$ hosts (*E. coli* HB101) and in *rec* A$^+$ hosts (*E. coli* C600 and X1776). The *rec* A$^-$ host was preferred as it was believed that it would be more likely to accurately propagate the cloned sequence. However, the single error observed thus far is in a sequence maintained exclusively in the *rec* A$^-$ host.

III. ALTERNATIVE METHODS FOR CLONING mRNA SEQUENCES WHICH DO NOT UTILIZE THE SELF-PRIMING PROPERTIES OF THE cDNA SYNTHESIZED WITH REVERSE TRANSCRIPTASE

Several laboratories[52-54] have cloned nucleotide sequences derived from mRNAs by methods which are fundamentally different from that discussed above in that they do not utilize the fact that cDNAs made with AMV reverse transcriptase are self-priming.

A. Artificial Priming on Sequences Generated by Terminal Transferase

The first of these procedures is that of Rougeon, et al.[52] It is shown in Figure 2 and can be summarized as follows. A cDNA copy is made from globin mRNA and terminal transferase is subsequently used to add poly T tails to the 3′ end of the copy. Oligo A is used as a primer to initiate second-strand synthesis by DNA polymerase I on the poly T tails. Finally, any remaining single stranded regions are trimmed from the du-

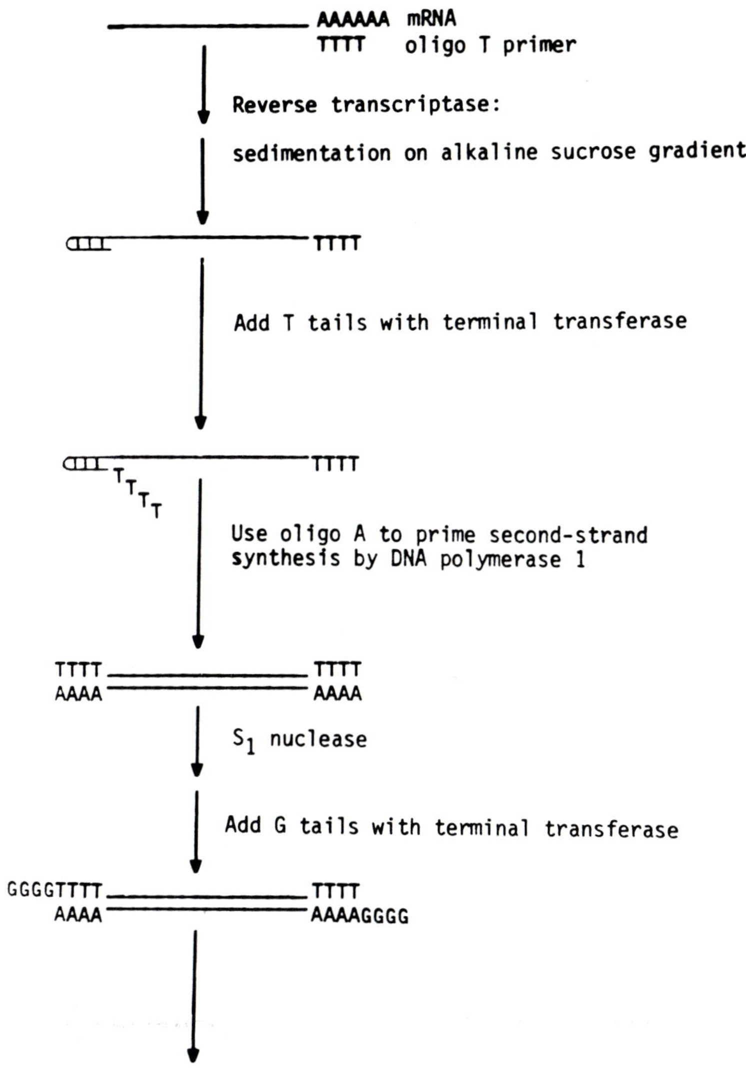

FIGURE 2. cDNA Cloning utilizing artificial priming of second-strand synthesis on sequences generated by terminal transferase. See text for details.

plex gene copy by digestion with S1 nuclease, and terminal transferase is used to add poly G tails prior to joining to a poly C-tailed plasmid vector.

Under the reaction conditions used by Rougeon et al., it is known that terminal transferase prefers to add to single-stranded 3′ ends, adding only with difficulty to base-paired 3′ ends. From our earlier observation that globin cDNAs bear a short duplex hook at the 3′ end, which makes them self-priming,[11] it can be predicted that the addition of poly T tails to such a substrate might have been inefficient and autocatalytic, giving a highly asynchronous addition of tails. In this case, it is possible that priming the second-strand synthesis reaction by oligo A would have been inefficient. If so, then perhaps most of the second-strand synthesis was by the natural self-priming inherent in the cDNA[11] to yield duplex hairpin structures. The second-strand synthesis was followed by an S1 nuclease digestion to remove single-stranded ends. The conditions of the digestion are not given in enough detail to know if it was sufficient to open any hairpin structures present. If such structures resulting from the natural self-priming were opened, it might be expected that they would have been cloned in the subsequent steps along with those molecules in which the oligo A priming off a poly T tail was successful. The efficiency of cloning obtained by Rougeon et al. is about 100 times lower than that reported in Table 3.1, possibly due to these complications.

B. Use of Poly T-tailed Plasmids as Primer for Globin cDNA Synthesis

Rabbits[54] used an approach in which poly T tails previously added to linearized plasmid molecules using terminal transferase were used as the primer for synthesis of globin cDNA. As shown in Figure 3, this should yield a plasmid with globin cDNA tails at each end. To obtain circular complexes which could self-replicate, Rabbits made a further addition of poly T sequences with terminal transferase to the 3′ end of the globin cDNA sequences. Due to the duplex hooks expected at the ends of the cDNA sequences, 3′ termini should be base paired and and would ordinarily be poor substrates for terminal transferase (just as in the experiments of Rougeon et al. discussed above). Rabbits, however, carried out the terminal transferase reaction with a cobalt buffer presently known[16,55,56] to abolish this enzyme's requirement that the substrate be a protruding (unpaired) 3′ terminus.

To obtain circular complexes, the plasmid-globin hybrid sequences prepared in this way were mixed with other plasmid molecules which had been cleaved with EcoR1 and tailed with poly A. This mixture was subsequently denatured and reannealed. One of the four major products expected from this reannealing is shown in Figure 3. In this product, one strand of the poly A-tailed plasmid has hybridized to the complementary strand of the poly T-tailed molecule containing a globin cDNA linked to the plasmid. The association of the poly A and T tails should make this a circular complex whose single-stranded regions can be filled in by the repair enzymes within *E. coli* to give a viable plasmid carrying the globin gene sequence. Rabbits gives insufficient data to permit one to compute the efficiency of the transformation process. It is therefore impossible to say whether the efficiencies are comparable to those obtained in methods utilizing the natural self-priming ability of the cDNA or to much lower efficiencies reported by Rougeon et al.[52] The data presented indicates that only a minority of the transformants obtained (perhaps as few as 4 out of 64) contain globin sequences of appreciable size. In this respect, the results are similar to those presented by Rougeon et al. (one positive out of five transformants), and both contrast with the results obtained by the methods described in Section II,[13-19] in which a majority of the transformants obtained are found to contain inserted sequences.

C. Direct Ligation of a Globin mRNA/Globin cDNA Heteroduplex to EcoR1-Cleaved Plasmid DNA

Wood and Lee[53] have used a still different approach in which short poly dA tails are added to a globin mRNA/cDNA heteroduplex molecule. The resulting structures have a poly dA tail at the 3′ end of the cDNA strand. These are ligated directly to EcoR1 treated plasmid molecules using bacteriophage T4 ligase, which is capable of joining RNA and DNA molecules.[57] There are a number of unexplained features in this procedure. The poly A tails should have only two nucleotides of complimentarity with the EcoR1 site; therefore, the use of an annealing step prior to the ligation would seem superfluous. However, it was found to be essential.[58] The presence of DNA polymerase I during the ligation raises other questions. This enzyme could have elongated the 3′ termini of the various molecules to give blunt ends. The DNA concentrations used in the ligation are not given; consequently, it is difficult to say whether the molar concentrations of ends would have been high enough to favor blunt-end ligation.

Of the transformants obtained, 40 were screened to determine the size of the plas-

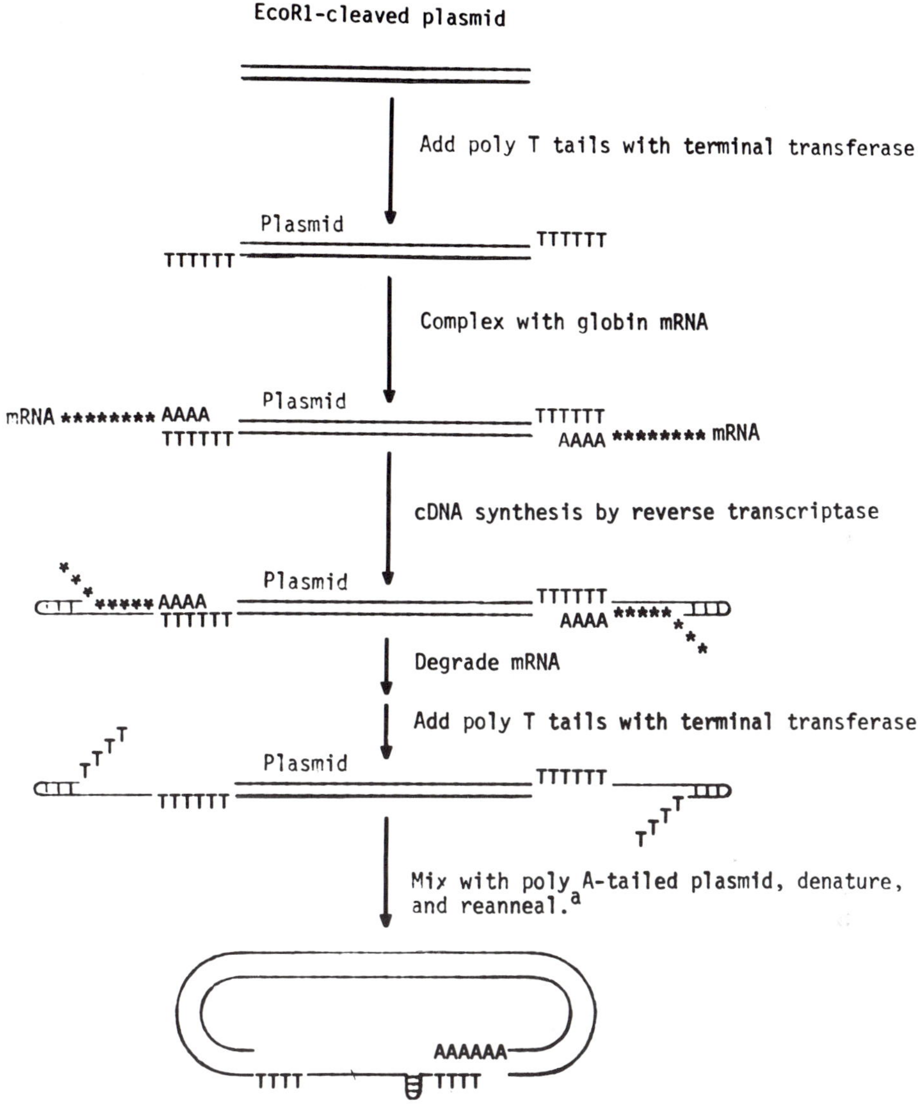

FIGURE 3.   cDNA Cloning utilizing poly T-tailed plasmids as primer for globin cDNA synthesis. See text for details. ª Only one of the possible products is shown.

mids. Two (colE₁-RG1 and -RG2) were shown by hybridization tests to contain globin gene sequences. However, they were not selected at random as stated in the original paper; rather, they were randomly chosen from those plasmids containing the largest inserts.[58] Therefore, it is difficult to determine the success rate, although the total number of transformants obtained in *E. coli* C600 (70/µg of total input DNA, 93/µg of globin heteroduplex) equals roughly 40 transformants per picamole of globin gene copy or 1200 transformants per picamole of plasmid vector. The efficiency per picamole of plasmid vector is about 9-fold less than that in Table 1 and approximately 100-fold more than in Reference 50. However, the efficiency computed per picamole of globin gene copy is about 300-fold less than in Table 1 and 3-fold less than in Reference 50.

The three approaches discussed above provide alternatives which may be useful in special circumstances. They may become more generally useful if refinements lead to increased cloning efficiencies. Unfortunately, none of the clones produced by these three methods have received a detailed analysis by restriction mapping and nucleotide sequencing, which has now become routine with clones produced using the natural self-priming ability of the cDNA.[16,17,50,51,82] It is only through such detailed analysis that it will be possible to compare the relative ease of obtaining complete gene copies by the different approaches.

IV. VARIATIONS ON THE BASIC PROCEDURE

Section II discussed the basic procedures which we have to clone mRNA sequences from a variety of sources. Modifications of these procedures can frequently extend the usefulness of the technique, e.g., it is possible to clone sequences from ribosomal RNA or bacterial and other messengers which lack poly A by using enzymes which can add homopolymer tails to RNA in vitro.[59,60] This section will consider the steps of the approach in greater depth, indicating possible complications and alternatives which may be useful in special cases.

A. Synthesis of Complete cDNA Copies

Under many experimental conditions, AMV reverse transcriptase makes only partial cDNA copies of an mRNA sequence. When the cDNA copies are analyzed by acrylamide-gel electrophoresis in denaturing conditions, one frequently sees partial copies in distinct bands which suggest that there are certain preferential termination points in a particular mRNA sequence. The fact that it is more difficult to obtain full-size copies with some mRNAs than with others also suggests that the enzyme tends to terminate at specific nucleotide sequences or specific features of mRNA base pairing. A number of laboratories have conducted detailed studies of conditions which encourage the synthesis of complete cDNA copies.[61-63] The use of high deoxynucleoside triphosphate concentrations is helpful; we find that 1000 μM concentrations of each deoxynucleoside triphosphate consistently give about 15% more synthesis than the 500 μM concentrations recommended by Efstratiadis et al.[61] It is also useful to carry out the cDNA synthesis at an elevated temperature as suggested by Monahan et al.[62] The conditions which we find most effective (see Section II.D) are different from theirs[62] in several other respects. Kacian[63] has recently proposed a set of still different conditions to maximize the extent of cDNA synthesis and size of the product. We have compared our high temperature-low actinomycin conditions with those suggested by Kacian and with various combinations of the two sets of conditions shown in Table 11. The size distribution of the products made in each set of conditions from a chick globin mRNA template was monitored by autoradiography after electrophoresis on formamide-acrylamide gels. All of the different combinations of conditions gave similar size distributions in which most of the product appeared to be full sized; however, the yield was highest with our high temperature-low actinomycin conditions — about twofold higher than for those specified in Figure 2D of Kacian.[63]

There is currently ample evidence to suggest that all cDNAs made by AMV reverse transcriptase may be self-priming templates for DNA polymerase I. This has been demonstrated by direct cloning experiments which utilized this self-priming characteristic for both the alpha- and beta-globin mRNAs of the rabbit,[17,50,51] chick[83] and mouse.[13] Nor is it limited to globin mRNAs, as it has been successfully used in cloning mouse kappa- and gamma-chain immunoglobulin mRNAs,[84] ovalbumin mRNA,[43] and representatives of a mixed mRNA population from *Dictyostelium discoidum*.[85] Moreover, the available evidence suggests that the short, duplex hook is produced wherever reverse transcriptase terminates synthesis; therefore, cDNAs which have been prematurely terminated are cloned about as efficiently as full-length cDNAs.

TABLE 2

Conditions used	Relative yield of cDNA
High temperature, low actinomycin[a]	1.00
"Kacian and Myers" conditions[b]	0.55
High temperature, low actinomycin, 0.2 mM dGTP	1.10
High temperature, low actinomycin, 0.2 mM dGTP, 4 mM PPI	0.96
High temperature, low actinomycin, 0.2 mM of each dNTP	0.97
High temperature, low actinomycin, 0.2 mM of each dNTP, 4 mM PPi	1.06
High temperature, low actinomycin, 0.2 mM of each dNTP, 4 mM PPi, actinomycin D omitted.	1.16
High temperature, low actinomycin, 0.2 mM dGTP, 4 mM PPi, 50 mM KCl	1.34
High temperature, low actinomycin, 0.2 mM dGTP, 4 mM PPi, incubated at 37°C	0.86
High temperature, low actinomycin, incubated at 37°C	0.89

[a] These conditions were developed by Paddock, Woo, Higuchi and Salser (unpublished observations and References 43 and 83) and are derived from the conditions described by Higuchi et al.[15] and Salser et al.[16] The conditions are as follows — 50 mM Tris-HCl pH 8.3, 20 mM beta-mercaptoethanol, 10 mM MgCl₂, 1.0 mM concentrations of each of the four deoxynucleoside triphosphates, 30 μg of actinomycin D per milliliter, 20 μg oligo dT₁₂₋₁₈ per milliliter, 100 μg globin mRNA per milliliter and 120 units reverse transcriptase per milliliter. Synthesis is carried out for 15 min at 46°C.

[b] These are the conditions described by Kacian and Myers[63] and are as follows — 50 mM Tris-HCl pH 8.3, 0.4 mM dithiothreitol, 8 mM MgCl₂, 0.2 mM concentrations of each of the four deoxynucleoside triphosphates, 50 mM KCl, 4 mM Na pyrophosphate, 5 μg oligo dT₁₂₋₁₈ per milliliter, 50 μg globin mRNA per milliliter and 45 units reverse transcriptase per milliliter. The reaction is carried out for 2 hr at 37°C. The remaining sets of conditions listed represent modifications of our "high temperature, low actinomycin" conditions by changing one or more factors to equal the synthesis conditions used by Kacian and Myers. In only one case (addition of 50 mM KCl) did this seem to result in a substantial further increase in synthesis. The products from each reaction were further analyzed by electrophoresis on 7% polyacrylamide gels to determine the size distribution of the products. The results showed that with all of the sets of conditions used there was an excellent yield of material which appeared to be full sized; there were no substantial differences in the size distributions obtained.

From Heindell, H., Ph.D. Thesis, University of California, Los Angeles.

B. Second-strand Synthesis and "Roadblocks"

We originally assumed that cloned cDNA sequences would almost always contain the sequence at the 3′ end of the mRNA next to poly A and run all or part of the way to the 5′ terminus, depending upon the ability of AMV reverse transcriptase to complete the first-strand synthesis. An analysis of cloned, rabbit alpha-globin sequences suggests that roadblocks in the synthesis of the second-strand by DNA polymerase I are the most serious impediments to obtaining a full-size gene copy with this mRNA. Liu et al.[51] have recently published an analysis of rabbit alpha-globin plasmid pHb72 which shows that it contains the 5′ end of the structural gene intact, but lacks the entire 3′ untranslated sequence and approximately 75 nucleotides from the 3′ end of the structural gene. This is the result which would be expected if second-strand synthesis prematurely terminated at the sequence corresponding to amino acid residue 117, or if the terminal transferase had tailed at a nick. A preliminary analysis of the other rabbit alpha globin clones suggests that this is not an isolated phenomenon.[95a]

If DNA polymerase I does not give completed second-strand synthesis, one might use other enzymes such as reverse transcriptase[18] or T4 DNA polymerase[64] for this step. O'Malley et al.[18] used reverse transcriptase; however, the results were not particularly encouraging. The efficiency of cloning was low, and the inserts obtained were very small.[65] In subsequent experiments using DNA polymerase I to synthesize the second strand, much larger inserts were obtained;[43] however, the improved results might be due to any of several differences.

C. Attempts to Obtain cDNA Gene Copies of Maximal Size

Where it is important to obtain plasmid inserts of maximal size, the S_1-treated duplex gene copies may be fractionated by electrophoresis on nondenaturing polyacrylamide gels; subsequently, the band of full-sized gene copies is located by autoradiography, eluted, tailed, and cloned. In this case, the duplex gene copies should be labeled with ^{32}P rather than ^3H; it is inconvenient to use ^3H fluorography because of difficulty in recovering the desired bands. The ^{32}P specific activity should be chosen with some care. It should be high enough so that the autoradiography will not take a prohibitively long time and low enough so that the ^{32}P label will not interfere with the quantitation of the length of homopolymer tails added in the next step. For instance, if the second-strand synthesis is carried out with nucleoside triphosphates having an average specific activity of 0.1 Ci/mM, then $\frac{1}{3}$ μg of duplex gene copy will contain about 0.5×10^5 cpm. If this material is run on a single 1-cm track of an acrylamide gel of 1-mm thickness, the products can be clearly visualized by autoradiography for less than 1 hr at 4°C.* Tailing such gene copies with ^3H nucleoside triphosphates at a specific activity of 50 Ci/mM gives an approximately 100-fold excess of ^3H dpm over ^{32}P dpm so that accurate quantitation of the tailing reaction is straightforward. The ratio of $^{32}P/^3H$ was computed under the assumption that homopolymer tails 50 nucleotides in length are to be added to gene copies 500 base pairs in size.

Where it is not necessary to obtain very accurately sized material for cloning, the procedures described above may be unnecessary. However, even a relatively small physical quantity of very low molecular weight fragments may outnumber the full-size copies on a molar basis and thus contribute a disproportionate number of the final clones. In this situation, fractionation on a neutral sucrose gradient provides a convenient way of eliminating the majority of the incomplete gene copies.

It is obviously highly desirable to select for full-size clones for a complete mRNA sequence analysis. For many other purposes, however, it may be undesirable to carry out a stringent size selection. To provide specific probes for assaying the message or cloning the sequences surrounding the gene in chromosomal DNA, it is of little importance that the gene copy be complete. Stringent sizing may be a drastic error when cloning a mixture of mRNAs, since blocks in either the first- or second-strand synthesis may prevent any of the gene copies of some messengers from attaining full size. This may explain why Maniatis et al.[17] obtained only rabbit beta-globin mRNA clones and no alpha-gene sequences while we obtained 9 alpha-globin mRNA clones out of a series of 35 which were examined in detail.[51]

Even where the complete mRNA sequence is desired, it may be better to use a well-characterized partial clone as a probe to clone the corresponding gene from the chromosomal DNA rather than strive for full-size cDNA copy. The placque hybridization technique developed by Benton and Davis[66] should permit the rapid isolation of clones containing single-copy mammalian genes as long as a pure probe is available. Since it is essential that such probes be free of contaminating mRNA or ribosomal RNA sequences, cDNA cloning may provide the best means for obtaining the desired probes. The obvious advantage of this route is that using even a partial cDNA clone as probe one may obtain not only the full-size gene copy, but the surrounding nucleotide sequences as well.

---

*   Note that the use of intensification screens is avoided since this requires freezing the gel. Frozen 4 to 8% gels, unlike 20% gels, become seriously distorted when thawed, interfering with recovery of the desired band.

D. Alternative Conditions for the Use of Polynucleotide Terminal Transferase

Under the conditions of Lobban and Kaiser,[21] terminal transferase preferentially adds nucleotides to single-strand 3′ termini. This avoids the tailing of nicks; however, it poses a problem since termini produced by restriction enzymes, especially those such as EcoR1 which leave a depressed 3′ terminus (i.e., the 5′ terminus extends beyond the 3′ terminus) are also poor substrates. The addition of a few nucleotides to a nick or depressed 3′ terminus should make it become a good substrate for further additions so that the reaction is autocatalytic and asynchronous, producing some tails much longer than others. Lobban and Kaiser avoided this by the use of bacteriophage λ exonuclease which will degrade 5′ termini and leave the 3′ termini exposed.

In their original purification and characterization of the enzyme, Chang and Bollum[31] found that the enzyme showed greater activity in cobalt buffers than under other conditions, i.e., those used by Lobban and Kaiser.[21] We have used cobalt[28] routinely in adding poly T tails since we found that this reaction goes slowly in the buffer specified by Lobban and Kaiser. Under these conditions, however, the treatment with λ exonuclease is unnecessary,[16] suggesting that cobalt abolishes the need for a single-stranded 3′ terminus. Similar observations have been independently reported in greater detail by Roychoudhury et al.[55] and Brutlag et al.[56] We now carry out tailing reactions with any of the four bases using cobalt buffers (or the conditions of Reference 56 in the case of poly A tailing) and omitting the use of bacteriophage λ exonuclease.

Cobalt buffers encourage tailing of nicks; however, tailing at nicks can also happen under the conditions of Lobban and Kaiser. When molecules to be cloned are tailed at internal nicks as well as at the termini, annealing with plasmid vectors bearing the complementary tails can result in topologically complex structures in which, for instance, several plasmids are complexed with one insert. That this may cause subtle problems of analysis is suggested by observations of Hogness and collaborators.[67] It was noted in one experiment that individual transformants gave rise to a variety of different clones containing inserts of different amounts of *Drosophila* DNA. Heteroduplex analysis showed that the different inserts derived from a single transformant were related to each other and appeared to be stable during further growth. Electron microscopy of the annealed DNA used to transform the cells in this particular preparation was remarkable in revealing the presence of a preponderance of topologically complex structures. It was known that the DNA used in the preparation was nicked, and the structures seen were as expected if, for instance, the addition of homopolymer tails at nicks permitted one *Drosophila* DNA fragment to associate with several plasmid molecules. When tailing at nicks is avoided, the structures seen are predominantly circles or lariats.

These complex results could be explained if it were imagined that the complex multiarmed structures seen in the electron microscope could be replicated with difficulty in vivo. There would be a strong selection pressure for any molecule in which one of the interconnected circles was eliminated. Such simplification could occur in several ways and give rise to a variety of products, each having deleted a different part of the parent structure. These simplified structures would replicate stably during further growth. Although Hogness[67] emphasizes that this hypothesis is purely conjectural, it does serve to emphasize the complications which may result when one uses either badly nicked DNA or a preparation of terminal transferase which has its own contaminating endonucleases, as is typical of some commercial preparations of the enzyme. Bender and Davidson,[32] for instance, have shown that one of the commercial preparations of terminal transferase contains sufficient nicking activity to efficiently add multiple poly T tails to SV40 DNA even when the DNA is supplied as unnicked supercoils.

E. Plasmid Vectors Certified for EK-2 use

The use of an EK-2 host-vector system is required for many experiments. Candidate

EK-2 systems are scrutinized by the National Recombinant DNA Committee which makes recommendations for certification to the Director of the National Institutes of Health, based upon the Federal Guidelines for Recombinant DNA Research.[25] Each system requires a specific host-vector combination. Presently, all plasmid-host systems require use of the *E. coli* host X1776.[35] The first plasmids approved for EK-2 use with X1776 were pSC101[36] and pCR1.[37] The National Recombinant DNA Committee has subsequently recommended certification of plasmids pMB9,[38] pBR313,[39] and pBR322[39] for certification for EK-2 use with the same host. A brief summary of some of the characterisitcs of each of these plasmids is given below. The number of base pairs has been estimated from the molecular weight using the ratio established by the direct sequences analysis of plasmid pBR345: 0.7×10^6 daltons, 1216 base pairs.[95] This gives values about 10% lower than some other common estimates.

1. pSC101 — 5.8×10^6 daltons: 10 kb pairs, tetracycline resistance selective marker, cloning possible using insertions at EcoR1 site with use of tetracycline resistance to select for transformants. The plasmid does not amplify in chloramphenicol so the yield is manyfold lower than that possible with the other plasmids listed below.
2. pCR1 — 8.9×10^6 daltons: 15.5 kb pairs, kanomycin and colicin resistance selective markers, cloning possible using insertion at EcoR1 site or at Pst sites (with loss of Pst fragment).
3. pMB9 — 3.5×10^6 daltons: 6 kb pairs, tetracycline and colicin resistance selective markers, cloning possible using insertions at EcoR1 site with selection for either resistance marker or at the HindIII, BamH1, or SalI sites in the *tet* gene using the colicin resistance.
4. pBR313 — $5.8 \times$ daltons: 10 kb pairs, tetracycline, ampicillin, and colicin resistance selective markers; cloning possible at EcoR1 site using any of the three resistance markers to select for transformants. Cloning at the HindIII, BamH1, and SalI sites in the *tet* gene is more convenient than with pMB9, since ampicillin rather than colicin resistance can be used for selection of transformants.
5. pBR322 — 2.6×10^6 daltons, 4.5 kb pairs, similar to pBR313 except that all Pst sites in the plasmid have been eliminated except for that located in the gene for ampicillan resistance. Consequently, this plasmid can be used in the same ways as pBR313 and has the ability to clone at the Pst site while using tetracycline resistance to select for transformants.

ampicillin resistance. Consequently, this plasmid can be used in the same ways as

Since cDNA cloning involves homopolymer tailing, the choice of restriction sites available in the plasmids listed above may seem unimportant. In fact, one may wish to choose a particular site because it permits convenient recovery of the inserted sequence, if one wishes the insert to be actively transcribed in the bacterial host, or to prevent the site from being transcribed and possibly disrupting the bacterial host thereby losing the plasmid. The proper choice of insertion sites and selection techniques can also permit the experimentor to eliminate the background due to plasmids which do not carry inserted gene copies. Even in the most efficient cDNA cloning systems, the efficiency of transformation with the desired complexes is approximately 100-fold less than with plasmid-twisted circles (see Table 1). Thus, when one inserts into pMB9 at the EcoR1 site, a 1% background level of pMB9 unopened by the EcoR1 treatment can give more tetracycline-resistant transformants lacking inserts than those containing the eukaryotic gene copy. This problem can be avoided by assaying the tailed vector preparation for its ability to transform. If the background is high, a second EcoR1 digestion will usually solve the problem. An alternative approach is to use a vector such as pBR322 which carries two antibiotic resistance markers. If, for instance, the cDNA gene copy is inserted at the Pst site in the gene for ampicillin resistance, the plasmids carrying no insert can be eliminated by using cycloserine to kill transformants growing in the presence of ampicillin. The intact gene for tetracycline resistance allows a positive selection of the transformants. By reversing the scheme of positive and negative selections, it is possible to obtain only those transformants carrying gene insertions in the gene for tetracycline resistance.

F. Transformation in the Biocontainment Host X1776

E. coli X1776 has been modified extensively to insure that it will be unable to propagate in a natural environment. Unfortunately, these modifications also make it more difficult to use in the laboratory. It is less easily transformed than *E. coli* C600 and gives variable results unless the conditions are carefully controlled. Some of this variability has been ascribed by Curtiss[69] to this strain's extreme sensitivity to traces of detergents and other interfering substances which may be present in the DNA preparation or otherwise introduced into the transformation. A complete description of the characteristics of X1776 is beyond the scope of this presentation; however, it may be useful to describe a transformation protocol which we have found to be dependable and which has worked when alternative methods have failed.[47] Our protocol has been used with good results by a number of other laboratories for cloning in X1776: One example is the cloning of the rat insulin gene.[81]

The procedure is a modification of conditions developed by Roy Curtiss III and collaborators.[68] Several of the steps are added as safety precautions to insure the integrity of the biological containment. An efficiency of about 26,000 transformants per microgram of pMB9 plasmid DNA has been reproducibly obtained, while recombinant DNA clones have been obtained at an efficiency of 1800 to 2900 transformants per microgram of cDNA gene copy.

Cultures of X1776 are prepared as described in Section II.I. and incubated overnight. If the control cultures show no contamination, the overnight culture is diluted into 50 mℓ of L-broth with 10mg/mℓ diaminopimelic acid (DAP) and 4 mg/mℓ thymidine (THD) to give a turbidity (A_{600}) of 0.05. This culture is incubated at 37°C until the turbidity has reached an A_{600} of 0.2 and is then placed on ice until use. Lower transformation efficiencies are obtained if cells are allowed to grow into late log or stationary phase. Cells (50 mℓ) are centrifuged at 2°C in a sterile polypropylene tube. The pellet is resuspended in 20 mℓ of ice-cold 10 mM NaCl and centrifuged again. This pellet is resuspended in 20 mℓ of an ice-cold solution containing the following: 10 mM Tris-HCl, pH 7.0; 140 mM NaCl; and 75 mM CaCl$_2$. The resuspended cells are kept on ice for 15 min and then centrifuged and resuspended in 2.5 mℓ of the same buffer and kept on ice while awaiting use. During a recombinant DNA cloning experiment, a 10-$\mu\ell$ aliquot should be added to a 10-mℓ unsupplemented L-broth culture and incubated for at least 2 days at 37°C. If this control shows growth, all transformants obtained should be considered contaminants and promptly autoclaved if the experiment requires EK-2 biocontainment. Assays are carried out in autoclaved 1.5-mℓ plastic microfuge tubes; 100 $\mu\ell$ of the concentrated calcium-shocked cells prepared as described above are added to the plasmid-insert complexes, which are concentrated (if necessary) by ethanol precipitation and dissolved in 50 $\mu\ell$ of an ice-cold solution containing the following: 0.01 M Tris, pH 8.0; 0.1 M CaCl; and 1 mM EDTA (this latter solution is the same buffer used for reannealing the inserts and plasmid for cloning). Plasmid DNA concentrations of 0.2 or 2.0 μg/mℓ were used in these assays.

Experiments carried out under somewhat different conditions[86] show that strains X1776 and X1849 (another strain prepared by Curtiss et al.), as with other *E. coli* strains, transform at decreased efficiencies when higher DNA concentrations are used. The mixture is kept on ice for 15 min, heat-shocked at 25°C for 4 min, and returned to ice for 30 min. Higher heat-shock temperatures reduce the number of viable transformants. The transformation mixture is plated on 1.5% Agar L-broth plates supplemented with DAP, THD, and 12.5 μg/mℓ tetracycline. Plates are incubated at 32°C for 2 to 3 days. Testing the individual transformant colonies on tetracycline plates lacking DAP and THD provides an additional precaution against transformation into contaminants. Care should be used throughout the procedure to avoid glassware or plasticware which might be contaminated with traces of detergent or other interfering substances. Special care is required to eliminate detergents from the plasmid DNA preparation.

G. Recovery of Inserted Sequences

One advantage of recombinant DNAs constructed by ligation of restriction fragments, such as those made with EcoR1, is the ease of recovering the eukaryotic insert free from plasmid DNA sequences. This is frequently helpful in restriction mapping, sequence analysis, and in providing amplified sequences in pure form for a variety of purposes. It is more difficult to obtain the inserts when cDNA sequences are cloned using the homopolymer tailing technique. However, a variety of techniques have been developed which either cleave specifically at the poly dA-dT joints or artificially create restriction sites at the ends of the inserted sequence.

1. Cleavage of dA-dT Joints with Nuclease S_1

Hofstetter et al.[69] were able to recover inserts by digesting the molecules with the single-strand specific S_1 nuclease under conditions which destabilized the base pairing of the dA-dT joints. The plasmid used for these experiments had AT joints of about 100 and 200 base pairs. Others have found it impossible to excise an insert bounded by AT joints of only 30 and 60 base pairs by similar methods,[70] and it may be that an efficient excision of inserts cannot be achieved if the dA-dT joints are as small as 30 to 60 nucleotides.

2. Reconstruction of Restriction Sites

Theoretically, restriction sites can be recreated in a number of ways so that inserts joined by homopolymer tailing can be excised using the appropriate restriction enzyme. Rougeon et al.[52] attempted to recreate EcoR1 sites which have the sequence 5′ GAATTC 3′ and are cleaved to yield termini with protruding 5′ ends

$$. . . G\ 3'$$
$$. . . CTTAA\ 5'$$

In such a case, the first step must be to fill in the 3′ terminus to give

$$. . . GAATT\ 3'$$
$$. . . CTTAA\ 5'$$

This may be accomplished with DNA polymerases or reverse transcriptase; Rougeon et al. used reverse transcriptase. Secondly, it is necessary to add homopolymer tails which will supply the missing CG pair which completes the restriction site. This was accomplished by adding poly C to the vector and poly G to the cDNA gene copy which was to be inserted. The resulting recombinant DNAs were found to be sensitive to EcoR1; however, this was expected since the sequences inserted (rabbit globin gene copies) had already been shown to contain internal EcoR1 sites.[4] Subsequent analysis has shown that the reconstruction of the EcoR1 restriction sites proceeded with poor efficiency in this experiment, and plasmids with excisable inserts were the exception rather than the rule.[71]

Ohtsuka[72] has devised an alternative procedure for the reconstruction of cleavage sites for the enzyme Pst which appears to be more efficient. Pst is unique among restriction enzymes as it produces a staggered cleavage with a protruding 3′ terminus. Consequently, it is not necessary to fill in nucleotides with DNA polymerase or reverse transcriptase. The addition of a single G residue to the 3′ terminus should recreate the restriction site. This is achieved by using terminal transferase to tail the Pst-cleaved plasmid vector with poly G and the insert with poly C. It should be possible to obtain efficient joining with tails only 10 or so base pairs in length due to the high stability of GC base pairs. We find that tailing with G proceeds with variable efficiencies. In part, this may be related to the tendency of poly G to form triple-stranded complexes. For instance, Burd and Wells[73] found that poly G addition reproducibly stopped when the tails had reached about 15 nucleotides in length, as would be expected if sequences greater than this length formed aggregates and withdrew them from further reaction. However, we have been able to add much longer tails in some reactions and found it difficult to add more than a few nucleotides in others. One strategy for avoiding these problems is to add a few G residues to reconstruct the Pst site and subsequently tail with poly A or poly T to carry out a conventional AT joining reaction.

3. Use of Chemically Synthesized Restriction Enzyme Recognition Sites

A variety of restriction enzyme recognition sites have been chemically synthesized, including sites for EcoR1, BamI and HindIII.[74,75,89] These sites may be directly joined to duplex DNA molecules having flush ends by blunt-end ligation with T4 DNA ligase.[76-78] These methods have not been used for cloning gene copies made from cDNAs, and it remains to be demonstrated that the S1 nuclease treatments give gene copies with flush ends which will enable efficient blunt end ligation. Somewhat related problems have been solved by Sheller et al.;[77] however, this approach may be difficult where the cDNA gene copy is in short supply and high cloning efficiencies are desired.

V. CHARACTERIZATION OF CLONES AND REDUCTION OF BIOCONTAINMENT REQUIREMENTS

A. Certification of Individual Globin cDNA Clones for Use under P2 + EK-1 Containment

The Federal Guidelines for Recombinant DNA Research[25] specify that "when a cloned DNA recombinant has been rigorously characterized and there is sufficient evidence that it is free of harmful genes, then experiments involving this recombinant DNA can be carried out under P1 + EK1 conditions if the inserted DNA is from a species that exchanges genes with *E. coli* and under P2 + EK1 conditions if not." A footnote to these same guidelines indicates that it is necessary to obtain approval from the National Institutes of Health for the use of these reduced containment levels, such approval being

contingent upon data concerning: (a) the absence of potentially harmful genes (e.g., sequences contained in indigenous tumor viruses or which code for toxic substances), (b) the relation between the recovered and desired sequences (e.g., hybridization and restriction endonuclease fragmentation analysis where applicable), and (c) maintenance of the biological properties of the vector.

Fortunately, when one starts with mRNAs for proteins of known amino acid sequence, it is very easy to meet the above requirements for cDNA clones constructed by the homopolymer tailing approach. The crucial point is that this method insures that only one mRNA sequence can be inserted per plasmid. Consequently, if one can conclusively demonstrate that the desired mRNA sequence is present in a monomeric plasmid, it may be confidently assumed that no other mRNA sequence is present. If the desired mRNA is known to be monocistronic, a demonstration of the presence of a desired nonharmful gene can help rule out the possibility that potentially harmful genes are present. As reported by Liu et al.,[51] we have used this rationale in successful petitions to the National Institutes of Health for permission to use P2 and EK1 conditions for globin plasmids which we appropriately characterized.[79] These include the rabbit beta globin plasmids sequenced by Browne et al.[50] and the rabbit alpha globin plasmids described by Liu et al.[51] Approvals were based upon the presentation of extensive nucleotide sequence data which showed conclusively that the desired gene sequences were present. This was corroborated by RNA-DNA hybridization data and restriction enzyme mapping studies. Finally, the required proof of the "maintenance of the biological properties of the vector" included: (1) a demonstration of the maintenance of the expected antibiotic resistance, (2) a confirmation that the appropriate plasmids retained their ability to amplify in chloramphenicol, and (3) a detailed comparison of the globin-gene-carrying plasmids with the original vector by restriction fragmentation analysis using HaeIII and other restriction endonucleases.

B. Sequencing as Part of General Criteria for Reducing Containment Requirements for Plasmids with Globin Gene Inserts

In addition to the approvals for reduced containment of specific globin plasmids, I felt that it would be desirable to establish general principles for the use of nucleotide sequence analysis to provide criteria for reducing containment requirements for plasmids with globin gene inserts. Accordingly, I petitioned[79] that

it be established as a general principle for all globin genes which are non-primate in origin that P2 + EK1 conditions can be employed if i) the plasmids were constructed by cDNA cloning so that proof of insertion of the desired gene can be taken as reasonable proof that undesired genes are not present; *and* ii) presence of the desired globin gene is indicated by *both* colony hybridization and by the unambiguous establishment of a sequence of more than 30 nucleotides which agree with known globin amino acid sequences from the same organism.

In addition it is necessary to demonstrate the maintenance of the biological properties of the vector as discussed in Section V.A and to show that the size of the recombinant DNA molecule is no greater than expected from the combination of the original vector plus a full-size copy of the messenger RNA.

In this petition, which was approved by the National Recombinant DNA Committee, it was argued that the agreement of a 30 nucleotide sequence with the appropriate known globin amino acid sequence clearly establishes that the desired gene is present. Since there is convincing evidence that globin mRNAs are monocistronic and the cDNA cloning method described cannot insert more than one sequence per plasmid, this is sufficient to rule out the possibility that other potentially hazardous genes are present. The probability of finding a sequence of 30 nucleotides which randomly agrees with some part of the expected globin amino acid sequence may be estimated as less than one in 10^9.

C. Demonstrating the Absence of Potentially Hazardous Genes by Comparisons with "Type" Plasmids

Once a particular gene has been cloned and unambiguously identified by detailed nucleotide sequence analysis or other methods, it becomes a "type" plasmid which makes it much easier to identify other clones which carry the same gene sequence. The power of restriction enzyme analysis permits rapid comparisons to be made between this type plasmid and any other plasmid thought to carry the desired gene sequence.

For example, digestion with both EcoR1 and BamH1 restriction enzymes yields a characteristic DNA fragment 70 base pairs in size from plasmids carrying that portion of the rabbit beta globin gene which codes for amino acid residues 98 through 122.[50] To test for presence of the rabbit beta globin gene sequence, plasmids are digested with EcoR1 plus BamH1 on 8% acrylamide gels alongside similar digests of the rabbit beta globin type plasmid. When the comparison is carried out in this way, the presence of the 70 base pair fragment can accurately be distinguished from other fragments which differ in size by a few base pairs. Confirmation of the assignment can be made by using other enzymes to digest the type plasmid and the plasmid to be tested.

VI. CONCLUSION

The cDNA cloning technique has been successfully used in many laboratories, and the available amplified gene sequences are finding an increasing number of uses. Since this approach makes available specific gene sequences in quantities far greater than could be obtained from cDNA synthesis by reverse transcriptase alone, the usefulness of the approach is hardly surprising. The nature of the cloning technique makes it especially useful when applied to mixtures of mRNAs which cannot be conveniently resolved into pure single species in any other way. The availability of cloned cDNA sequences will facilitate any approach for which cDNA was formerly used. However, these techniques which could formerly be used with only a few easily purified mRNAs may now be applied to many different species. This chapter has already emphasized the advantages for sequence analysis. In addition, cDNA clones open up a whole new range of possibilities which require duplex DNA rather than cDNA. The rapid nucleotide sequence analysis possible when duplex gene copies are available in quantity has also been emphasized. Chromatin studies, i.e., making duplex DNA affinity columns to search for specific binding proteins, and studies of expression of eukaryotic genes in bacterial cells are also important. A few additional examples of immediate technical importance are discussed below.

A. Purification of mRNAs and mRNA Precursors

The milligram quantities of mammalian gene copy which may be easily obtained by growing plasmids can be used to make denatured DNA affinity columns suitable for the large-scale isolation of mRNA or mRNA precursors. Since the cloned sequence is pure and the discriminating power of RNA-DNA hybridization techniques can be very great, it should be possible to recover material of especially high purity by such approaches. This is particularly important in the rapid isolation of an mRNA present in small amount or an RNA precursor which amounts to only a tiny fraction of the RNA sequences in the nucleus.

B. "Southern" Gels

Chromosomal DNA is cleaved with restriction enzymes in this approach and the fragments are separated by gel electrophoresis and subsequently transferred to a nitrocellulose filter by a blotting technique. The location of specific fragments is revealed by hybridizing a radioactive RNA probe for a specific gene followed by autoradiographic analysis. While this approach has proved extremely valuable, it has been difficult to apply it to mammalian single-copy genes because of problems of both the sensitivity of the technique and the presence of impurities in the probes. The availability of cDNA clones solves both of these problems. A great increase in sensitivity may be obtained by the use of cDNA plasmids labeled by nick translation with ^{32}P nucleoside triphosphates. Not only does an individual plasmid contain more radioactivity; additionally, there appears to be a cascade effect in which many copies of the plasmid sequence are bound at the proper site. Equally important is the complete elimination of impurities, especially low levels of ribosomal RNA fragments which would otherwise produce strong, spurious signals.

C. Gene Enrichment Techniques

There are many ways in which cloned cDNA sequences can be used to isolate the corresponding genes along with the surrounding sequences from the chromosomal DNA. The triplexing or R-loop technique will be mentioned here. In this approach, an RNA probe forms a triple-strand complex with the complementary gene sequence. The difficulty is that such triplexes are only stable in a very narrow range of conditions which may vary widely from gene to gene. Having the gene available in large quantity as a cloned cDNA simplifies the establishment of the proper R-loop conditions which may subsequently be used to isolate the corresponding chromosomal gene.

One of the most important uses of the cDNA clones may be simply to provide pure probes for the rapid screening of chromosomal DNA clones. As screening techniques become more rapid, it appears that researchers will soon be able to easily clone the chromosomal DNA sequences surrounding almost any gene for which a pure probe is available. The availability of such cloned sequences should lead to a very rapid increase in our understanding of the complexities of the structure and regulation of the eukaryotic genome.

REFERENCES

1. **Verma, I. M., Temple, G. F., Fan, H., and Baltimore, D.,** *In vitro* synthesis of DNA complementary to rabbit reticulocyte 10S RNA, *Nature New Biol.,* 235, 163, 1972.
2. **Kacian, D. L., Spiegelman, S., Bank, A., Terada, M., Metafora, S., Dow, L., and Marks, P. A.,** *In vitro* synthesis of DNA components of human genes for globins, *Nature New Biol.,* 235, 167, 1972.
3. **Ross, J., Aviv, H., Scolnick, E., and Leder, P.,** In vitro synthesis of DNA complementary to purified rabbit globin mRNA, *Proc. Natl. Acad. Sci. U.S.A.,* 69, 264, 1972.
4. **Poon, R., Paddock, G. V., Heindell, H., Whitcome, P., Salser, W., Kacian, D., Bank, A., Gambino, R., and Ramirez, F.,** Nucleotide sequence analysis of RNA synthesized from rabbit globin complementary DNA, *Proc. Natl. Acad. Sci. U.S.A.,* 71, 3502, 1974.
5. **Paddock, G. V., Poon, R., Heindell, H. C., Isaacson, J., and Salser, W.,** Rabbit globin mRNA: analysis of T1 RNase digestion fragments, *J. Biol. Chem.,* 252, 3446, 1977.
6. **Salser, W., Fry, K., Brunk, C., and Poon, R.,** Nucleotide sequencing of DNA: preliminary characterization of the products of specific cleavages at guanosine, cytosine or adenine residues, *Proc. Natl. Acad. Sci. U.S.A.,* 69, 238, 1972.
7. **Paddock, G. V., Heindell, H. C., and Salser, W.,** Deoxysubstitution in RNA by RNA polymerase *in vitro*: a new approach to nucleotide sequence determinations, *Proc. Natl. Acad. Sci. U.S.A.,* 71, 5017, 1974.
8. **Salser, W., Bowen, S., Browne, D., El Adli, F., Fedoroff, N., Fry, K., Heindell, H., Paddock, G. V., Poon, R., Wallace, B., and Whitcome, P.,** Investigation of the organization of mammalian chromosomes at the DNA sequence level, *Fed. Proc. Fed. Am. Soc. Exp. Biol.,* 35, 23, 1976.
9. **Maxam, A. and Gilbert, W.,** A new method for sequencing DNA, *Proc. Natl. Acad. Sci. U.S.A.,* 74, 560, 1977.
10. **Sanger, F. and Coulson, A. R.,** A rapid method for determining sequences in DNA by primed synthesis with DNA polymerase, *J. Mol. Biol.,* 94, 441, 1975.
11. **Salser, W.,** DNA sequencing techniques, *Annu. Rev. Biochem.* 43, 923, 1974. (See p. 948—949).
12. **Efstratiadis, A., Kafatos, F., Maxam, A. M., and Maniatis, T.,** Enzymatic *in vitro* synthesis of globin genes, *Cell,* 7, 279, 1976.
13. **Rougeon, F. and Mach, B.,** Stepwise biosynthesis *in vitro* of globin genes from globin mRNA by DNA polymerase of avian myeloblastosis virus, *Proc. Natl. Acad. Sci. U.S.A.,* 73, 3418, 1976.

14. Higuchi, R., Paddock, G. V., Wall, R., and Salser, W., Insertion of rabbit globin sequences into *E. coli* plasmids pSC101 and pMB9, *Fed. Proc. Abstr.*, 35 (7), 1369, 1976.

15. Higuchi, R., Paddock, G. V., Wall, R., and Salser, W., A general method for cloning eukaryotic structural gene sequences, *Proc. Natl. Acad. Sci. U.S.A.*, 73, 3146, 1976.

16. Salser, W., Browne, J., Clarke, P., Heindell, H., Higuchi, R., Paddock, G. V., Roberts, J., Studnicka, G., and Zakar, P., Determination of globin mRNA sequences and their insertion into bacterial plasmids, *Prog. Nucleic Acids Res. Mol. Biol.*, 19, 177, 1976.

17. Maniatis, T., Kee, S. G., Efstratiadis, A., and Kafatos, F. C., Amplification and characterization of a beta globin gene synthesized *in vitro, Cell*, 8, 163, 1976.

18. O'Malley, B. W., Woo, S. L. C., Monahan, J. J., McReynolds, L., Harris, S. E., Tsai, M. J., Tsai, S. Y., and Means, A. R., The synthesis, isolation, amplification and transcription of the ovalbumin gene, in *Molecular Mechanisms in the Control of Gene Expression*, Nierlich, D. P., Rutter, W. J., and Fox, C. F., Eds., Academic Press, New York, 1976, 309.

19. Maniatis, T., Efstratiadis, A., Kee, S. G., and Kafatos, F. C., *In vitro* synthesis and molecular cloning of eukaryotic structural genes, in *Molecular Mechanisms in the Control of Gene Expression*, Nierlich, D. P., Rutter W. J., and Fox, C. F., Eds., Academic Press, New York, 1976, 513.

20. Jackson, D. A., Symons, R. H., and Berg, P., Biochemical method for inserting new genetic information into DNA of simian virus 40: circular SV40 DNA molecules containing lambda phage genes and the galactose operon of *Escherichia coli, Proc. Natl. Acad. Sci. U.S.A.*, 69, 2904, 1972.

21. Lobban, P. E. and Kaiser, A. D., Enzymatic end-to-end joining of DNA molecules, *J. Mol. Biol.*, 78, 453, 1973.

22. Wensink, P. C., Finnegan, D. J., Donelson, J. E., and Hogness, D. S., A system for mapping DNA sequences in the chromosomes of *Drosophila melanogaster, Cell*, 3, 315, 1974.

23. Mandel, M. and Higa, A., Calcium-dependent bacteriophage DNA infection, *J. Mol. Biol.*, 53, 159, 1973.

24. Berg, P., Baltimore, D., Brenner, S., Roblin, R., and Singer, M., Asilomar Conference on Recombinant DNA Molecules, *Science*, 188, 991, 1975.

25. Federal Guidelines for Recombinant DNA Research, *Fed. Regist.*, 41 (131), 27917, 1976.

26. Kacian, D. L. and Spiegelman, S., Purification and detection of reverse transcriptase in viruses and cells, in *Nucleic Acids and Protein Synthesis*, Grossman, L. and Moldave, K., Eds., **Methods in Enzymology**, 29E, Academic Press, New York, 1974, 150.

27. Richardson, C., Schildkrout, C., Aposhian, H., and Kornberg, A., Enzymatic synthesis of deoxyribonucleic acid, *J. Biol. Chem.*, 239, 222, 1964.

28. Vogt, V. M., Purification and further properties of single-strand specific nuclease from *Aspergillis oryzae, Eur. J. Biochem.*, 33, 192, 1973.

29. Little, J. W., Lehman, I. R., and Kaiser, A. D., An exonuclease induced by bacteriophage lambda, *J. Biol. Chem.*, 242, 672, 1967.

30. Personal communication from R. Wall and M. Kamaromy as well as observations in my own laboratory.

31. Chang, L. M. S. and Bollum, F. J., Deoxynucleotide-polymerizing enzymes of calf thymus gland. V. Homogeneous terminal deoxynucleotidyl transferase, *J. Biol. Chem.*, 246, 909, 1971.

32. Bender, W. and Davidson, N., Mapping of poly (A) sequences in the electron microscope reveals unusual structure of type C oncornavirus RNA molecules, *Cell*, 7, 595, 1976.

33. Boyer, H. W. and Roulland, D. D., A complementation analysis of the restriction and modification of DNA in *Escherichia coli, J. Mol. Biol.*, 41, 459, 1969.

34. Appleyard, R. K., Segregation of new lysogenic types during growth of a doubly lysogenic strain derived from *Escherichia coli* K12, *Genetics*, 39, 440, 1954.

35. Curtis, R., Pereira, D. A., Hsu, J. C., Hull, S. C., Clark, J. E., Maturin, L. J., Goldschmidt, R., Moody, R., Inoue, M., and Alexander, L., Biological containment: the subordination of *Escherichia coli* K-12, in *Recombinant Molecules: Impact on Science and Society*, Beers, R. F. and Bassett, E. G., Eds., Raven Press, New York, 1977, 45.

36. Cohen, S. N., Chang, A. C. Y., Boyer, H. W., and Helling, R. B., Construction of biologically functional bacterial plasmids *in vitro, Proc. Natl. Acad. Sci. U.S.A.*, 70, 3240, 1973.

37. Covey, C., Richardson, D., and Carbon, J., A method for the deletion of restriction sites in bacterial plasmid DNA, *Mol. Gen. Genet.*, 145, 155, 1976.

38. Rodrieguez, R., Bolivar, F., Goodman, H., Boyer, H., and Betlach, M., Construction and Characterization of Cloning Vehicles, p. 471 in *Molecular Mechanisms in the Control of Gene Expression*, Academic Press, New York.

39. Bolivar, F., Rodrieguez, R., Betlach, M., and Boyer, H., Construction and characterization of new cloning vehicles: ampicillin resistant derivatives of the plasmid pMB9, *Gene*, in press (1977).

39a. Bolivar, F., Rodrieguez, R., Green, P., Betlàch, M., Heynecker, H., Boyer, H., Crosa, J., and Falkow, S., Construction and characterization of new cloning vehicles: ampicillin resistant derivatives of the plasmid pMB9, *Gene*, in press (1977).

40. **Gilbert, J. M. and Anderson, W. F.,** Cell-free hemoglobin synthesis. II. Characteristics of the transfer ribonucleic acid-dependent assay system, *J. Biol. Chem.,* 245, 2342, 1970.
41. **Lipman, S., Toth, K., Fedoroff, N., and Wall, R.,** A general method for the large scale isolation of polysomes and messenger RNA applied to MOPC21 mouse myeloma tumours, *Anal. Biochem.,* in press (1977).
42. **Palmiter, R. D.,** Magnesium precipitation of ribonucleoprotein complexes. Expedient techniques for the isolation of undegraded polysomes and messenger ribonucleic acid, *Biochemistry,* 13 (17), 3606, 1974.
43. **McReynolds, L. A., Monahan, J. J., Bendure, D. W., Woo, S. L. C., Paddock, G. V., Salser, W., Dorson, J., Moses, R. E., and O'Malley, B. W.,** The ovalbumin gene: insertion of ovalbumin gene sequences in chimeric bacterial plasmids, *J. Biol. Chem.,* 252, 1840, 1977.
44. **Shenk, T. E., Rhodes, C., Rigby, P., and Berg, P.,** Biochemical method for mapping mutational alterations in DNA with S_1 nuclease: the location of deletions and temperature-sensitive mutations in Simian Virus 40, *Proc. Natl. Acad. Sci. U.S.A.,* 72, 989, 1975.
45. **Maniatis, T., Jeffrey, A., and van de Sande, H.,** Chain length determination of small double and single stranded DNA molecules by polyacrylamide gel electrophoresis, *Biochemistry,* 14, 3787, 1975.
46. **Cohen, S. N., Chang, A. C. Y., and Hsu, C. L.,** Nonchromosomal antibiotic resistance in bacteria: genetic transformation of *Escherichia coli* by R-factor DNA, *Proc. Natl. Acad. Sci. U.S.A.,* 69, 992, 1972.
47. **Manske, C., Wall, R., and Salser, W.,** Conditions for Transformation of *E. coli* Strain X1776, Nucleic Acid Recombinant Scientific Memoranda NAR-46, September, 1976.
48. **Grunstein, M. and Hogness, D. S.,** Colony hybridization: a method for the isolation of cloned DNAs that contain a specific gene, *Proc. Natl. Acad. Sci. U.S.A.,* 72, 3961, 1975.
49. **Battula, N. and Loeb, L.,** The infidelity of avian myeloblastosis virus deoxyribonucleic acid polymerase in polynucleotide replication, *J. Biol. Chem.,* 249, 4086, 1974.
50. **Browne, J., Paddock, G. V., Liu, A., Clarke, P., Heindell, H., and Salser, W.,** Nucleotide sequences from the rabbit beta globin gene inserted into *Escherichia coli* plasmids, *Science,* 195, 389, 1977.
51. **Liu, A., Paddock, G. V., Heindell, H., and Salser, W.,** Nucleotide sequences from a rabbit alpha globin gene inserted in a chimeric plasmid, *Science,* 196, 192, 1977.
52. **Rougeon, F., Kourilsky, P., and Mach, B.,** Insertion of a rabbit beta globin gene sequence into an *E. coli* plasmid, *Nucleic Acids Res.,* 2, 2365, 1975.
53. **Wood, I. K. and Lee, J. C.,** Integration of synthetic globin genes into an *E. coli* plasmid, *Nucleic Acids Res.,* 3, 1961, 1976.
54. **Rabbits, T. H.,** Bacterial cloning of plasmids carrying copies of rabbit globin messenger RNA, *Nature,* 260, 221, 1976.
55. **Roychoudhury, R., Jay, E., and Wu, R.,** Terminal labeling and addition of homopolymer tracts to duplex DNA fragments by terminal deoxynucleotidyl transferase, *Nucleic Acids Res.,* 3, 101, 1976.
56. **Brutlag, D., Fry, K., Nelson, T., and Hung, P.,** Synthesis of hybrid bacterial plasmids containing highly repeated satellite DNA, *Cell,* 10, 509, 1977.
57. **Fareed, G., Wilt, E., and Richardson, C.,** Enzymatic breakage and joining of deoxynucleic acid. VIII. Hybrids of ribo- and deoxyribonucleotide homopolymers as substrates for polynucleotide ligase of bacteriophage T4, *J. Biol. Chem.,* 246, 925, 1971.
58. **Lee, J.,** personal communication.
59. **Hell, A., Young, B. D., and Birnie, G. D.,** Synthesis of DNAs complementary to human ribosomal RNAs polyadenylated *in vitro, Biochim. Biophys. Acta,* 442, 37, 1976.
60. **Mans, R. J. and Huff, N. J.,** Utilization of ribonucleic acid and deoxyoligomer primers for polyadenylic acid synthesis by adenosine triphosphate: polynucleotidylexotransferase from Maize, *J. Biol. Chem.,* 250, 3672, 1975.
61. **Efstratiadis, A., Maniatis, T., Kafatos, F. C., Jeffrey, A., and Vournakis, J. N.,** Full length and discrete partial reverse transcripts of globin and chorion mRNAs, *Cell,* 4, 367, 1975.
62. **Monahan, J. J., Harris, S. E., Woo, S. L. C., Robberson, D. L., and O'Malley, B. W.,** The synthesis and properties of the complete complementary DNA transcript of ovalbumin mRNA, *Biochemistry,* 15, 223, 1976.
63. **Kacian, D. L. and Myers, J. C.,** Synthesis of extensive, possibly complete, DNA copies of polio virus RNA in high yields and at high specific activities, *Proc. Natl. Acad. Sci. U.S.A.,* 73, 2191, 1976.
64. **Kruger, L. J., Benbow, R. M., Frauss, M. R., Caryk, T. M., and Anderson, W. F.,** Enzymatic synthesis of full size globin genes, in press, 1977.
65. **Woo, S.,** personal communication.
66. **Benton, W. D. and Davis, R. W.,** Screening lambda gt recombinant clones by hybridization to single plaques *in situ, Science,* 196, 180, 1977.
67. **Hogness, D.,** personal communication.
68. **Curtiss, R., III,** personal communication.

69. **Hofstetter, H., Schambock, A., Van den Berg, J., and Weissmann, C.,** Specific excision of the inserted DNA segment from hybrid plasmids constructed by the poly (dA)-poly(dT) method, personal communication.
70. **Roberts, J.,** personal communication.
71. **Mach, B.,** personal communication.
72. **Ohtsuka, A.,** Regeneration of Pst I restriction sites during addition of deoxynucleotide tails, submitted to *Gene*, 1977.
73. **Burd, J. F. and Wells, R. D.,** Synthesis and characterization of the Duplex Block Polymer $d(C_{15}A_{15}) \cdot d(T_{15}G_{15})$, *J. Biol. Chem.,* 249, 7094, 1974.
74. **Sheller, R. H., Dickerson, R. E., Boyer, H., Riggs, A., and Itakura, K.,** Chemical synthesis of restriction enzyme recognition sites useful for cloning, *Science,* 196, 177, 1977.
75. **Bahl, C. P., Marians, K. J., Wu, R., Stowinski, J., and Narang, S. A.,** A general method for inserting specific DNA sequences into cloning vehicles, *Gene,* 1, 81, 1976.
76. **Sgaramella, V., van de Sande, J. H., and Khorana, H. G.,** Studies on polynucleotides, C. A novel joining reaction catalyzed by the T4 polynucleotide ligase, *Proc. Natl. Acad. Sci. U.S.A.,* 67, 1468, 1970.
77. **Sheller, R. H., Thomas, T., Lee, A., Klein, W., Hiles, W., Britten, R., and Davidson, E.,** Clones of individual repetitive sequences from sea urchin DNA constructed with synthetic EcoR1 sites, *Science,* 196, 197, 1977.
78. **Heyneker, H. L., Schine, J., Goodman, H. M., Boyer, H. W., Rosenberg, J., Dickerson, R. E., Narang, S. A., Itakura, K., Lin, S. Y., and Riggs, A. D.,** Synthetic lac operator DNA is functional *in vivo, Nature,* 263, 748, 1976.
79. Petitions Salser, W., considered and approved by the National Recombinant DNA Committee at their Miami meeting in January, 1977.
80. **McFall, Z. Pardee, A., and Stent, G.,** Effects of radiophosphorus decay on some synthetic capacities of bacteria, *Biophys. Biochem. Acta,* 27, 282, 1958.
81. **Ullrich, A., Shine, J., Chirgwin, J., Pictet, R., Tischer, E., Rutter, W. J., and Goodman, H. M.,** Rat insulin genes: construction of plasmids containing the coding sequences, *Science,* 196, 1313, 1977.
82. Manuscripts in preparation by **Cummings, I., Liu, A., Heindell, H., Padayatty, J., Paddock, G., and Salser, W.** include sequence data which complete the entire rabbit alpha globin mRNA sequence and bring the chicken alpha and beta globin mRNA sequences to within more than 97% and 85% completion, respectively.
83. **Padayatty, J., Cummings, I., Manske, C., Higuchi, R., Woo, S., and Salser, W.,** manuscript in preparation.
84. **Paddock, G., Higuchi, R., Salser, W., and Wall, R.,** unpublished observations.
85. **Higuchi, R., Paddock, G., Firtel, R., and Salser, W.,** unpublished observations.
86. **Manske, C.,** unpublished results.
87. **Heindell, C. and Salser, W.,** unpublished results.
88. **Studnicka, G. and Salser, W.,** unpublished results.
89. **Bahl, C. P., Marians, K. J., and Wu, R., Stawinsky, J., and Narang, S.,** A general method for inserting specific DNA sequences into cloning vehicles, *Gene,* 1, 81, 1976.
90. **Southern, E. M.,** Detection of specific sequences among DNA fragments separated by gel electrophoresis, *J. Mol. Biol.,* 98, 503, 1975.
91. **Botchan, M., Topp, W., and Sambrook, J.,** The arrangement of Simian virus sequences in the DNA of transformed cells, *Cell,* 9, 269, 1976.
92. **Thomas, M., White, R. L., and Davis, R. W.,** Hybridization of RNA to double-stranded DNA: formation of R-loops, *Proc. Natl. Acad. Sci. U.S.A.,* 73, 2294, 1976.
93. **White, R. L. and Hogness, D. S.,** R-loop mapping of the 18S and 28S sequences in the long and short repeating units of *Drosophila melanogaster* rDNA, *Cell,* 10, 177, 1977.
94. **Casey, J. and Davidson, N.,** Rates of formation and thermal stabilities of RNA:DNA duplexes at high concentrations of formamide, in press, 1977.
95. **Boyer, H.,** personal communication.
95a. **Liu, A., Heindell H., Paddock, G., and Salser, W.,** unpublished results.
96. **Heindell, H.,** Ph.D. Thesis, University of California, Los Angeles.

Chapter 4

PLASMID CLONING VECTORS

J. G. Sutcliffe and F. M. Ausubel

TABLE OF CONTENTS

INTRODUCTION

The ability to obtain large amounts of specific DNA segments allows an experimentally deliberate and molecularly precise approach to the construction of genetic variants, the detailed probing of protein-DNA contacts, and the analysis of the organization of eukaryotic genes. DNA cloning is a means to propagate DNA of specific sequence in a highly enriched form. Specific DNA sequences may be cloned either because of their intrinsic interest, because of their usefulness as hybridization probes, or because the DNA codes for desired proteins. The coordinate development of in vitro DNA cloning techniques, the use of restriction enzymes as tools for generating unique DNA fragments, and direct and rapid DNA sequencing techniques has greatly widened the scope of biochemical genetics, allowing a more detailed analysis of molecular biological processes.

Figure 1 presents a simple schematic view of a plasmid DNA cloning experiment. A small DNA genome which is capable of autonomous replication in a bacterial host has its physical continuity interrupted. The continuity is then restored in such a way that a new segment of DNA is carried as a passenger at the site of the interruption. The resealed DNA, with its passenger or "insert", is then propagated in bacteria. This

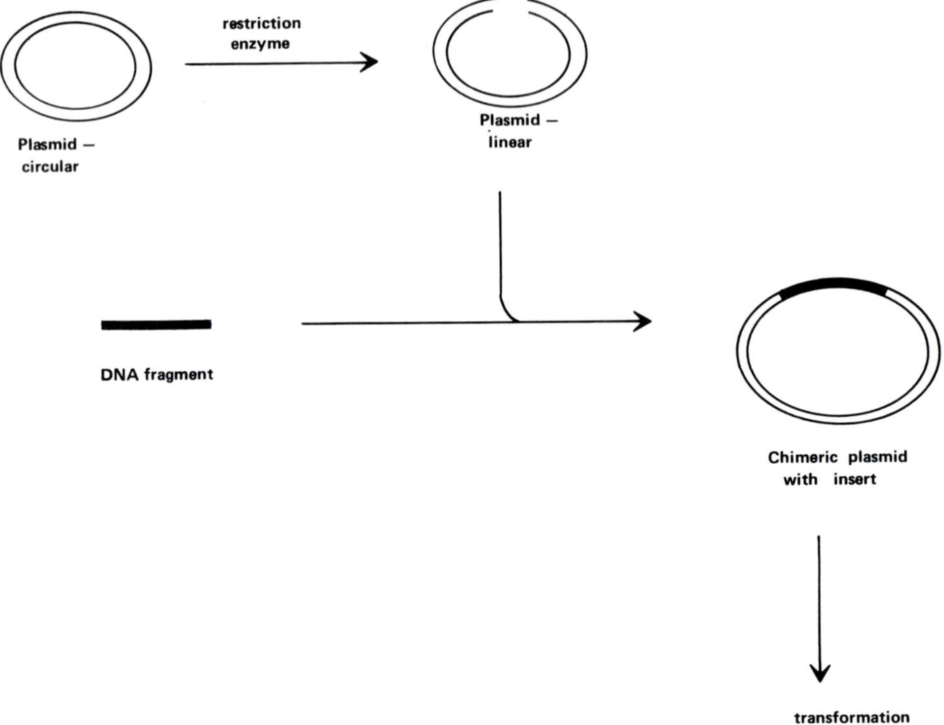

FIGURE 1.   A cloning experiment.

review intends to describe the development of plasmid cloning techniques. The topic naturally divides itself into several relatively distinct areas: plasmid vectors, the source of insert DNA, in vitro DNA manipulation, propagation of the recombinant plasmid, and insert analysis.

II. PLASMID VECTORS

The development of new concepts in cloning methodology has led to the construction of increasingly versatile plasmid vectors, as well as vectors having specialized uses. The plasmids originally used in cloning experiments are comparatively "primitive" and are now interesting primarily from an historical view. Detailed discussions will be reserved for the most recent and most highly engineered plasmids — those which are the closest to what are considered to be ideal vectors.

DNA cloning is a means of enriching for particular DNA sequences. The feature that most distinguishes "recombinant DNA technology" from the most advanced forms of in vivo genetic engineering is the use of restriction enzymes to precisely define where recombination or deletion events occur. The experimenter may thus play an active role in determining which new genetic arrangements are generated for molecular studies. The use of restriction enzymes also extends the range of "recombinants" which can be studied. DNA from different species can be made to coexist in combinations which might be hard to detect in the wild.

Cohen et al.[1] were the first to utilize this approach when they observed that the type II restriction enzyme EcoR1 cleaved the plasmid pSC101 at a single site. EcoR1 cleaves DNA at a specific symmetric hexanucleotide sequence and leaves a four-nucleotide 5' extension.[2] The ends left by EcoR1, therefore, are unique and "sticky" (complementary to other EcoR1 ends). The researchers found they could open the circular pSC101

DNA molecule with EcoR1 to form a linear molecule that could recircularize via its cohesive EcoR1 ends and regain its biological activity. pSC101 could be detected in a bacterium because it confers resistance to the antibiotic tetracycline.[3,4] When EcoR1-opened pSC101 DNA was mixed with EcoR1 restriction fragments from plasmid R6-5 DNA and this mixture was incubated with DNA ligase, molecules were formed that were chimeric: that is, some molecules contained all of the pSC101 DNA and also a segment of pR6-5 DNA inserted at the EcoR1 site.[1] These chimeric plasmids were biologically active. Both tetracycline resistance (from the pSC101 portion) and kanamycin resistance (from the pR6-5 portion) were expressed in transformed bacteria.

These experiments were the prototypes of "passive cloning". If the enzyme EcoR1 produces a suitable restriction fragment containing the gene of interest, then the fragment may be propagated by inserting it at the unique EcoR1 site of pSC101 and examining tetracycline-resistant bacterial transformants for a plasmid which carries the desired insert. If the insert confers on the recipient bacteria a scorable or selectable phenotype, then the new plasmid can be fished out of the background of other unwanted plasmids by simple genetic manipulations. When the plasmid DNA is isolated from the host, the inserted fragment is greatly enriched in relation to the sequences of its original chromosomal context. The insert can be removed from pSC101 simply by cutting the new plasmid with EcoR1 and separating the insert from the pSC101 DNA by agarose or acrylamide gel electrophoresis.

Analogous cloning experiments were performed by Hershfield et al. using the plasmid ColE1.[5] This plasmid produces the colicin factor E1 and also confers immunity to the colicin.[6] Like pSC101, it, too, contains a single EcoR1 restriction site which is located within the colicin gene. Insertion of DNA at this site does not affect the replication properties of the plasmid. Hershfield et al. were able to clone the tryptophan biosynthetic genes of *E. coli* in the EcoR1 site of ColE1. Such plasmids allowed bacteria which were tryptophan auxotrophs to become tryptophan independent when they carried the plasmid.

A feature of colE1 that renders it extremely valuable as a cloning vector is its mode of replication. pSC101 is under "stringent control", and is found in two to five copies per cell.[7,8] Its replication is coordinate with host replication. On the other hand, ColE1 is found in about 25 to 30 copies per cell under normal growth conditions.[9] Moreover, it replicates under "relaxed control", unlinked to the host. During amino acid starvation or other inhibition of protein synthesis (such as by chloramphenicol treatment), cell growth stops, but the replication of ColE1 continues for several hours until there are 1000 to 3000 copies per cell.[5,9-11] The yield of insert DNA in ColE1 per liter of bacterial culture can be two to three orders of magnitude greater than that of an insert cloned in a stringently controlled plasmid. Furthermore, because the gene dosage is high, the level of insert expression is also high. Hershfield et al.[5] observed that the level of tryptophan synthetase was 18-fold higher whene its gene was carried as a ColE1 insert than when the gene was chromosomal. A similar increase in trp mRNA was also measured. A plasmid which expresses its insert not only enriches the amount of insert DNA but also increases the levels of its products.

Another useful feature of ColE1 which makes it a valuable EcoR1 cloning vector is the fact that the EcoR1 site is located within the colicin gene. When a DNA fragment is inserted in the EcoR1 site, the chimeric plasmid confers on its host the phenotype colicin E1$^-$, ImmE1$^+$. The immunity gene provides a positive selection for the plasmid, and the ColE1$^-$ phenotype a scorable marker for a plasmid carrying an insert at the EcoR1 site. This is called "insertional inactivation" and removes the requirement that the fragment carry a scorable marker of its own.

Colicin immunity was found to be inconvenient as a selective marker. Consequently, several research groups responded by constructing hybrid plasmids combining the positive features of ColE1 and pSC101. Those features are

1. A strong positive selective marker for the plasmid
2. A unique restriction site in which to clone an insert
3. A plasmid that replicates in the relaxed mode

So et al.[12] constructed one such plasmid, pSF2124, by genetically transferring an ampicillin-resistance gene onto ColE1. This was accomplished by mating plasmid R1drd, which carries a transposable ampicillin resistance, TnA,[13] with a cell containing ColE1. pSF2124 still carries the unique EcoR1 site of ColE1 and replicates in the relaxed mode. Inserts at the EcoR1 site inactivate colicin E1 expression.

Betlach et al.[14] constructed a similar vector by transposing TnA from pR1drd to pMB1 to produce pMB3. pMB1 is an isolate from nature similar to ColE1, except that it carries the gene for the restriction enzyme EcoR1, as well as colicin factor E1 and colicin immunity. pMB1 also has a relaxed mode of replication.

Both pSF2124 and pMB3 were improvements, but they shared two shortcomings. TnA is a transposon and therefore is somewhat unstable. TnA is also rather large (3.2 $\times 10^6$ daltons),[15] and thus the vectors pSF2124 and pMB3 are unnecessarily bulky. Not only does extra size increase the complexity of the DNA sequence of the plasmid and hinder restriction mapping, but it also lowers copy number per cell and hence insert DNA yield.

A plasmid which circumvents the disadvantages of TnA is pMB9.[16] It was made by a series of molecular scrambling events (called EcoR1* rearrangements) involving the combination of the relaxed replicator of pMB3 with the tetracycline resistance of pSC101. EcoR1* (R1 star) activity refers to cleavage by the enzyme at an altered recognition site: various in vitro conditions (such as high salt or manganese ion) perturb the normally observed kinetic properties of the enzyme EcoR1 (and other enzymes as well), such that the entire sequence 5′-GAATTC-3′ is no longer required for cleavage; instead, only the tetranucleotide core sequence 5′-AATT-3′ is necessary.[17,18] In EcoR1* rearrangements, the continuity of the product plasmid DNA may not directly correspond to that of its parent. Limited digestion of pMB3 with EcoR1* activity, followed by ligase treatment, yielded a tiny plasmid (1.7 $\times 10^6$ daltons) called pMB8 that retained enough genetic information to replicate but no longer carried ampicillin resistance. pMB8 contained a unique EcoR1 site that allowed fusion with pSC101. This double plasmid was reduced in size by EcoR1* rearrangement to form pMB9.[16] pMB9 has proved to be a very useful cloning vector. Its relatively small size (3.5 $\times 10^6$ daltons), high copy number (20 per cell), relaxed replicator, stable antibiotic resistance providing a strong positive selection, and unique EcoR1 site make it very popular.

Not all regions of DNA will have conveniently positioned EcoR1 sites, and inevitably some genes will be cut into parts by EcoR1. Since it was desirable to be able to choose among several restriction endonucleases to flank inserts, Hamer and Thomas[19] constructed vectors that could be used for cloning fragments generated by SalI or BamHI which, like EcoR1, recognize symetrical hexanucleotide sequences and leave four-nucleotide cohesive ends.[20] A higher number of available cloning sites in the plasmid (i.e., different restriction enzyme single recognition sites) increases the likelihood of being able to insert a restriction fragment that carries the interesting gene.

By screening several of the plasmids already discussed in this article, Hamer and Thomas[19] discovered that both SalI and BamHI had single sites within the tet^r gene of pSC101, such that inserts cloned at these sites inactivated the tet^r gene. pSF2124 was shown to contain no SalI sites. A joint plasmid, pGM706, was constructed by ligating pSF2124 and pSC101 at their single EcoR1 sites. pGM706 is a 13 $\times 10^6$ dalton plasmid that expresses amp[r] and tet[r], has a unique SalI site for insertional inactivation of the tet[r] phenotype, and replicates in the relaxed mode from the pSF2124 (ColE1) replicator. The plasmid also carries the stringent pSC101 replicator and all of TnA.

Another plasmid, pGM16, is a joint vector formed by Hamer and Thomas from pSC101 and pML21. pML21 carries the kanamycin resistance of pR6-5 ligated by its EcoR1 ends to mini-ColE1, a deletion variant of ColE1 which still has a unique EcoR1 site.[21] pML21 has two EcoR1 sites, carries kan[r], and has no BamHI sites. Partial digestion of pML21 by EcoR1 followed by ligation to EcoR1-digested pSC101 yielded pGM16: kan[r], tet[r], unique BamHI site for tet[r] insertional inactivation, 12×10^6 daltons, ColE1 relaxed replicator, and pSC101 stringent replicator.[19] pGM706 and pGM16 each has two positive selective markers, one of which is insertionally inactivated, so that clones containing plasmids which carry an insert in the tet[r] gene can be detected by replica plating. These two plasmids provided two new features: (1) an expanded repertoire of cloning sites and (2) an insertional inactivation marker which was simple to assay. However, both plasmids are rather large and specialized.

A plasmid with universal application which combines the positive features of pGM706 and pGM16 and the convenience of pMB9 is pBR322, currently the most comprehensive and useful vector. Bolivar et al.[22] transposed TnA from pSF2124 onto pMB9 to form a large plasmid, pBR312, which was amp[r], tet[r]. This was decreased in size by EcoR1* rearrangement to form pBR313. Precise deletion of pBR313 led to plasmid pBR322.[24] Figure 2 presents the lineage of pBR322. This plasmid carries DNA derived from three different naturally occurring plasmids. The amp[r] gene is of type RTEM1 and derives from the parent of pR1drd, pR7268, which was isolated in London in 1963 from *Salmonella paratyphi* B.[24,25] The tet[r] gene is derived from pSC101, also a natural *Salmonella* plasmid.[4] Finally, the origin of replication comes from pMB1, a natural isolate from *Eschericia coli*.[14]

pBR322 has the following notable features:

1. It is small — 2.8×10^6 daltons.[23,26]
2. It has a relaxed mode of replication.[23]
3. It carries two strong positive selective markers — amp[r] and tet[r].[23]
4. It has unique sites for several restriction enzymes — EcoR1, HindIII, BamHI, SalI, AvaI, and PstI.[23,27]
5. Insertional inactivation is possible in both antibiotic resistance genes. Inserts cloned in the Pst I site inactivate amp[r]. Inserts in the BamHI or SalI sites and some inserts in the HindIII site inactivate tet[r].[24] The HindIII site is in the tet[r] promoter.[29] Presumably, inserts which do not cause loss of tet[r] can provide some information which allows the promoter to function, or else substitute another promoter. The above properties make pBR322 as useful as the earlier plasmids in every respect. The following qualities greatly increase its versatility over previous plasmids.
6. There is direct selection for insertions (Figure 3).
     Any insert which inactivates tetracycline resistance can be selected for by cycloserine enrichment.[24] Tetracycline is a bacteriostatic antibiotic — that is, it stops cell growth by inhibiting protein synthesis, but does not kill. The amino acid analogue cycloserine is lethal if incorporated into proteins.[29] Therefore, in the presence of tetracycline, cycloserine kills tet[r] but not tet[s] cells. Cells which survive a cycle of cycloserine treatment are greatly enriched for tet[s], and several successive cycles multiply the enrichment. Selection for cells that are amp[r] followed by cycloserine enrichment for cells that are tet[s] yields cells containing plasmids with inserts in the tet[r] gene.

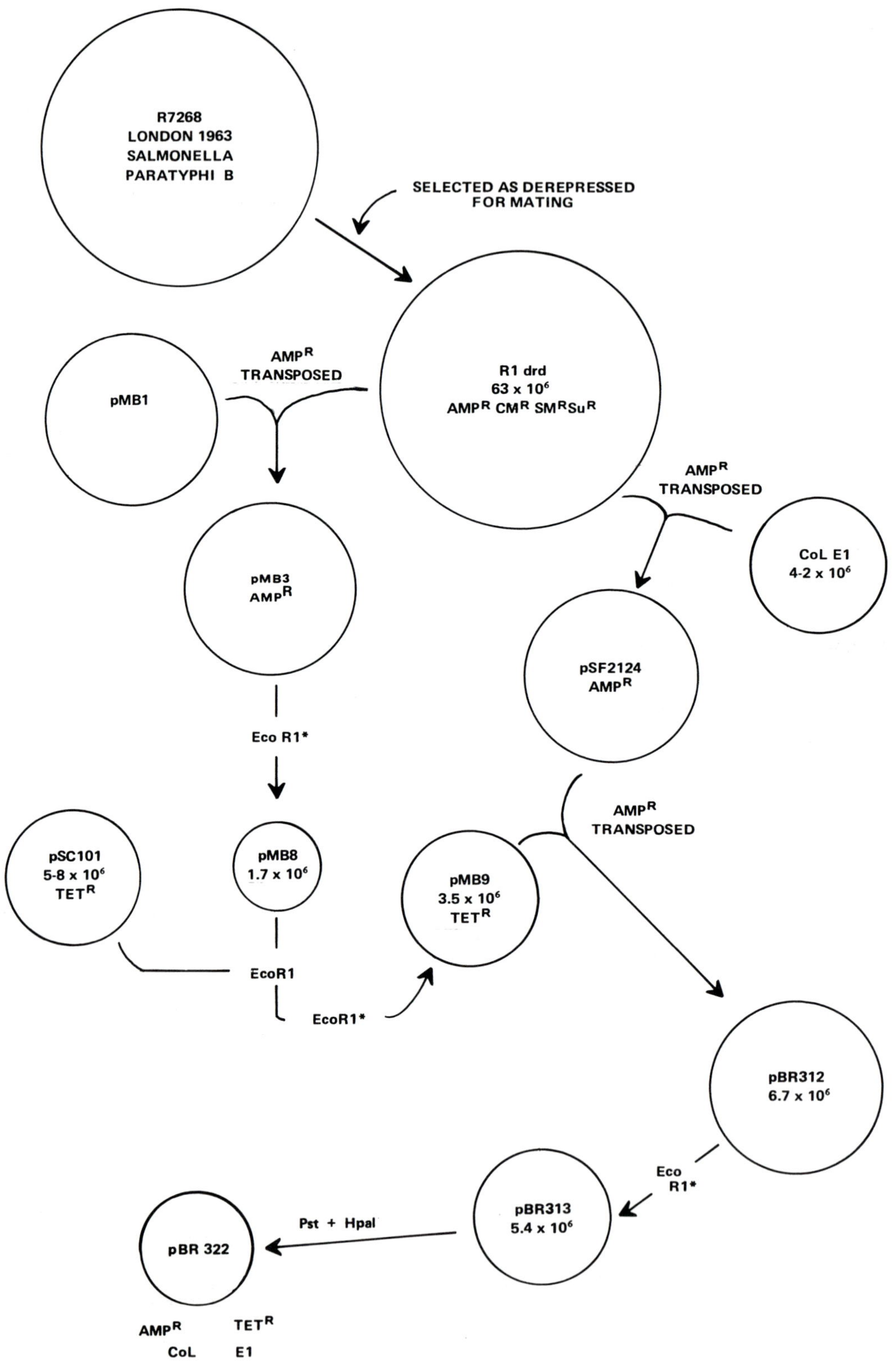

FIGURE 2.   Lineage of pBR322.

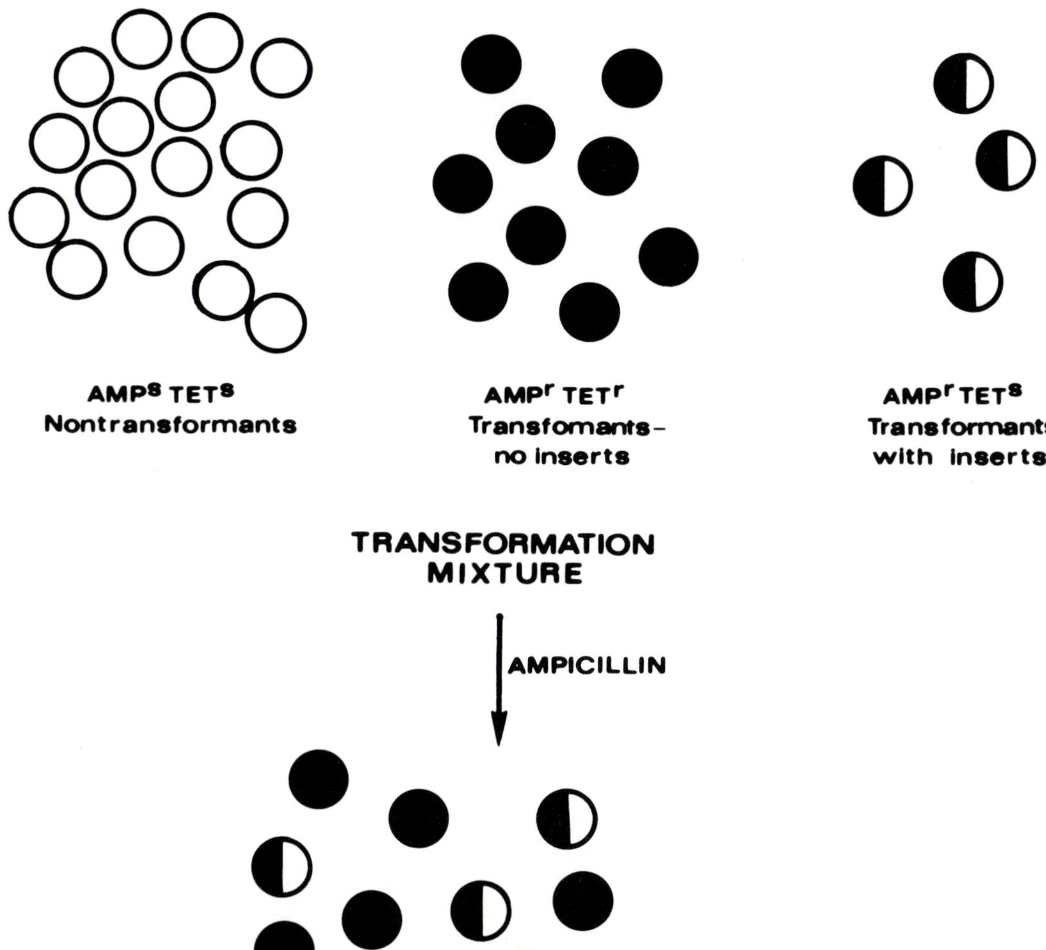

FIGURE 3.   Cycloserine enrichment.

7.   Blunt-end ligation is possible (Figure 4).

T4 DNA ligase will join flush-ended DNA fragments.[30,31]. pBR322 provides a HindII flush cut site in the amp^r gene. There are two HindII sites on the molecule. One site is common with the unique SalI site.[23] SalI recognizes 5′-GTCGAC-3′ and makes a staggered cut. HindII recognizes 5′-GTPyPuAC-3′ and makes a flush cut.[20]. If pBR322 is first cut by SalI and then by HindII, two fragments are produced, each with one flush end and one staggered end. At low DNA concentra-

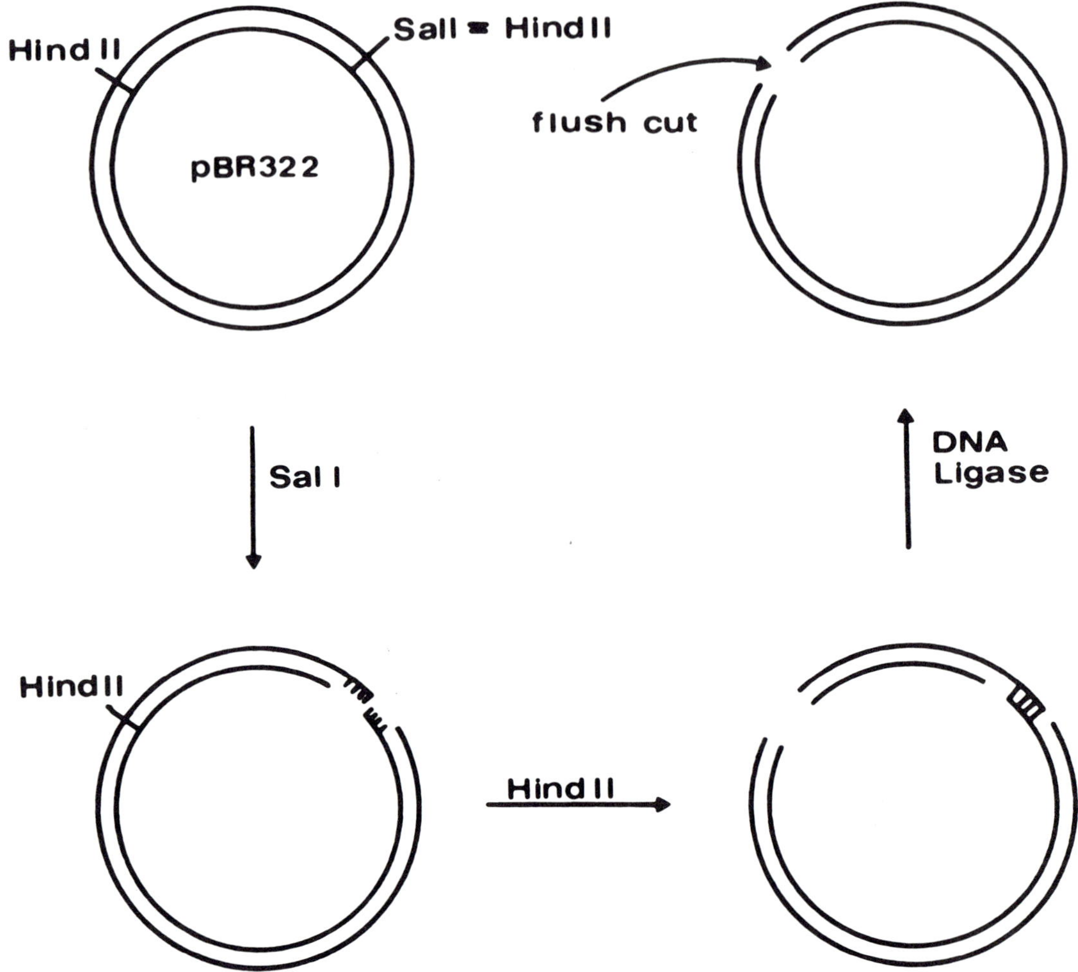

FIGURE 4.   Blunt end site.

tions, ligase will reseal only the SalI staggered ends, leaving a linear molecule with a flush cut in the amp[r] gene. If flush-ended DNA fragments are added at high concentration to the ligation mixture, then inserts can be incorporated at the HindII site in the amp[r] gene. Such plasmids will confer the phenotype tet[r], amp[s] and can be detected by replica plating.

8.   A site exists for terminal transferase tailing (Figure 5).

In some cloning experiments it is advantagous to use the terminal transferase tailing procedure rather than using cohesive single-stranded ends generated by restriction endonucleases. This is particularly true in cloning cDNA fragments produced by reverse transcriptase. The preferred terminal transferase substrate is a single-stranded 3′-OH, such as that left by cleavage with the enzyme PstI. pBR322 contains a single PstI site located in the amp[r] gene. PstI cleaves 5′-CTGCAG-3′ leaving a 4 nucleotide 3′-extension.[32] Terminal transferase adds a polynucleotide tail onto the 3′ end of the DNA fragment.[33] If the 3′ end of the PstI cut plasmid is extended with polydG, and the potential insert DNA is tailed with polydC, then the insert may be annealed to the plasmid by virtue of G-C base pairing of the tails. The insert may later be retrieved from the chimeric plasmid by PstI cleavage because the poly dG tailing recreates the recognition sequence.[23]

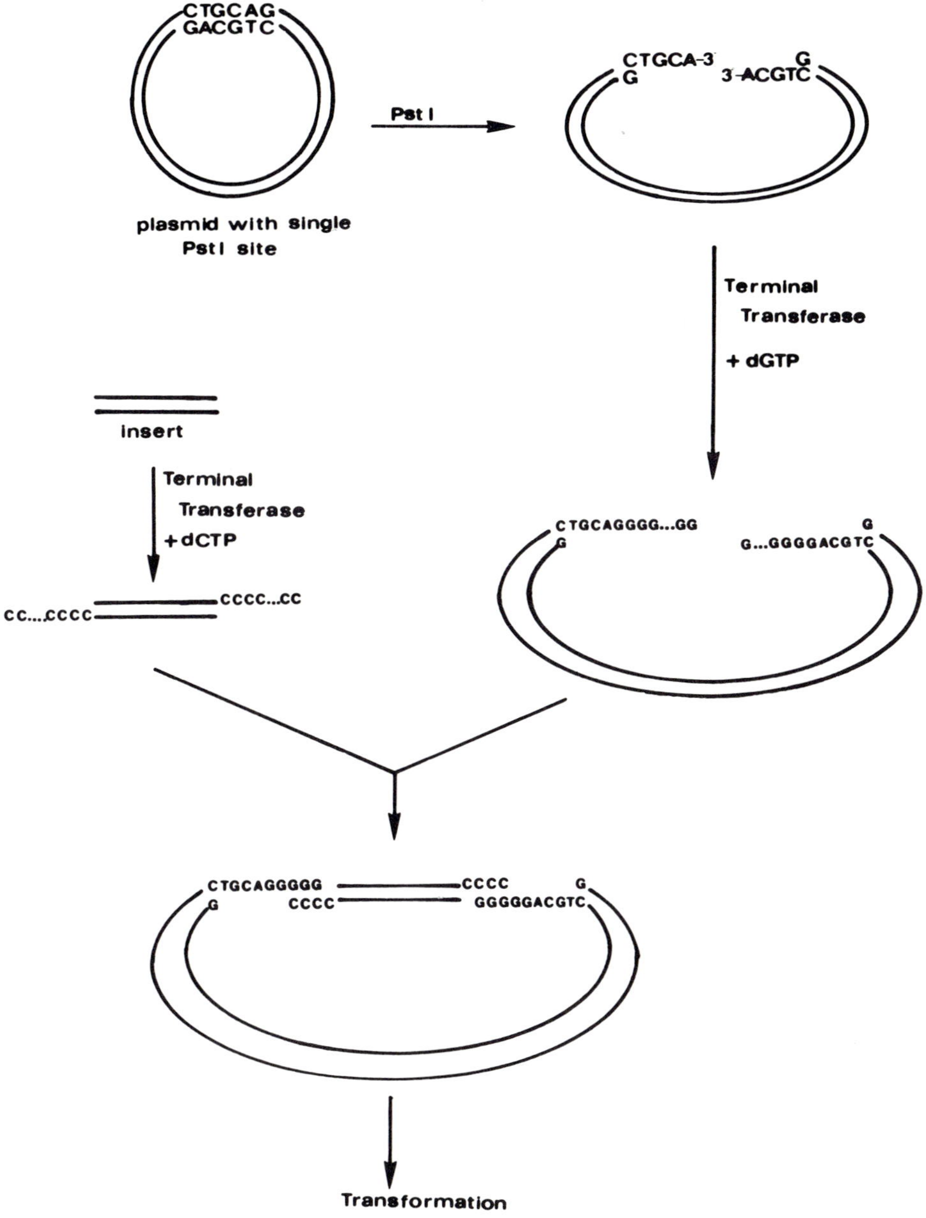

FIGURE 5.   Tailing in the PstI site.

9.   Insert analysis is simplified (Figure 6).

The entire pBR322 nucleotide sequence has been determined.[26] These data greatly simplify the analysis of inserted fragments. It usually is not necessary to purify the insert away from the vector DNA in order to map the restriction cuts within the insert; in fact, it is preferable not to do so. Because the exact size and location of every pBR322 restriction fragment have been derived directly from the sequence,[34] a restriction digest of a chimeric plasmid will contain internal markers of known size. When displayed by gel electrophoresis, the pBR322 fragments should allow very accurate sizing of digestion products. Whenever direct examination of the insert is a necessary step in plasmid selection, the pBR322 sequence should greatly simplify the task.

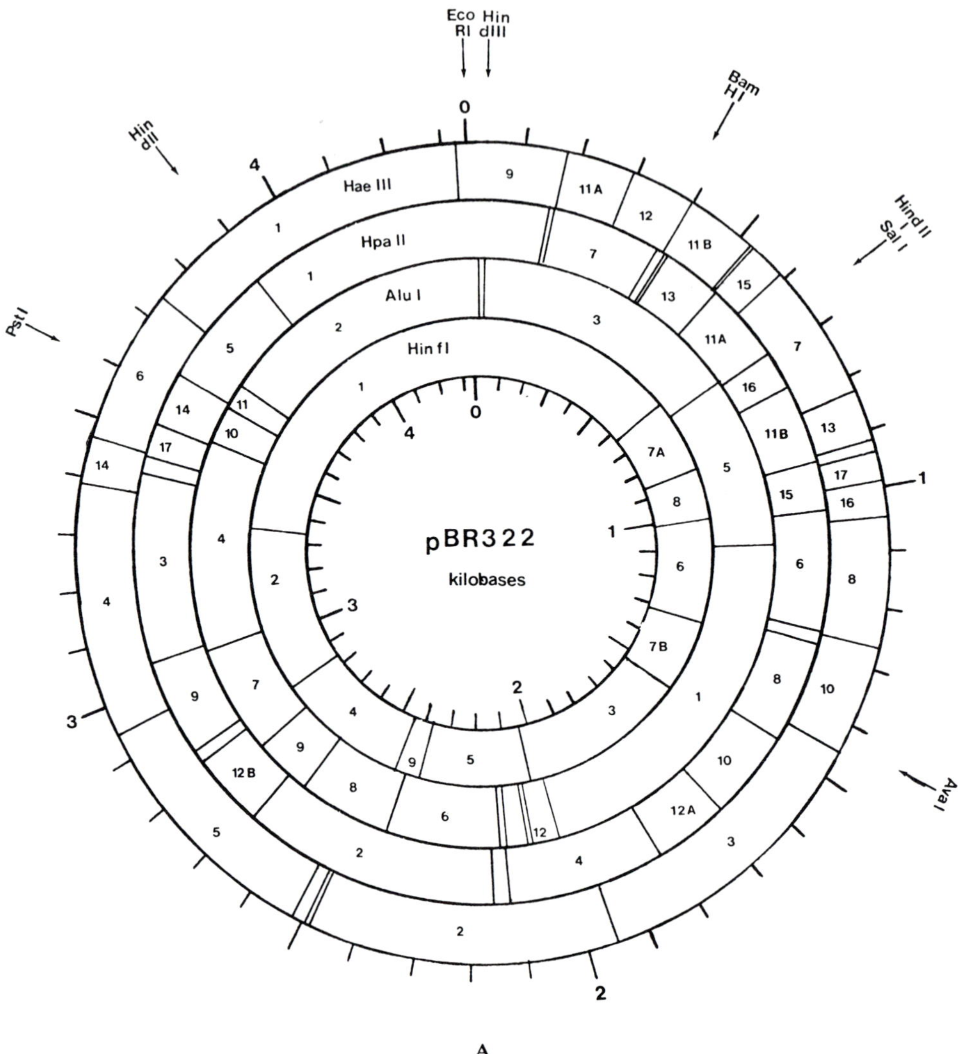

FIGURE 6.   (A,B) Restriction maps for pBR322. (C) Nucleotide sequence of pBR322.

10. Much is known about the biology of pBR322.

The three components of pBR322 have been characterized by several groups. The amp^r gene codes for a β-lactamase of 263 amino acids which inactivates penicillins and cephalosporins.[35,36] The β-lactamase is synthesized with a 23-residue hydrophobic signal peptide which allows the enzyme to traverse the cell membrane.[35] The signal peptide is cleaved from the proenzyme as it is excreted into the periplasm. The structural and functional parameters of the enzyme are currently being investigated extensively.[37,38]

Tetracycline resistance is more complex. At least two proteins appear to be responsible for the high-level tetracycline resistance conferred by pMB9 and, presumably, by pBR322.[28,39] Expression of the larger of the two proteins is controlled by a promoter proximal to the EcoR1 site. This protein is inactivated by cloning in the SalI site. Examination of the DNA sequence of pBR322 reveals that the gene for the larger protein ($\sim$36,000 daltons) probably overlaps the gene for the smaller protein ($\sim$17,000 daltons).[26] The mechanism of tetracycline resistance is

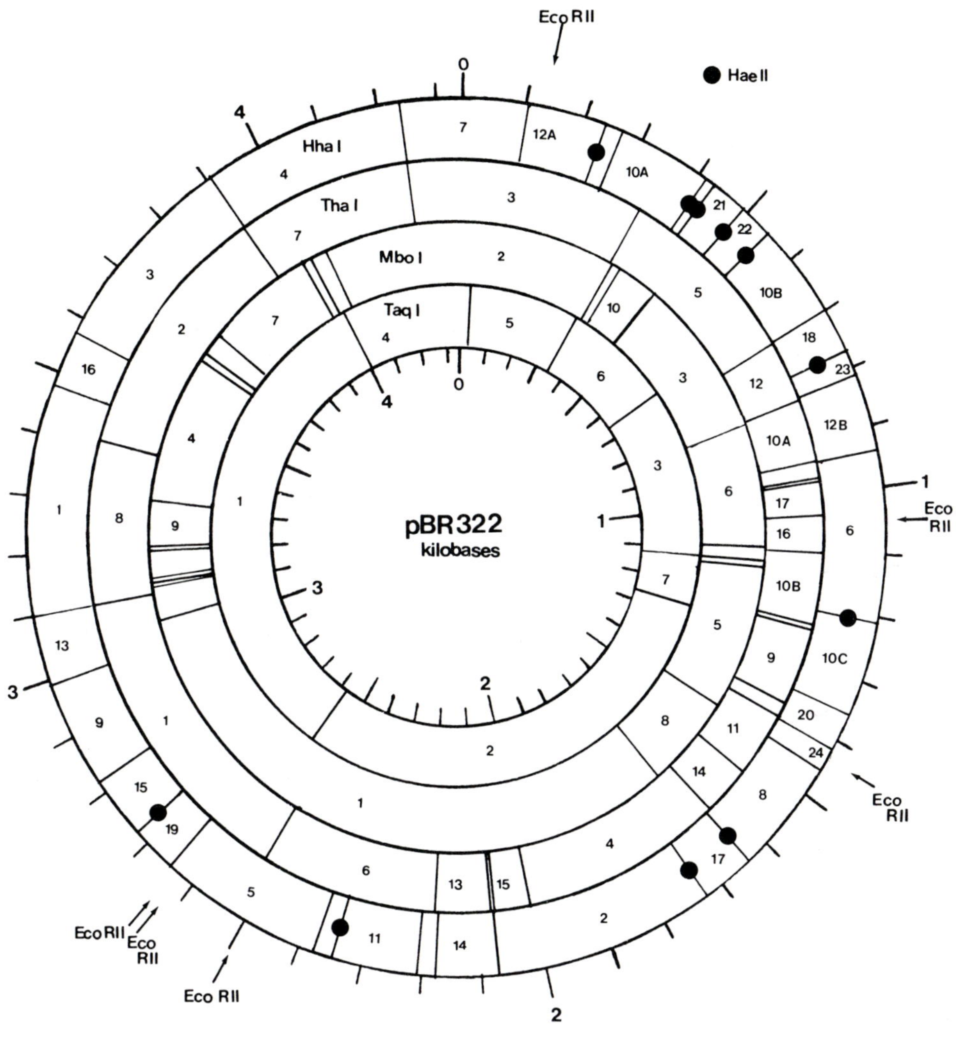

FIGURE 6B.

somewhat baffling because resistance takes many forms. Tetracycline is an inhibitor of protein synthesis. Tetracycline-resistant cells have been variously reported to resist tet uptake, to exclude tet by an energy-dependent mechanism, and to be able to resist elevated intracellular tet concentrations.[39-41] Some plasmids which confer tet resistance when isolated in minicells elaborate several proteins.[39,42] The existence of plasmids which encode only a few of these proteins will simplify the sorting out of various tet-resistance phenotypes.

Sequence analyses reveal that the origins of replication of pBR322 and ColE1 are quite similar.[26,43,44] The replication of ColE1 has been characterized in vivo and in vitro.[9,10,44-47] The ColE1 replication cycle is initiated by the synthesis by RNA polymerase of an RNA primer. DNA polymerase I is required for ColE1 replication.

11. pBR322 is a certified NIH EK2 vector.[48]

pBR322 is only one of several versatile cloning vectors, and there are undoubtedly experimental scenarios for which the others are superior. Chang and Cohen[49] have constructed two plasmids similar in design to pBR322. These plasmids utilize the replication region of p15A, a miniplasmid apparently related ancestrally to ColE1 and

```
(GAA)TTCTCATGTTTGACAGCTTATCATCGATAAGCTTTAATGCGGTAGTTTATCACAGTTAAATTGCTAACGCAGTCAGGCACCGTGTATGAAATCTAACAAT
     .                   .                   .                   .                   .             100.
     AAGAGTACAAACTGTCGAATAGTAGCTATTCGAAATTACGCCATCAAATAGTGTCAATTTAACGATTGCGTCAGTCCGTGGCACATACTTTAGATTGTTA

     GCGCTCATCGTCATCCTCGGCACCGTCACCCTGGATGCTGTAGGCATAGGCTTGGTTATGCCGGTACTGCCGGGCCTCTTGCGGGATATCGTCCATTCCG
     .                   .                   .                   .                   .             200.
     CGCGAGTAGCAGTAGGAGCCGTGGCAGTGGGACCTACGACATCCGTATCCGAACCAATACGGCCATGACGGCCCGGAGAACGCCCTATAGCAGGTAAGGC

     ACAGCATCGCCAGTCACTATGGCGTGCTGCTAGCGCTATATGCGTTGATGCAATTTCTATGCGCACCCGTTCTCGGAGCACTGTCCGACCGCTTTGGCCG
     .                   .                   .                   .                   .             300.
     TGTCGTAGCGGTCAGTGATACCGCACGACGATCGCGATATACGCAACTACGTTAAAGATACGCGTGGGCAAGAGCCTCGTGACAGGCTGGCGAAACCGGC

     CCGCCCAGTCCTGCTCGCTTCGCTACTTGGAGCCACTATCGACTACGCGATCATGGCGACCACACCCGTCCTGTGGATCCTCTACGCCGGACGCATCGTG
     .                   .                   .                   .                   .             400.
     GGCGGGTCAGGACGAGCGAAGCGATGAACCTCGGTGATAGCTGATGCGCTAGTACCGCTGGTGTGGGCAGGACACCTAGGAGATGCGGCCTGCGTAGCAC

     GCCGGCATCACCGGCGCCACAGGTGCGGTTGCTGGCGCCTATATCGCCGACATCACCGATGGGGAAGATCGGGCTCGCCACTTCGGGCTCATGAGCGCTT
     .                   .                   .                   .                   .             500.
     CGGCCGTAGTGGCCGCGGTGTCCACGCCAACGACCGCGGATATAGCGGCTGTAGTGGCTACCCCTTCTAGCCCGAGCGGTGAAGCCCGAGTACTCGCGAA

     GTTTCGGCGTGGGTATGGTGGCAGGCCCGTGGCCGGGGGACTGTTGGGCGCCATCTCCTTGCATGCACCATTCCTTGCGGCGGCGGTGCTCAACGGCCTC
     .                   .                   .                   .                   .             600.
     CAAAGCCGCACCCATACCACCGTCCGGGCACCGGCCCCCTGACAACCCGCGGTAGAGGAACGTACGTGGTAAGGAACGCCGCCGCCACGAGTTGCCGGAG

     AACCTACTACTGGGCTGCTTCCTAATGCAGGAGTCGCATAAGGGAGAGCGTCGACCGATGCCCTTGAGAGCCTTCAACCCAGTCAGCTCCTTCCGGTGGG
     .                   .                   .                   .                   .             700.
     TTGGATGATGACCCGACGAAGGATTACGTCCTCAGCGTATTCCCTCTCGCAGCTGGCTACGGGAACTCTCGGAAGTTGGGTCAGTCGAGGAAGGCCACCC

     CGCGGGGCATGACTATCGTCGCCGCACTTATGACTGTCTTCTTTATCATGCAACTCGTAGGACAGGTGCCGGCAGCGCTCTGGGTCATTTTCGGCGAGGA
     .                   .                   .                   .                   .             800.
     GCGCCCCGTACTGATAGCAGCGGCGTGAATACTGACAGAAGAAATAGTACGTTGAGCATCCTGTCCACGGCCGTCGCGAGACCCAGTAAAAGCCGCTCCT




     CCGCTTTCGCTGGAGCGCGACGATGATCGGCCTGTCGCTTGCGGTATTCGGAATCTTGCACGCCCTCGCTCAAGCCTTCGTCACTGGTCCCGCCACCAAA
     .                   .                   .                   .                   .             900.
     GGCGAAAGCGACCTCGCGCTGCTACTAGCCGGACAGCGAACGCCATAAGCCTTAGAACGTGCGGGAGCGAGTTCGGAAGCAGTGACCAGGGCGGTGGTTT

     CGTTTCGGCGAGAAGCAGGCCATTATCGCCGGCATGGCGGCCGACGCGCTGGGCTACGTCTTGCTGGCGTTCGCGACGCGAGGCTGGATGGCCTTCCCCA
     .                   .                   .                   .                   .             1000.
     GCAAAGCCGCTCTTCGTCCGGTAATAGCGGCCGTACCGCCGGCTGCGCGACCCGATGCAGAACGACCGCAAGCGCTGCGCTCCGACCTACCGGAAGGGGT

     TTATGATTCTTCTCGCTTCCGGCGGCATCGGGATGCCCGCGTTGCAGGCCATGCTGTCCAGGCAGGTAGATGACGACCATCAGGGACAGCTTCAAGGATC
     .                   .                   .                   .                   .             1100.
     AATACTAAGAAGAGCGAAGGCCGCCGTAGCCCTACGGGCGCAACGTCCGGTACGACAGGTCCGTCCATCTACTGCTGGTAGTCCCTGTCGAAGTTCCTAG

     GCTCGCGGCTCTTACCAGCCTAACTTCGATCACTGGACCGCTGATCGTCACGGCGATTTATGCCGCCTCGGCGAGCACATGGAACGGGTTGGCATGGATT
     .                   .                   .                   .                   .             1200.
     CGAGCGCCGAGAATGGTCGGATTGAAGCTAGTGACCTGGCGACTAGCAGTGCCGCTAAATACGGCGGAGCCGCTCGTGTACCTTGCCCAACCGTACCTAA

     GTAGGCGCCGCCCTATACCTTGTCTGCCTCCCCGCGTTGCGTCGCGGTGCATGGAGCCGGGCCACCTCGACCTGAATGGAAGCCGGCGGCACCTCGCTAA
     .                   .                   .                   .                   .             1300.
     CATCCGCGGCGGGATATGGAACAGACGGAGGGGCGCAACGCAGCGCCACGTACCTCGGCCCGGTGGAGCTGGACTTACCTTCGGCCGCCGTGGAGCGATT

     CGGATTCACCACTCCAAGAATTGGAGCCAATCAATTCTTGCGGAGAACTGTGAATGCGCAAACCAACCCTTGGCAGAACATATCCATCGCGTCCGCCATC
     .                   .                   .                   .                   .             1400.
     GCCTAAGTGGTGAGGTTCTTAACCTCGGTTAGTTAAGAACGCCTCTTGACACTTACGCGTTTGGTTGGGAACCGTCTTGTATAGGTAGCGCAGGCGGTAG

     TCCAGCAGCCGCACGCGGCGCATCTCGGGCAGCGTTGGGTCCTGGCCACGGGTGCGCATGATCGTGCTCCTGTCGTTGAGGACCCGGCTAGGCTGGCGGG
     .                   .                   .                   .                   .             1500.
     AGGTCGTCGGCGTGCGCCGCGTAGAGCCCGTCGCAACCCAGGACCGGTGCCCACGCGTACTAGCACGAGGACAGCAACTCCTGGGCCGATCCGACCGCCC

     GTTGCCTTACTGGTTAGCAGAATGAATCACCGATACGCGAGCGAACGTGAAGCGACTGCTGCTGCAAAACGTCTGCGACCTGAGCAACAACATGAATGGT
     .                   .                   .                   .                   .             1600.
     CAACGGAATGACCAATCGTCTTACTTAGTGGCTATGCGCTCGCTTGCACTTCGCTGACGACGACGTTTTGCAGACGCTGGACTCGTTGTTGTACTTACCA
```

FIGURE 6C.

```
CTTCGGTTTCCGTGTTTCGTAAAGTCTGGAAACGCGGAAGTCAGCGCCCTGCACCATTATGTTCCGGATCTGCATCGCAGGATGCTGCTGGCTACCCTGT
         .                   .                   .                   .                   .        1700.
GAAGCCAAAGGCACAAAGCATTTCAGACCTTTGCGCCTTCAGTCGCGGGACGTGGTAATACAAGGCCTAGACGTAGCGTCCTACGACGACCGATGGGACA

GGAACACCTACATCTGTATTAACGAAGCGCTGGCATTGACCCTGAGTGATTTTTCTCTGGTCCCGCCGCATCCATACCGCCAGTTGTTTACCCTCACAAC
         .                   .                   .                   .                   .        1800.
CCTTGTGGATGTAGACATAATTGCTTCGCGACCGTAACTGGGACTCACTAAAAAGAGACCAGGGCGGCGTAGGTATGGCGGTCAACAAATGGGAGTGTTG

GTTCCAGTAACCGGGCATGTTCATCATCAGTAACCCGTATCGTGAGCATCCTCTCTCGTTTCATCGGTATCATTACCCCCATGAACAGAAATTCCCCCTT
         .                   .                   .                   .                   .        1900.
CAAGGTCATTGGCCCGTACAAGTAGTAGTCATTGGGCATAGCACTCGTAGGAGAGAGCAAAGTAGCCATAGTAATGGGGGTACTTGTCTTTAAGGGGGAA

ACACGGAGGCATCAAGTGACCAAACAGGAAAAAAACCGCCCTTAACATGGCCCGCTTTATCAGAAGCCAGACATTAACGCTTCTGGAGAAACTCAACGAGC
         .                   .                   .                   .                   .        2000.
TGTGCCTCCGTAGTTCACTGGTTTGTCCTTTTTTGGCGGGAATTGTACCGGGCGAAATAGTCTTCGGTCTGTAATTGCGAAGACCTCTTTGAGTTGCTCG

TGGACGCGGATGAACAGGCAGACATCTGTGAATCGCTTCACGACCACGCTGATGAGCTTTACCGCAGCTGCCTCGCGCGTTTCGGTGATGACGGTGAAAA
         .                   .                   .                   .                   .        2100.
ACCTGCGCCTACTTGTCCGTCTGTAGACACTTAGCGAAGTGCTGGTGCGACTACTCGAAATGGCGTCGACGGAGCGCGCAAAGCCACTACTGCCACTTTT

CCTCTGACACATGCAGCTCCCGGAGACGGTCACAGCTTGTCTGTAAGCGGATGCCGGGAGCAGACAAGCCCGTCAGGGCGCGTCAGCGGGTGTTGGCGGG
         .                   .                   .                   .                   .        2200.
GGAGACTGTGTACGTCGAGGGCCTCTGCCAGTGTCGAACAGACATTCGCCTACGGCCCTCGTCTGTTCGGGCAGTCCCGCGCAGTCGCCCACAACCGCCC

TGTCGGGGCGCAGCCATGACCCAGTCACGTAGCGATAGCGGAGTGTATACTGGCTTAACTATGCGGCATCAGAGCAGATTGTACTGAGAGTGCACCATAT
         .                   .                   .                   .                   .        2300.
ACAGCCCCGCGTCGGTACTGGGTCAGTGCATCGCTATCGCCTCACATATGACCGAATTGATACGCCGTAGTCTCGTCTAACATGACTCTCACGTGGTATA

GCGGTGTGAAATACCGCACAGATGCGTAAGGAGAAAATACCGCATCAGGCGCTCTTCCGCTTCCTCGCTCACTGACTCGCTGCGCTCGGTCGTTCGGCTG
         .                   .                   .                   .                   .        2400.
CGCCACACTTTATGGCGTGTCTACGCATTCCTCTTTTATGGCGTAGTCCGCGAGAAGGCGAAGGAGCGAGTGACTGAGCGACGCGAGCCAGCAAGCCGAC


CGGCGAGCGGTATCAGCTCACTCAAAGGCGGTAATACGGTTATCCACAGAATCAGGGGATAACGCAGGAAAGAACATGTGAGCAAAAGGCCAGCAAAAGG
         .                   .                   .                   .                   .        2500.
GCCGCTCGCCATAGTCGAGTGAGTTTCCGCCATTATGCCAATAGGTGTCTTAGTCCCCTATTGCGTCCTTTCTTGTACACTCGTTTTCCGGTCGTTTTCC

CCAGGAACCGTAAAAAGGCCGCGTTGCTGGCGTTTTTCCATAGGCTCCGCCCCCCTGACGAGCATCACAAAAATCGACGCTCAAGTCAGAGGTGGCGAAA
         .                   .                   .                   .                   .        2600.
GGTCCTTGGCATTTTTCCGGCGCAACGACCGCAAAAAGGTATCCGAGGCGGGGGGGACTGCTCGTAGTGTTTTTAGCTGCGAGTTCAGTCTCCACCGCTTT

CCCGACAGGACTATAAAGATACCAGGCGTTTCCCCCTGGAAGCTCCCTCGTGCGCTCTCCTGTTCCGACCCTGCCGCTTACCGGATACCTGTCCGCCTTT
         .                   .                   .                   .                   .        2700.
GGGCTGTCCTGATATTTCTATGGTCCGCAAAGGGGGACCTTCGAGGGAGCACGCGAGAGGACAAGGCTGGGACGGCGAATGGCCTATGGACAGGCGGAAA

CTCCCTTCGGGAAGCGTGGCGCTTTCTCAATGCTCACGCTGTAGGTATCTCAGTTCGGTGTAGGTCGTTCGCTCCAAGCTGGGCTGTGTGCACGAACCCC
         .                   .                   .                   .                   .        2800.
GAGGGAAGCCCTTCGCACCGCGAAAGAGTTACGAGTGCGACATCCATAGAGTCAAGCCACATCCAGCAAGCGAGGTTCGACCCGACACACGTGCTTGGGG

CCGTTCAGCCCGACCGCTGCGCCTTATCCGGTAACTATCGTCTTGAGTCCAACCCGGTAAGACACGACTTATCGCCACTGGCAGCAGCCACTGGTAACAG
         .                   .                   .                   .                   .        2900.
GGCAAGTCGGGCTGGCGACGCGGAATAGGCCATTGATAGCAGAACTCAGGTTGGGCCATTCTGTGCTGAATAGCGGTGACCGTCGTCGGTGACCATTGTC

GATTAGCAGAGCGAGGTATGTAGGCGGTGCTACAGAGTTCTTGAAGTGGTGGCCTAACTACGGCTACACTAGAAGGACAGTATTTGGTATCTGCGCTCTG
         .                   .                   .                   .                   .        3000.
CTAATCGTCTCGCTCCATACATCCGCCACGATGTCTCAAGAACTTCACCACCGGATTGATGCCGATGTGATCTTCCTGTCATAAACCATAGACGCGAGAC

CTGAAGCCAGTTACCTTCGGAAAAAGAGTTGGTAGCTCTTGATCCGGCAAACAAACCACCGCTGGTAGCGGTGGTTTTTTTGTTTGCAAGCAGCAGATTA
         .                   .                   .                   .                   .        3100.
GACTTCGGTCAATGGAAGCCTTTTTCTCAACCATCGAGAACTAGGCCGTTTGTTTGGTGGCGACCATCGCCACCAAAAAAACAAACGTTCGTCGTCTAAT

CGCGCAGAAAAAAAGGATCTCAAGAAGATCCTTTGATCTTTTCTACGGGGTCTGACGCTCAGTGGAACGAAAACTCACGTTAAGGGATTTTGGTCATGAG
         .                   .                   .                   .                   .        3200.
GCGCGTCTTTTTTTTCCTAGAGTTCTTCTAGGAAACTAGAAAAGATGCCCCAGACTGCGAGTCACCTTGCTTTTGAGTGCAATTCCCTAAAAACCAGTACTC
```

FIGURE 6C (continued).

```
ATTATCAAAAAGGATCTTCACCTAGATCCTTTTTAAATTAAAAATGAAGTTTTAAATCAATCTAAAGTATATATGAGTAAACTTGGTCTGACAGTTACCAA
         .                   .                   .                   .                   .          3300.
TAATAGTTTTTCCTAGAAGTGGATCTAGGAAAATTTAATTTTTACTTCAAAATTTAGTTAGATTTCATATATACTCATTTGAACCAGACTGTCAATGGTT

TGCTTAATCAGTGAGGCACCTATCTCAGCGATCTGTCTATTTCGTTCATCCATAGTTGCCTGACTCCCCGTCGTGTAGATAACTACGATACGGGAGGGCT
         .                   .                   .                   .                   .          3400.
ACGAATTAGTCACTCCGTGGATAGAGTCGCTAGACAGATAAAGCAAGTAGGTATCAACGGACTGAGGGGCAGCACATCTATTGATGCTATGCCCTCCCGA

TACCATCTGGCCCCAGTGCTGCAATGATACCGCGAGACCCACGCTCACCGGCTCCAGATTTATCAGCAATAAACCAGCCAGCCGGAAGGGCCGAGCGCAG
         .                   .                   .                   .                   .          3500.
ATGGTAGACCGGGGTCACGACGTTACTATGGCGCTCTGGGTGCGAGTGGCCGAGGTCTAAATAGTCGTTATTTGGTCGGTCGGCCTTCCCGGCTCGCGTC

AAGTGGTCCTGCAACTTTATCCGCCTCCATCCAGTCTATTAATTGTTGCCGGGAAGCTAGAGTAAGTAGTTCGCCAGTTAATAGTTTGCGCAACGTTGTT
         .                   .                   .                   .                   .          3600.
TTCACCAGGACGTTGAAATAGGCGGAGGTAGGTCAGATAATTAACAACGGCCCTTCGATCTCATTCATCAAGCGGTCAATTATCAAACGCGTTGCAACAA

GCCATTGCTGCAGGCATCGTGGTGTCACGCTCGTCGTTTGGTATGGCTTCATTCAGCTCCGGTTCCCAACGATCAAGGCGAGTTACATGATCCCCCATGT
         .                   .                   .                   .                   .          3700.
CGGTAACGACGTCCGTAGCACCACAGTGCGAGCAGCAAACCATACCGAAGTAAGTCGAGGCCAAGGGTTGCTAGTTCCGCTCAATGTACTAGGGGGTACA

TGTGCAAAAAAGCGGTTAGCTCCTTCGGTCCTCCGATCGTTGTCAGAAGTAAGTTGGCCGCAGTGTTATCACTCATGGTTATGGCAGCACTGCATAATTC
         .                   .                   .                   .                   .          3800.
ACACGTTTTTTCGCCAATCGAGGAAGCCAGGAGGCTAGCAACAGTCTTCATTCAACCGGCGTCACAATAGTGAGTACCAATACCGTCGTGACGTATTAAG

TCTTACTGTCATGCCATCCGTAAGATGCTTTTCTGTGACTGGTGAGTACTCAACCAAGTCATTCTGAGAATAGTGTATGCGGCGACCGAGTTGCTCTTGC
         .                   .                   .                   .                   .          3900.
AGAATGACAGTACGGTAGGCATTCTACGAAAAGACACTGACCACTCATGAGTTGGTTCAGTAAGACTCTTATCACATACGCCGCTGGCTCAACGAGAACG

CCGGCGTCAACACGGGATAATACCGCGCCACATAGCAGAACTTTAAAAGTGCTCATCATTGGAAAACGTTCTTCGGGGCGAAAACTCTCAAGGATCTTAC
         .                   .                   .                   .                   .          4000.
GGCCGCAGTTGTGCCCTATTATGGCGCGGTGTATCGTCTTGAAATTTTCACGAGTAGTAACCTTTTGCAAGAAGCCCCGCTTTTGAGAGTTCCTAGAATG


CGCTGTTGAGATCCAGTTCGATGTAACCCACTCGTGCACCCAACTGATCTTCAGCATCTTTTACTTTCACCAGCGTTTCTGGGTGAGCAAAAACAGGAAG
         .                   .                   .                   .                   .          4100.
GCGACAACTCTAGGTCAAGCTACATTGGGTGAGCACGTGGGTTGACTAGAAGTCGTAGAAAATGAAAGTGGTCGCAAAGACCCACTCGTTTTTGTCCTTC

GCAAAATGCCGCAAAAAAGGGAATAAGGGCGACACGGAAATGTTGAATACTCATACTCTTCCTTTTTCAATATTATTGAAGCATTTATCAGGGTTATTGT
         .                   .                   .                   .                   .          4200.
CGTTTTACGGCGTTTTTTCCCTTATTCCCGCTGTGCCTTTACAACTTATGAGTATGAGAAGGAAAAAGTTATAATAACTTCGTAAATAGTCCCAATAACA

CTCATGAGCGGATACATATTTGAATGTATTTAGAAAAATAAACAAATAGGGGTTCCGCGCACATTTCCCCGAAAAGTGCCACCTGACGTCTAAGAAACCA
         .                   .                   .                   .                   .          4300.
GAGTACTCGCCTATGTATAAACTTACATAAATCTTTTTATTTGTTTATCCCCAAGGCGCGTGTAAAGGGGCTTTTCACGGTGGACTGCAGATTCTTTGGT

TTATTATCATGACATTAACCTATAAAAATAGGCGTATCACGAGGCCCTTTCGTCTTCAAGAA(TTC)
         .                   .                   .                   .
AATAATAGTACTGTAATTGGATATTTTTATCCGCATAGTGCTCCGGGAAAGCAGAAGTTCTT
```

FIGURE 6C (continued)

pMB1.[50,51] It can be amplified with chloramphenicol, requires DNA polymerase I for replication, and has DNA homology with ColE1 in the replicator region. However, p15A or one of its derivatives (such as pACYC177 or pACYC184) can cohabit the same cell as a ColE1- or pMB1-derived plasmid. Normally any two ColE1-type plasmids are mutually incompatible in the same host.[52] Compatible vectors make it possible to construct strains stably carrying two different cloned fragments.

As is the case for pBR322, pACYC177 and pACYC184 contain segments from several different ancestral plasmids.[49] Restriction maps are shown in Figure 7. DNA fragments with cohesive ends generated by PstI, HindIII, and XhoI can be cloned directly into the corresponding cleavage sites in pACYC177. Blunt-end ligation is possible in the HindII or SmaI sites. Insertions in one of these sites inactivate either the ampicillin or kanamycin resistance genes. In the case of pACYC184, cloned EcoR1 fragments can be detected by insertional inactivation of the chloramphenicol gene. The tet' gene is derived from pSC101, so BamHI, SalI, and some HindIII inserts inactivate tet'.

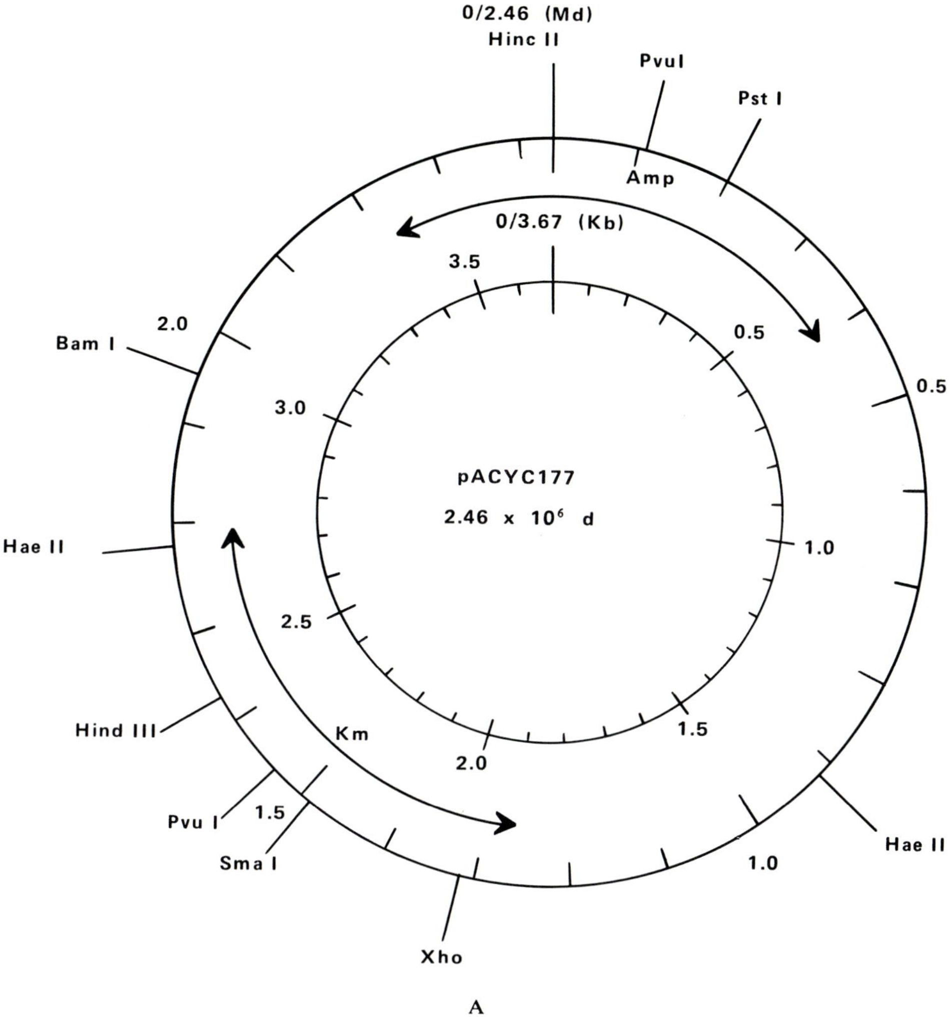

FIGURE 7.   (A) Restriction map for pACYC177. (B) Restriction map for pACYC184.

Another plasmid which is compatible with ColE1-type plasmids is RSF1030, a 5.6×10^6 dalton vector which has a relaxed replicator and confers ampicillin resistance.[15] RSF1030 contains a single BamHI site suitable for cloning inserts. ColE1 and RSF1030 share no DNA homology and will not recombine with each other in *rec⁺ E. coli* unless they contain homologous inserts.[61]

The properties of a number of different plasmid vectors that have been discussed are summarized by Table 1. pBR322, pACYC177, and pACYC184 are extremely versatile vectors and are the prime candidates to utilize for a modern cloning experiment. Figure 8 represents a condensation of the ideas of many authors[28,53] as to what characteristics the ideal general-purpose plasmid (at the current conceptual level) should possess. These three plasmids satisfy most of the requirements. However, for some special cloning purposes, vectors of less general application are more useful.

A. Mini-plasmids

Whenever the desired insert allows a positive selection, it is unnecessary for the vector itself to carry a selectable phenotype. It is also unnecessary to have insertional inactivation as an available screen, since all candidates which acquire the positive phen-

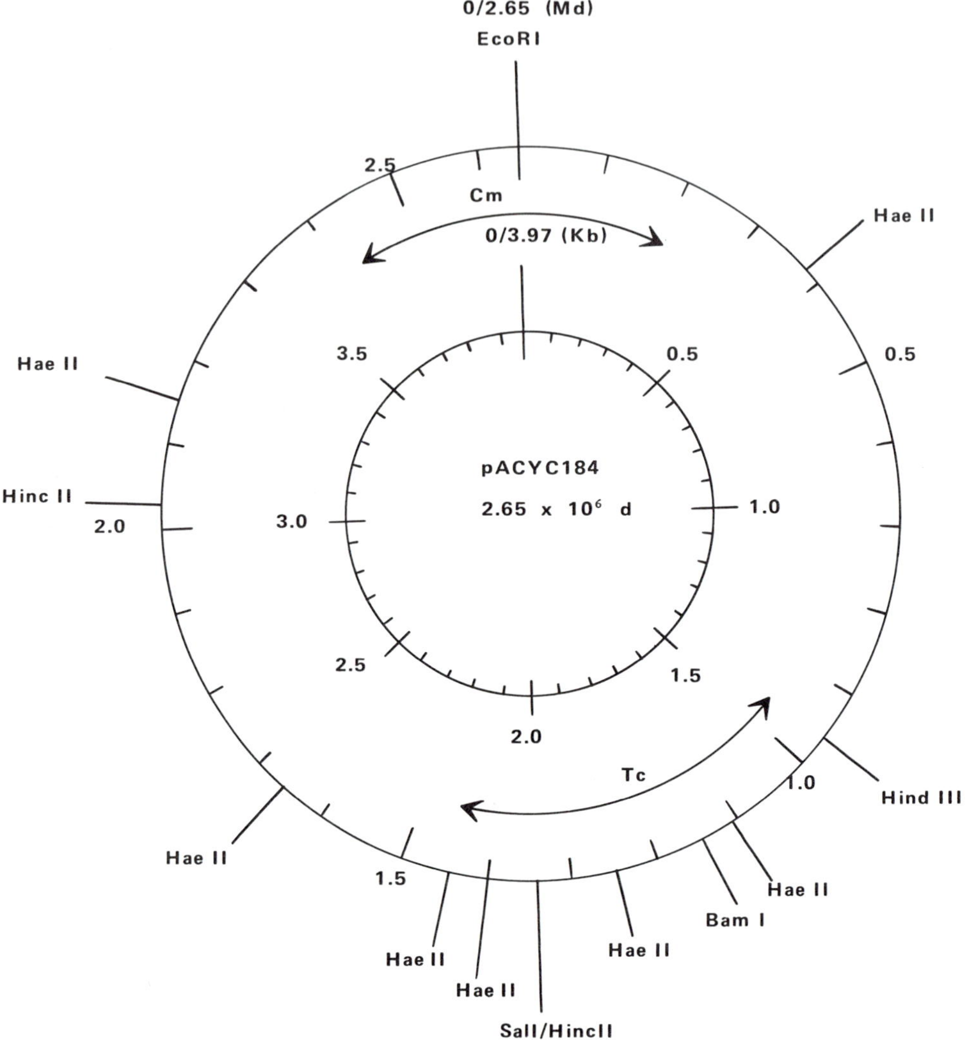

FIGURE 7B.

otype must carry an insert. It is possible, therefore, to discard most of the extra baggage and to utilize a vector which supplies only a relaxed replicator. Such plasmids provide an effective enrichment of fourfold in the insert-to-vector ratio. Three plasmids of this sort are mini-colE1[21] and pMB8,[16] which were previously mentioned, and pBR345.[54]

pBR345 carries two very small inserts of the lac operator at its EcoR1 site. These operators, when present at 20 or so copies per cell, compete with the host's chromosomal operators for lac repressor molecules. Lac⁺ cells that contain pBR345 (or other relaxed plasmids carrying the lac operator) become constitutive for lactose operon expression and, in particular, for β-galactosidase production. Such cells can be detected because they form blue colonies on X-gal indicator plates.[55] The formation of blue colonies is the only phenotypic marker of this plasmid.[54]

When pBR345 is opened with EcoR1, the lac operator inserts are released. The remaining plasmid is only 1160 basepairs long and is suitable for cloning any EcoR1 insert for which a positive selection exists. The complete nucleotide sequence of

TABLE 1

Plasmid Vectors

Plasmid	Size md	Size Kb	Copy number	Amplifiable with CAM	Replicon	Compatibility group	Phenotype conferred by plasmid	Single-site restriction enzymes	Insertional inactivation phenotypes	Ref.
pSC101	5.8	8.9	2—5	—	pSC101	pSC101	Tc^r	EcoR1	—	3, 4
ColE1	4.2	6.5	25—30	+	ColE1	ColE1	EI immunity Colicin production	EcoR1	loss EI production	5
pSF2124	7.4	11.4	20	+	ColE1	ColE1	Ap^r EI immunity Colicin production	BamHI EcoR1	loss EI production loss EI production	12
pCR1	8.7	13.4	16	+	ColE1	ColE1	Km^r EI immunity Colicin production	EcoR1 SalI PstI HindIII	loss EI production — — Km'	100
pTK16	2.8	4.2		+	ColE1	ColE1	Tc^r Km^r	EcoR1 BamHI SalI	— Tc' Tc'	97
pMB9	3.5	5.4	60	+	pMB1	ColE1	Tc^r E1 immunity	EcoR1 HindIII SalI BamHI	— Tc' Tc' Tc'	16
pBR313	5.8	8.9	60	+	pMB1	ColE1	Ap^r Tc^r	EcoR1 HpaI XmaI HindIII SalI BamHI	— — — Tc' Tc' Tc'	22
pBR322	2.8	4.3	60	+	pMB1	ColE1	Ap^r Tc^r	EcoR1 PstI SalI BamHI HindIII AvaI	— Ap' Tc' Tc' Tc' ?	23, 26
pACYC177	2.46	3.54		+	p15A	p15A	Ap^r Km^r	HincII PstI BamI HindIII	Amp' Amp' — Km'	49

TABLE 1 (continued)

Plasmid Vectors

Plasmid	Size md	Size Kb	Copy number	Amplifiable with CAM	Replicon	Compatibility group	Phenotype conferred by plasmid	Single-site restriction enzymes	Insertional inactivation phenotypes	Ref.
pACYC184	2.65	3.9		+	p15A	p15A	Cm^r Tc^r	Xho SmaI EcoR1 SalI BamI HindIII	Km^r Km^r Cm^r Tc^r Tc^r (Tc^r)	49, 99 104
RSF1030	5.6	8.6	20—40	+	RSF1030	RSF1030	Amp^r	BamHI	—	15
pMK20	2.4	3.7	70	+	ColE1	ColE1	Km^r E1 immunity	HindIII SamI EcoR1 PstI	Km^r Km^r — —	97
pMK2004	3.2	4.9	50	+	pMB1	ColE1	Km^r Tc^r Ap^r	SmaI SalI BamHI PstI HpaI	Km^r Tc^r Tc^r Ap^r —	97
pDF41	8.5	13.1	1—2	—	F	F	trpE	EcoR1 HindIII SalI BamHI	— — — —	98
pML21	6.6	10.2	30	+	ColE1	ColE1	Km^r E1 immunity	SalI	—	21
pGM706	13.1	20.2		+	ColE1 + PSC101	ColE1 + pSC101	Tc^r Ap^r	SalI	Tc^r	19
pGM16	12.8	19.7		+	ColE1 + pSC101	ColE1 + pSC101	Tc^r Km^r	BamHI	Tc^r	19
Mini-ColE1	2.2		200	+	ColE1	ColE1	E1 immunity	EcoR1		21
pBR345	0.8	1.2		+	pMB1	ColE1	*LacI*$^-$	EcoR1	—	54
RK2	37.0	57		—	RK2	P	Tc^r Km^r Ap^r	HindIII EcoR1 BamHI		53
pRK248	5.6	8.6		—	RK2	P	Tc^r	SalI EcoR1	Tc^r —	91

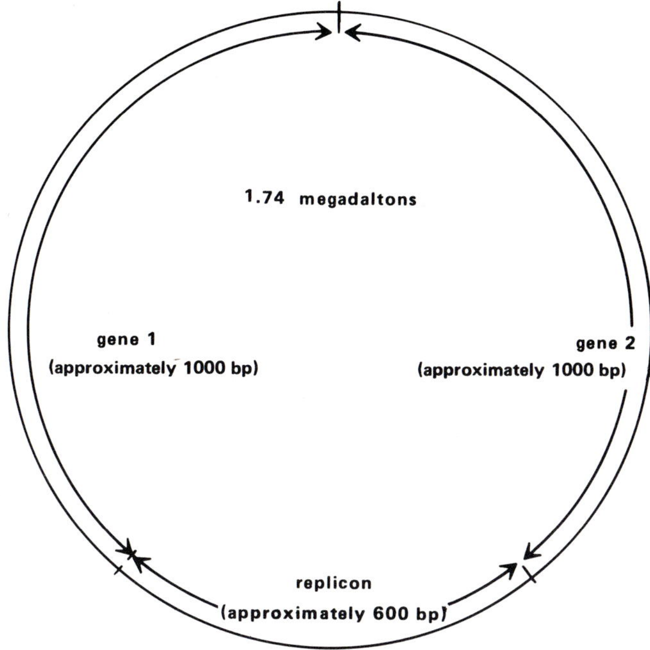

(approximately 600 bp)

FIGURE 8.   The ideal plasmid. Desired features:

1.   Two marker genes
   A.   Resistance to bacteriostatic drug (e.g., tetracyclene)
   B.   Gene which satisfies auxotrophy in host bacterium
        (e.g., amino acid biosynthesis gene)

2.   A relaxed amplifiable replicon with a broad host range

3.   Several single-site restriction enzymes which
   A.   Do not cut replicon
   B.   Generate $5'PO_4^-$ single-stranded ends (substrates for
        BAP)
   C.   Generate $3'OH$ single-stranded ends (substrates for
        terminal transferase)
   D.   Generate blunt ends
   E.   Result in insertional inactivation of marker gene
        when cloned DNA sequence is ligated
   F.   Cut adjacent to promoter and ribosome binding site
        of inactivated gene

4.   Known nucleotide sequence

5.   Safety features
   A.   Decreased transmissibility in presence of conjugative
        plasmid
   B.   Temperature-sensitive replication at mammalian
        body temperatures

pBR345 has been determined and these data provide the same advantages which are afforded by the pBR322 sequence.[54]

B. Specialized Vectors

Some cloning endeavors require that the vector play an active role in the expression of the insert's phenotype. The vector may be required to provide the transcriptional promoter or ribosome binding site, either because the insert's signals are absent or too weak or because the insert is of eukaryotic origin and has no such signals which will function in a bacterium. It may be desirable to provide control over insert expression. If it is desired that a protein be secreted by the bacterium, then it may be necessary to splice a secretion signal onto its amino terminus. In a somewhat extreme case, an expressed protein may be unstable in the cell and require a covalent polypeptide extension in order to survive for analysis.

Two types of promotor plasmids have been developed. The so-called "portable promotors" are short fragments containing lac promotors flanked by EcoR1 sites. When an EcoRI site is filled out with DNA polymerase and then joined by blunt-end ligation to either a HaeIII or AluI restriction fragment, the EcoR1 site is regenerated. In order to construct a portable lac promotor, Fuller inserted a 203-basepair HaeIII fragment carrying a catabolite activator-independent lac promotor (UV5), the lac z ribosome binding site, and the first eight lac z codons into the filled-out EcoR1 site of pMB9.[56] The chimeric plasmid, pOP203-1, is detectable in lac⁺ colonies because, like pBR345, it makes them constitutive for β-galactosidase synthesis and, hence, blue on X-gal plates. Backman et al.[31] cut the promoter from pOP203-1 with EcoRI and inserted it into an EcoRI site in front of the λcI gene, which codes for the λ immunity repressor. This hybrid plasmid, pKB252, produces λ repressor under the control of the "portable" lac promotor.

Johnsrud constructed a portable promotor plasmid starting with a 95-basepair AluI fragment from the UV5 lac operator region.[57] This plasmid, pLJ3, has two aligned UV5 lac promoters cloned into the EcoR1 site in pMB9. This insert, flanked by EcoR1 sites, may be cut out by EcoR1 such that the lac z ribosome binding site immediately precedes the cut. Backman and Ptashne moved the pLJ3 portable promotor in front of the λcI gene to form pKB280.[58] In this plasmid, the lac z ribosome binding site functions to initiate cI synthesis. The initiator methionine codon is provided by the cI gene.

Another class of promotor plasmids, exemplified by pBGP120, pOP203-3, and pBH20, utilizes a "stationary" lac promotor pointing at a unique EcoR1 site. On the plasmid, transcription initiated at the lac promotor proceeds through the EcoR1 site. On pOP203-3 (a variant of pOP203-1)[56] and pBH20 (lac HaeIII-203 inserted in pBR322),[59] transcription must cover the first eight codons of the lac z gene before reaching the EcoR1 site; pBGP120 transcription must transverse considerably more of the z gene.[60] Another stationary promotor plasmid, pOP95 (constructed by Fuller), carries the pLJ3 lac Alu 95 fragment locked into pMB9. The EcoR1 site of pOP95 is preceded by the lac z ribosome binding site.[56]

Itakura et al.[59] cloned a chemically synthesized gene for the hormone somatostatin in the EcoR1 site of pBH20. They were unable to detect somatostatin synthesis in cells carrying the plasmid pSomI and concluded that the hormone, which is only 14 amino acids long, was being degraded in the bacterium. By fusing the somatostatin insert into an EcoR1 site inside the lac z gene, Itakura et al. detected synthesis of a fused β-galactosidase — somatostatin peptide. Apparently, the amino terminus of β-galactosidase was sufficient to stop degradation of the peptide hormone.

III. SOURCES OF DNA

The plasmid vector is only a means to achieve an end. The important ingredient of a cloning experiment is the insert DNA. In most cases, that DNA is a scarce commodity; therefore, much work has gone into engineering the other facets of cloning in order to optimize insert formation. A few details of the preparation of DNA for cloning will be briefly mentioned here.

Any method of truncating large DNA molecules is a potential tool for insert preparation. It is possible to generate a complete array of DNA fragments covering any region of a chromosome by sonic shearing.[62] Such fragments can be cloned by using terminal transferase to extend their 3′-OH ends with a homopolynucleotide tail. These tailed fragments can then be annealed to opened plasmid vectors that carry tails of the complementary homopolynucleotide and sealed to the plasmid with DNA polymerase and ligase. Similarly, the cDNA synthesized by the action of reverse transcriptase using either total mRNA or a purified messenger as a template can be tailed with terminal transferase and cloned as described above.[63-65]

Fragments cloned by tailing are difficult to recover from the vector. Unless the insert has regenerated PstI sites, there will be no flanking restriction cuts. Even if there are PstI sites, these will be separated from the insert by a homopolymer stretch of sequence. If possible, it is more useful to clone fragments into sticky-ended restriction sites.

The cleavage site for a restriction endonuclease which recognizes a hexanucleotide sequence is expected to appear every 4096 basepairs, on the average. However, a given enzyme may not produce a fragment which carries the entire region of the interesting gene for every case. Several antibiotics, such as distamycin A, actinomycin D, and bleomycin, bind to DNA in a nonrandom fashion. In the presence of one of these antibiotics, a restriction enzyme cleaves its different sites at considerably different rates.[66,67] It is possible to generate partial restriction digests such that some sites have been completely cut and others barely cut at all. In this manner, it is sometimes possible to get a fragment that spans a gene that otherwise would have been cut by the enzyme. Distamycin A partial digestions have allowed the cloning of *Dictyostelium discoideium* rRNA genes[68] and the *Klebsiella pneumoniae* nitrogen fixation (nif) genes.[69] In the latter case, an EcoR1 fragment containing three internal EcoR1 cleavage sites was successfully cloned. These fragments were arranged in the same order and orientation on the plasmid as on the *Klebsiella* chromosome.[70] Of course, it is possible to utilize another restriction enzyme and avert this type of operation, but sometimes the restriction enzyme to be used is dictated by the vector or by a subsequent part of an experiment. Inserts obtained by this method will tend to be large, but a plasmid can accommodate relatively large inserts. A bacteriophage cloning vector, on the other hand, is limited in the absolute amount of DNA that can be packaged into a viable virion.

A shotgun experiment is one in which a total restriction digest of cellular DNA, or total cDNA prep, is cloned and, subsequently, interesting colonies are identified. If the genome of the cell is large, the percentage of inserts which are the desired ones will be very small. Unless a very strong selection for the particular insert exists, it is a very laborious task to find the plasmid that carries the correct insert. The genome of a mouse, for example, will contain about 10^6 EcoR1 restriction fragments. To make this a more manageable problem, techniques of enriching for desired inserts have been developed. These rely on the availability of a hybridization probe — usually cDNA made by reverse transcriptase from a purified messenger. A restriction digest of the whole genome is fractionated, for instance, by size on an agarose gel, and sequences that hybridize to the probe are identified and pooled. When two or more methods of separation, each of which relies on a different property of the DNA, are combined, a considerable enrichment can be achieved.[71] Besides agarose gel separation, RPC 5 chromatography and various CSCl and $CsSO_4$ equilibrium centrifugation techniques separate DNA fragments according to different chemical parameters.[72-74] When DNA enriched in such a manner is cloned, the problem of picking out the desired insert is enormously simplified.

IV. IN VITRO MANIPULATIONS

Three procedures are currently used to form recombinant plasmids between plasmid vectors and DNA fragments. The most efficient procedure is complementary homopolymeric tailing of vector and fragment with terminal transferase. When this procedure is used, neither the tailed plasmid vector nor the tailed DNA fragments can anneal with themselves. In order to circularize, a plasmid must incorporate an insert. Thus, almost all transformants contain plasmids with DNA inserts. In experiments utilizing blunt- or sticky-end ligation, the vast majority of transformants may contain self-ligated vectors rather than recombinant plasmids.

Dugaiczyk et al.[75] have made a systematic study of the optimal DNA concentrations that promote intermolecular ligations as opposed to circularization. They found that the relative rates of circularization vs. end-to-end joining of fragments and vector depend not only on the concentration of the fragments and the vector, but also on the fragment and vector lengths.

In experiments which utilize fragments with short complementary terminii, self-ligation of the vector can be minimized by removal of the terminal 5′ phosphates of the linearized cloning vector with calf intestine alkaline phosphatase.[65] This treatment effectively prevents self-ligation of the vector because the ligase reaction requires the presence of a 5′ phosphate. When nonphosphatased fragments and phosphatased vectors are mixed, partially ligated circular hybrid plasmids will be formed. This procedure can reduce the "background" transformation due to the vector itself by as much as 60-fold.[70]

Analagously, self-ligation of the fragments to be cloned can result in greatly reduced efficiency of hybrid plasmid formation in vitro. In some special cases, this problem can be circumvented. As more restriction enzymes are characterized, pairs of enzymes are being discovered that recognize different hexanucleotide sequences but generate the same single-stranded complimentary ends.[20] For example, the recognition sequences for BamHI and BglII are 5′-GGATCC-3′ and 5′-AGATCT-3′ respectively, but both enzymes leave a 5′-GATC-3′ single-stranded terminal sequence. This means that the ends of BamHI- and BglII-generated fragments will anneal with each other, but the resulting ligated hybrid hexanucleotide sequences will not form the recognition sequence for either enzyme. When BglII fragments are ligated into the BamHI site of a cloning vector in the presence of both BglII and BamHI, only BamHI-BglII hybrids will be stable. Any BglII-BglII or BamHI-BamHI joinings will be cleaved by the BglII or BamHI in the reaction mixture. In principle all transformants should carry inserts.[76,77]

V. PROPAGATION OF RECOMBINANT PLASMIDS

In vitro synthesized recombinant plasmids must be reintroduced ("transformed") into bacterial cells in order to be propagated. Cohen and Chang discovered that chilled, calcium-treated *E. coli* could be transformed with circular pSC101 DNA at a frequency of 10^4 to 10^5 tet^r transformants per microgram of DNA, or about one transformant per 10^6 to 10^7 plasmid molecules.[78] Subsequently, it was shown that this relatively poor transformation efficiency was due in part to the fact that only a small percentage of calcium-treated cells are competent for transformation.[79] The number of cells in this "transformable" pool (0.1 to 1%) can be determined empirically by transforming with saturating amounts of DNA.[80] Surprisingly, although the saturating DNA concentration is identical for different plasmids, the actual number of transformants can vary by an order of magnitude for different plasmids.[80] This variation of transforming ability appears to be an intrinsic property of each plasmid and not correlated with size.

Since the source of DNA that one desires to clone is often in short supply, the poor efficiency of transformation into calcium-treated cells can be a serious problem. A highly efficient transformation procedure devised by Suzuki and Szalay[101] utilizes temperature-sensitive peptidoglycan cell wall mutants which form spheroplasts at the nonpermissive temperature, but which regenerate cell walls and are 97 to 99% viable (form colonies) when returned to the permissive temperature. Suzuki and Szalay have shown that transformation of such spheroplasts is 500- to 800-fold more efficient than transformation of calcium-treated *E. coli*. Temperature-sensitive cell wall mutants of *Klebsiella pneumoniae* and *Agrobacterium tumefaciens* can be efficiently transformed, and it is likely that the procedure can be extended to other Gram-negative bacteria.

VI. INSERT IDENTIFICATION

When an insert confers a unique, selectable phenotype on a bacterium, there is no difficulty associated with identification of the desired clones. However, in many instances, particularly when the insert is of eukaryotic origin, it is not possible to pick out the wanted clones by genetic manipulations alone. If the size of the desired insert is known, then it should increment the size of the vector in a predictable manner and the desired chimeric plasmid can be identified by agarose gel electrophoresis. It is not necessary to purify the DNA, since the amount of plasmid DNA in a single colony is sufficient to see by ethidium bromide staining.[81]

Grunstein and Hogness developed a technique of hybridizing labeled probe to individual colonies.[82] Replica-plated clones are grown on a nitrocellulose filter. The DNA in these colonies is baked onto the filter, which is then incubated with radioactive probe, usually ^{32}P-cDNA or cRNA. Individual colonies containing sequences homologous to the probe can be detected by autoradiography.

Broome and Gilbert have developed a technique for identifying clones which express a particular protein.[83] Purified IgG directed against that protein will adsorb uniformly to a plastic disc. When this disc is placed over lysed colonies (replica-plated), any antigen associated with the clone is bound to the adsorbed antibody molecules. The plastic disc is then exposed to ^{125}I-IgG directed against the same protein. A "sandwich" complex is formed and the antigen is precisely located on the disc by autoradiography. The technique will detect 0.1 molecules of antigen per cell.

Once a clone has been identified, the insert must be characterized. In order to prepare DNA of a plasmid with a relaxed replicator, chloramphenicol is added to a late log-phase bacteria culture. Several hours later the cells are collected and gently lysed.[9,84] Small circular DNA is separated from bulk chromosomal DNA by low-speed centrifugation.[85,86] The soluble fraction or "cleared lysate" contains the plasmid, broken chromosomal DNA, RNA, and protein. When banded in an ethidium bromide-cesium chloride density gradient, closed circular plasmid DNA forms a band of slightly higher density than the nicked or broken DNA.[87]

A faster method of plasmid purification has recently been reported by Ohlsson et al.[88] After phenol extraction, cleared lysates are denatured, either by temperature or alkali, and are then phase partitioned in a PEG-dextran two-phase system. Closed circular DNA, because of its topology, reanneals very rapidly and is recovered in the upper phase, while other DNA does not reanneal and is found with the bulk of the RNA in the lower phase. After ethanol precipitation, plasmid DNA from the upper phase is suitable for restriction analysis. Total characterization is possible using the rapid DNA sequencing techniques of Maxam and Gilbert[89] or Sanger et al.[90]

VII. DEVELOPMENT OF OTHER HOST-VECTOR CLONING SYSTEMS

At the present time, almost all cloning experiments utilize *E. coli* K12 as the vector host. Only two other organisms, *Bacillus subtilis* and *Sacromyces cervisiae*, have been certified by the NIH recombinant DNA committee for use in recombinant DNA host-vector (HV) systems. As other organisms are certified as appropriate hosts for recombinant DNA projects, new plasmid vectors will have to be designed. *E. coli* vectors will probably be suitable for Gram-negative bacteria closely related to *E. coli*, such as *Salmonella typhimurium* and *Klebsiella pneumoniae*. The most versatile *E. coli* vectors, however, may prove to have an extremely limited host range and may not be replicated in more distantly related bacteria. For example, ColE1-related plasmids apparently fail to replicate in *Rhizobium, Agrobacterium*, and *Azotobacter*.[91]

It has been suggested that it may be possible to develop broad host-range cloning vehicles by utilizing the replication region of promiscuous conjugative plasmids such as RK2.[53] RK2 is a drug resistance transfer factor of 37×10^6 daltons which confers resistance to tetracycline, kanamycin, and ampicillin and belongs to the P compatibility group. The P group R factors have unusually wide host ranges within Gram-negative bacteria and are capable of mediating their transfer to most, if not all, Gram negatives.[92,93] Useful cloning vectors derived from RK2 have been constructed by reducing the plasmid in size in vitro with restriction endonucleases, thereby eliminating genes conferring some of the drug resistance properties and genes responsible for self-transmissibility.[53,91,94] One such plasmid, pRK248, is 5.6×10^6 daltons, confers tetracycline resistance, and contains single restriction sites for EcoR1 and SalI.[91]

One problem that may be encountered in the development of certain HV systems is the lack of an efficient transformation system to introduce recombinant plasmids into the host cell. One way to circumvent this difficulty is by constructing recombinant RK2 plasmids in *E. coli* and subsequently transferring them to the desired host with a conjugative helper plasmid. One such scheme involves a nonconjugative wide host-range vector, such as pRK248, and a conjugative plasmid that has the ability to mobilize the nonconjugative cloning to vector to various hosts but lacks the ability itself to replicate in other hosts.[91] This scheme minimizes the possibility of disseminating a recombinant plasmid by conjugation to undersirable recipients.

An alternative to using broad host-range vectors is the construction of cloning vectors from indigenous plasmids. One procedure, for example, would involve screening a number of different isolates of the species in question for small plasmids with relaxed replicators. It may be difficult to correlate a readily identifiable or selectable phenotype with the presence of a small indigenous plasmid. A selectable drug resistance plasmid could be transferred to plasmids in Gram-negative bacteria by using a promiscuous P-type plasmid to introduce a suitable resistance transposon to the cell. The detailed procedure would resemble that used by So et al.[12] in the construction of pSF2124.

VIII. A SCHEME FOR ISOLATING OVERLAPPING RESTRICTION FRAGMENTS

There are a number of experimental situations in which it may be desirable to clone a DNA fragment that overlaps and extends outside a fragment already cloned. Figure 9 diagrams a method devised by Bedbrook and Ausubel for genetically selecting recombinant plasmids carrying sequence homology to previously cloned DNA fragments.[61] The method is based on the observation that compatible, amplifiable plasmids recombine with each other only at regions of homology to form hybrid plasmids.[15,95] An essential part of the experimental strategy is a mutant derivative of ColE1 (ColE1[ts]) which fails to replicate at 42°C.[96] Fragment A, generated by restriction enzyme 1, is cloned in a kanamycin-resistant derivative of ColE1[s61] Whole chromosomal DNA cut with enzyme 2 is "shotgunned" into RSF1030, and recombinant plasmids containing a complete array of inserts are used to transform rec[+] *E. coli* harboring ColE1[ts] kan[+] (fragment A). Transformed cells are selected for a kan[+] Ap[r] phenotype at 42°C. Since RSF1030 will not recombine with ColE1[ts] kan[+] unless both plasmids contain homologous inserts, and since RSF1030 can only complement ColE1[ts] at 42°C, kan[r] Ap[r] transformants should contain (for the case illustrated in Figure 9) one of two hybrid plasmids: ColE1[ts]/A-RSF1030/B or ColE1[ts]/A-RSF1030/C. Since plasmid-plasmid

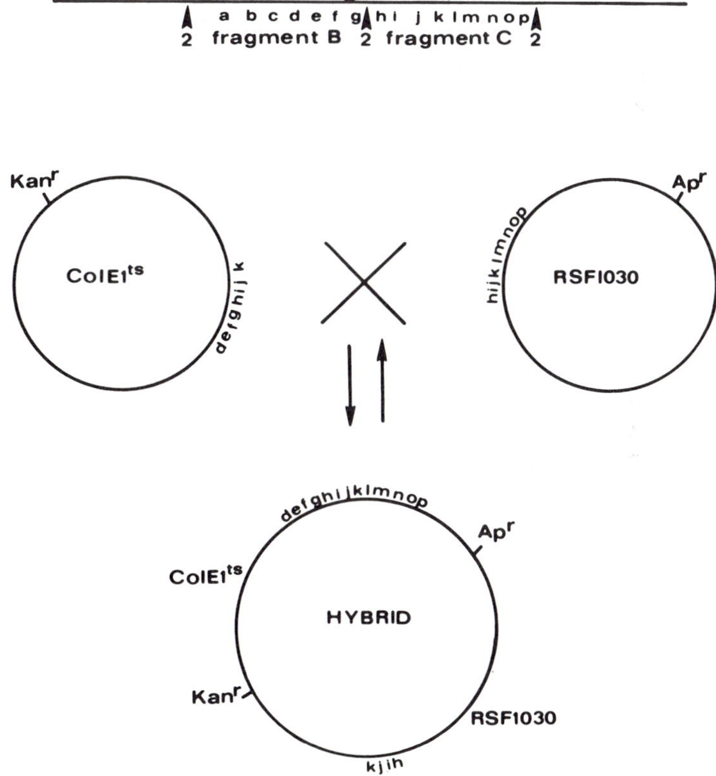

FIGURE 9.   Method for "walking down the chromosome".

Step

1.   Clone fragment A in ColE1[ts]kan[r]
2.   Cut genome with enzyme 2 and "shotgun entire genome into RSF130
3.   Transform RFF1030 hybrid plasmids into a rec[+] strain carrying ColE1[ts]kan[r] (fragment A)
4.   Select Ap[r] (RSF130) and kan[r] (ColE1[ts]) at 42°C. ColE1[ts]kan[r] will not replicate at this temperature unless it has recombined with RSF1030 (see below)
5.   Isolate plasmid DNA and separate plasmids by preparative agarose gel electrophoresis. Since plasmid-plasmid recombination is reversible, RSF130 (fragment B) and RSF130 (fragment C) should be present on the gels

(From Bedbrook, J. and Ausubel, F., unpublished data, 1977. With permission.)

recombination is reversible, these transformants should also contain RSF1030/B and RSF1030/C.[95] These latter plasmids can be isolated from closed circular DNA preparations by means of preparative agarose gel electrophoresis.

ACKNOWLEDGEMENTS

We wish to thank John Sims and Stephanie Bird for criticizing our manuscript. Some of the work cited here was supported by grants to F. Ausubel (NSF: PCM75-21435-A01) and to Walter Gilbert (NIHGM21514-3).

REFERENCES

1. **Cohen, S. N., Chang, A. C. Y., Boyer, H. W., and Helling, R. B.,** Construction of biologically functioning bacterial plasmids *in vitro*, *Proc. Natl. Acad. Sci. U.S.A.*, 70, 3240, 1973.
2. **Hedgpeth, J., Goodman, H. M., and Boyer, H. W.,** DNA nucleotide sequence restricted by the R1 endonuclease, *Proc. Natl. Acad. Sci. U.S.A.*, 69, 3448, 1972.
3. **Cohen, S. N. and Chang, A. C. Y.,** Recircularization and autonomous replication of a sheared R-factor DNA segment in *Escherichia coli* transformants, *Proc. Natl. Acad. Sci. U.S.A.*, 70, 1293, 1973.
4. **Cohen, S. N. and Chang, A. C. Y.,** Revised interpretation of the origin of the pSC101 plasmid, *J. Bacteriol.*, 132, 734, 1977.
5. **Hershfield, V., Boyer, H. W., Yanofsky, C., Lovett, M. A., and Helinski, D. R.,** Plasmid ColE1 as a molecular vehicle for cloning and amplification of DNA, *Proc. Nat. Acad. Sci. U.S.A.*, 71, 3455, 1974.
6. **Dewitt, W. and Helinski, D. R.,** Characterization of colicinogenic factor E1 from non-induced and a Mitomycin C-induced *Proteus* strain, *J. Mol. Biol.*, 13, 692, 1965.
7. **Timmis, K., Cabello, F., and Cohen, S. N.,** Utilization of two distinct modes of replication by a hybrid plasmid constructed *in vitro* from separate replicons, *Proc. Nat. Acad. Sci. U.S.A.*, 71, 4556, 1974.
8. **Helinski, D. R. and Clewell, D. B.,** Circular DNA, *Annu. Rev. Biochem*, 40, 899, 1972.
9. **Clewell, D. B. and Helinski, D. R.,** The effect of growth conditions on the formation of the relaxation complex of supercoiled ColE1 DNA and protein in *Escherichia coli*, *J. Bacteriol.*, 110, 1135, 1972.
10. **Clewell, D. B.,** Nature of ColE1 plasmid replication in *Escherichia coli* in the presence of chloramphenicol, *J. Bacteriol.*, 110, 667, 1972.
11. **Bazaral, M. and Helinski, D. R.,** Replication of a bacterial plasmid and an episome in *Escherichia coli*, *Biochemistry*, 9, 359, 1970.
12. **So, M., Gill, R., and Falkow, S.,** The generation of ColE1-Ap^r cloning vehicle which allows detection of inserted DNA, *Mol. Gen. Genet.*, 142, 239, 1976.
13. **Datta, N., Hedges, R., Shaw, E. J., Sykes, R., and Richmond, M. H.,** Properties of an R factor from *Pseudomonas aeroginosa*, *J. Bacteriol.*, 108, 1244, 1971.
14. **Betlach, M. C., Hershfield, V., Chow, L., Brown, W., Goodman, H. M., and Boyer, H. W.,** A restriction endonuclease analysis of the bacterial plasmid controlling the EcoR1 restriction and modification of DNA, *Fed. Proc. Fed. Am. Soc. Exp. Biol.*, 35, 2037, 1976.
15. **Heffron, F., Sublett, R., Hedges, R. W., Jacob, A., and Falkow, S.,** Origin of the TEM β-lactamase gene found on plasmids, *J. Bacteriol.*, 122, 250, 1975.
16. **Rodriquez, R. L., Bolivar, F., Goodman, H. M., Boyer, H. W., and Betlach, M.,** Construction and characterization of cloning vehicles, in *Molecular Mechanisms in the Control of Gene Expression*, Nierlich, D. P., Rutter, W. J., and Fox, C., Eds., Academic Press, New York, 1976, 471.
17. **Polisky, B., Green, P., Garfin, D. E., McCarthy, B. J., Goodman, H. M., and Boyer, H. W.,** Specificity of substrate recognition by the EcoR1 restriction endonuclease, *Proc. Nat. Acad. Sci. U.S.A.*, 72, 3310, 1975.
18. **Hsu, M. and Berg, P.,** Altering the specificity of restriction endonucleases: Effect of replacing Mg^{2+} with Mn^{2+}, *Biochemistry*, 17, 131, 1978.
19. **Hamer, D. H. and Thomas, C. A., Jr.,** Molecular cloning of DNA fragments produced by restriction endonucleases SalI and BamI, *Proc. Natl. Acad. Sci. U.S.A.*, 73, 1537, 1976.

20. **Roberts, R. J.**, Restriction endonucleases, in *DNA Insertion Elements, Plasmids and Episomes*, Bukhari, A. I., Shapiro, J. A., and Adhya, S. L., Eds., Cold Spring Harbor Laboratory, Cold Spring Harbor, N.Y., 1977.
21. **Hershfield, V., Boyer, H. W., Chow, L., and Helinski, D. R.**, Characterization of a mini-ColE1 plasmid, *J. Bacteriol.*, 126, 447, 1976.
22. **Bolivar, F., Rodriguez, R. L., Betlach, M. C., and Boyer, H. W.**, Construction and characterization of new cloning vehicles. I. Ampicillin-resistant derivatives of the plasmid pMB9, *Gene*, 2, 75, 1977.
23. **Bolivar, F., Rodriguez, R. L., Greene, P. J., Betlach, M. C., Heyneker, H. L., Boyer, H. W., Crosa, J. H., and Falkow, S.**, Construction and characterization of new cloning vehicles. II. A multipurpose cloning system, *Gene*, 2, 95, 1977.
24. **Datta, N. and Kontomichalou, P.**, Penicillinase synthesis controlled by infectious R factors in enterobacteriace, *Nature (London)*, 208, 239, 1965.
25. **Meynell, E. and Datta, N.**, Mutant drug resistance factors of high transmissibility, *Nature (London)*, 214, 885, 1967.
26. **Sutcliffe, J. G.**, The complete nucleotide sequence of pBR322, Cold Spring Harbor Symposium, Vol. 43, in press.
27. **Sutcliffe, J. G. and Church, G. M.**, The cleavage site of the restriction endonuclease Ava II, *Nucleic Acids Res.*, 5, 2313, 1978.
28. **Boyer, H. W., Betlach, M., Bolivar, F., Rodriquez, R. L., Heyneker, H. L., Shine, J., and Goodman, H. M.**, The construction of molecular cloning vehicles, in *Recombinant Molecules: Impact on Science and Society*, Beer, R. F. and Bassett, E. G., Eds., Raven Press, New York, 1977, 9.
29. **Curtiss, R., III, Charamella, L. J., Berg, C. M., and Harris, P. E.**, Kinetic and genetic analysis of D-cycloserine inhibition and resistance in *Escherichia coli*, *J. Bacteriol.*, 90, 1238, 1965.
30. **Sgaramella, V., van de Sande, J., and Khorana, H. G.**, A novel joining reaction catalyzed by the T4-poly-nucleotide ligase, *Proc. Natl. Acad. Sci. U.S.A.*, 67, 1468, 1970.
31. **Backman, K., Ptashne, M., and Gilbert, W.**, Construction of plasmids carrying the cI gene of bacteriophage lambda, *Proc. Natl. Acad. Sci. U.S.A.*, 73, 4174, 1976.
32. **Brown, N. L. and Smith, M.**, The mapping and sequence determination of the single site on φX174am3 RF DNA cleaved by Pst I, *FEBS Lett.*, 65, 284, 1976.
33. **Lobban, P. and Kaiser, A. D.**, Enzymatic end-to-end joining of DNA molecules, *J. Mol. Biol.*, 78, 453, 1973.
34. **Sutcliffe, J. G.**, pBR322 restriction map derived from DNA sequence; accurate DNA size markers up to 4361 nucleotide pairs long, *Nucleic Acids Res.*, 5, 2721, 1978.
35. **Sutcliffe, J. G.**, The nucleotide sequence of the ampicillin resistance gene of pBR322, *Proc. Natl. Acad. Sci., U.S.A.*, 75, 3737, 1978.
36. **Ambler, R. P. and Scott, G. K.**, The partial amino acid sequence of the penicillinase coded by the *E. coli* plasmid pR6K, *Proc. Natl. Acad. Sci., U.S.A.*, 75, 3732, 1978.
37. **Knox, J. R., Kelly, J. A., Moews, P. C., and Murthy, N. S.**, 5.5A crystallographic structure of penicillin β-lactamase and radius of gyration in solution, *J. Molec. Biol.*, 104, 865, 1978.
38. **Hall, A. and Knowles, J. R.**, Directed selective pressure on a β-lactamase to analyze changes involved in the development of enzyme function, *Nature (London)*, 264, 803, 1976.
39. **Tait, R. C. and Boyer, H. W.**, On the nature of tetracyline resistance controlled by the plasmid pSC101, *Cell*, 13, 73, 1978.
40. **Izaki, K., Kinichi, K., and Arima, K.**, Specificity and mechanism of tetracycline resistance in a multiple drug resistant strain of *E. coli*, *J. Bacteriol.*, 91, 628, 1966.
41. **Levy, S. B., McMurry, L., Onigman, P., and Saunders, R. M.**, Plasmid-mediated tetracycline resistance in *E. coli*, in *Topics in Infectious Diseases*, Drews and Hogenaur, Eds., Springer-Verlag, Vienna, 1978, 181.
42. **Meagher, R. B., Tait, R. C., Betlach, M., and Boyer, H. W.**, Protein expression in *Escherichia coli* mini cells by recombinant plasmids, *Cell*, 10, 509, 1977.
43. **Bolivar, F., Betlach, M. C., Heyneker, H. L., Shine, J., Rodriguez, R. L., and Boyer, H. W.**, Origin and replication of pBR345 plasmid DNA, *Proc. Natl. Acad. Sci. U.S.A.*, 74, 5265, 1977.
44. **Tomizawa, J., Ohmiri, H., and Bird, R. E.**, Origin of replication of colicin E1 plasmid DNA, *Proc. Natl. Acad. Sci. U.S.A.*, 74, 1865, 1977.
45. **Tomizawa, J., Sakakibara, Y., and Kakefunda, T.**, Replication of Colicin E1 plasmid DNA in cell extracts. Origin and direction of replication, *Proc. Natl. Acad. Sci. U.S.A.*, 71, 2260, 1974.
46. **Inselburg, J.**, Replication of Colicin E1 plasmid DNA in mini cells from a unique replication initiation site, *Proc. Natl. Acad. Sci. U.S.A.*, 71, 2256, 1974.
47. **Lovett, M. A., Guiney, D. G., and Helinski, D. R.**, Relaxation complexes of plasmids ColE1 and ColE2: unique site of the nick in open circular DNA of the relaxed complexes, *Proc. Natl. Acad. Sci. U.S.A.*, 71, 3854, 1974.
48. Recombinant DNA Tech. Bull. 1, (1), U.S. Department of Health, Education, and Welfare, 1977, 9.

49. **Chang, A. C. Y. and Cohen, S. N.,** Construction and characterization of amplifiable multicopy DNA cloning vehicles derived from the P15A cryptic miniplasmid, submitted for publication.

50. **Cozzarelli, N. R., Kelly, R. B., and Kornberg, A.,** A minute circular DNA from *Escherichia coli* 15*, *Proc. Natl. Acad. Sci. U.S.A.,* 60, 992, 1968.

51. **Kretschmer, P. J., Chang, A. C. Y., and Cohen, S. N.,** Indirect selection of bacterial plasmids lacking identifiable phenotypic properties, *J. Bacteriol.,* 124, 225, 1975.

52. **Cabello, F., Timmis, K., and Cohen, S. N.,** Replication control in a composite plasmid constructed by *in vitro* linkage of two distinct replicons, *Nature (London),* 259, 285, 1976.

53. **Helinski, D. R., Hershfield, V., Figurski, D., and Meyer, R. J.,** Construction and properties of plasmid cloning vehicles, in *Recombinant Molecules: Impact on Science and Society,* Beers, R. F. and Basset, E. G., Eds., Raven Press, New York, 1977, 151.

54. **Bolivar, F., Backman, K., Shine, J., Heyneker, H. L., Rodriguez, R. L., Tait, R. C., Yanofsky, S., and Boyer, H. W.,** in preparation.
Gen. Genet., 1978.

55. **Gilbert, W.,** personal communication, 1975.

56. **Fuller, F.,** unpublished data, 1975.

57. **Johnsrud, L.,** Contacts between *E. coli* RNA polymerase and a *lac* operon promotor, *Proc. Natl. Acad. Sci. U.S.A.,* in press.

58. **Backman, K. and Ptashne, M.,** Maximizing gene expression on a plasmid using recombination *in vitro, Cell,* 13, 65, 1978.

59. **Itakura, I., Hrose, T., Crea, R., Riggs, A. D., Heyneker, H. L., Bolivar, F., and Boyer, H. W.,** Expression in *E. coli* of a chemically synthesized gene for the hormone somatostatin, *Science,* 198, 1056, 1977.

60. **Polisky, B., Bishop, R. J., and Gelfand, D. H.,** A plasmid cloning vehicle allowing regulated expression of eukaryotic DNA in bacteria, *Proc. Natl. Acad. Sci. U.S.A.,* 73, 3900, 1976.

61. **Bedbrook, J. and Ausubel, F.,** unpublished data, 1977.

62. **Clark, L. and Carbon, J.,** Biochemical construction and selection of hybrid plasmids containing specific segments of the *E. coli* genome, *Proc. Natl. Acad. Sci. U.S.A.,* 72, 4361, 1975.

63. **Maniatas, T., Gekkee, S., Efstatiadis, A., and Kafatos, F. C.,** Amplification and characterization of a β-globin gene synthesized *in vitro, Cell,* 8, 1976.

64. **Higuchi, R., Paddock, G. V., Wall, R., and Salser, W.,** A general method for cloning eukaryotic structural gene sequences, *Proc. Natl. Acad. Sci. U.S.A.,* 73, 3146, 1976.

65. **Seeburg, P. H., Shine, J., Martial, J. A., Baxter, J. D., and Goodman, H. M.,** Nucleotide sequences and amplification in bacteria of structural gene for rat growth hormone, *Nature (London),* 270, 486, 1977.

66. **Nosikov, V. V., Braga, E. A., Narlishev, A. V., Shoze, A. L., and Poljanovskij, O. L.,** Protection of particular cleavage sites of restriction endonucleases by distamycin A and actinomycin D, *Nucleic Acid Res.,* 3, 2293, 1976.

67. **Kuroyedov, A. A., Grokhovsky, S. L., Zhyze, A. L., Nosikov, V. V. and Polyanovsky, O. L.,** Distamycin A and its analogs as agents for blocking of endo R EcoR1 activity, *Gene,* 1, 389, 1977.

68. **Maizels, N.,** Dictyostelium 17S, 25S and 5S rDNA lie within a 38,000 base pair repeated unit, *Cell,* 9, 431, 1976.

69. **Cannon, F. C., Riedel, G. E., and Ausubel, F. M.,** Recombinant plasmid that carries part of the nitrogen fixation (nif) gene cluster of *Klebsiella pneumoniae, Proc. Natl. Acad. Sci. U.S.A.,* 74, 2963, 1977.

70. **Riedel, G.,** personal communication, 1977.

71. **Tonegawa, S., Brack, C., Hozumi, N., and Schuller, R.,** Cloning of an immunoglobulin variable region gene from mouse embryo, *Proc. Natl. Acad. Sci. U.S.A.,* 74, 3518, 1977.

72. **Landy, A., Foeller, C., Reszelback, R., and Dudock, B.,** Preparative fractionation of DNA restriction fragments by high pressure column chromatography on RPC-5, *Nucleic Acids Res.,* 3, 2575, 1976.

73. **Handies, S. C. and Wells, R. D.,** Preparative fractionation of DNA restriction fragments by reversed phase-column chromatography, *Proc. Natl. Acad. Sci. U.S.A.,* 73, 3117, 1976.

74. **Tilghman, S. M., Tiemeier, D. C., Polsky, F., Edgell, M. H., Siedman, J. G., Leder, A., Enguist, L. W., Norman, B., and Leder, P.,** Cloning specific segments of the mammalian genome: bacteriophage λ containing mouse globin and surrounding gene sequences, *Proc. Natl. Acad. Sci. U.S.A.,* 74, 4406, 1977.

75. **Dugaiczyk, A., Boyer, H. W., and Goodman, H. M.,** Ligation of EcoR1 endonuclease-generated fragments into linear and circular structures, *J. Mol. Biol.,* 96, 171, 1975.

76. **Roberts, R. J.,** The role of restriction endonucleases in genetic engineering, in *Recombinant Molecules, Impact on Science and Society,* Beers, R. F. and Bassett, E. G., Eds., Raven Press, New York, 1977, 21.

77. **Broome, S.,** personal communication, 1977.
78. **Cohen, S. N., Chang, A. C. Y., and Hsu, C. L.,** Nonchromosomal antibiotic resistance in bacteria: genetic transformation of *E. coli* by R factor DNA, *Proc. Nat. Acad. Sci. U.S.A.,* 69, 2110, 1977.
79. **Kretschmer, P. J., Chang, A. C. Y., and Cohen, S. D.,** Indirect selection of bacterial plasmids lacking identifiable phenotypic properties, *J. Bacteriol.,* 124, 225, 1975.
80. **Bedbrook, J., Lehrach, H., and Ausubel, F.,** unpublished data, 1977.
81. **Barnes, W. M.,** Plasmid detection and sizing in single colony lysates, *Science,* 195, 393, 1977.
82. **Grunstein, M. and Hogness, D. S.,** Colony hybridization: a method for the isolation of cloned DNAs that contain a specific gene, *Proc. Natl. Acad. Sci. U.S.A.,* 72, 3961, 1975.
83. **Broome, S. and Gilbert, W.,** New immunological screening methods for specific translation products, *Proc. Nat. Acad. Sci. U.S.A.,* 1978.
84. **Clewell, D. B.,** Nature of ColE1 plasmid replication in *E. coli* in the presence of chloramphenicol, *J. Bacteriol.,* 110, 667, 1972.
85. **Hirt, B.,** Selective extraction of polyoma DNA from infected mouse cell cultures, *J. Mol. Biol.,* 26, 365, 1967.
86. **Clewell, D. B. and Helinski, D. R.,** Supercoiled circular DNA-protein complex in *E. coli:* purification and induced conversion to an open circular DNA form, *Proc. Natl. Acad. Sci. U.S.A.,* 62, 1159, 1969.
87. **Bauer, W. and Vinograd, J.,** The interaction of closed circular DNA with intercalative dyes, *J. Mol. Biol.,* 33, 141, 1968.
88. **Ohlsson, R., Hentschel, C. C., and Williams, J. G.,** A rapid method for the isolation of circular DNA using an aquaeous two-phase partition system, *Nucleic Acids Res.,* 5, 583, 1978.
89. **Maxam, A. M., and Gilbert, W.,** A new method for sequencing DNA, *Proc. Natl. Acad. Sci. U.S.A.,* 74, 560, 1977.
90. **Sanger, F., Nicklen, S., and Coulson, A. R.,** DNA sequencing with chain-termination inhibitors, *Proc. Natl. Acad. Sci. U.S.A.,* 74, 5463, 1977.
91. **Helinski, D.,** Plasmids as vectors for gene cloning, in *Genetic Engineering for Nitrogen Fixation,* Hollander, A., Ed., Plenum Press, New York, 1977, 19.
92. **Meyer, R., Figurski, D., and Helinski, D. R.,** Molecular vehicle properties of the broad host range plasmid RK2, *Science,* 190, 1226, 1976.
93. **Datta, N. and Hedges, R. W.,** Compatibility groups under FiR factors, *Nature (London),* 234, 222, 1976.
94. **Figurski, D., Meyer, R., Miller, D. S., and Helinski, D. R.,** Generation *in vitro* of deletions in the broad host range plasmid RK2 using phage Mu insertions and a restriction endonuclease, *Gene,* 1, 107, 1976.
95. **Bedbrook, J. R. and Ausubel, F. M.,** Recombination between bacterial plasmids leading to the formation of plasmid multimers, *Cell,* 9, 707, 1976.
96. **Collins, J. and Helinski, D. R.,** in preparation, 1978.
97. **Kahn, M. and Helinski, D. R.,** in preparation, 1978.
98. **Figurski, D. and Helinski, D. R.,** in preparation, 1978.
99. **Chang, S. and Cohen, S. N.,** *In vivo* site-specific genetic recombination promoted by EcoR1 restriction endonuclease, *Proc. Natl. Acad. Sci. U.S.A.,* 74, 4811, 1977.
100. **Covey, C., Richardson, D., and Carbon, J.,** A method for the deletion of restriction sites in bacteria plasmid DNA, *Mol. Gen. Genet.,* 145, 155, 1976.
101. **Suzuki, and Szalay, A.,** personal communication.

Chapter 5
RESTRICTION ENZYMES AND THEIR USES IN GENETIC ENGINEERING

K. Murray

TABLE OF CONTENTS

I. RESTRICTION AND MODIFICATION

The ability of a bacteriophage to grow in a particular bacterial host may depend upon the strain in which it was last propagated. If bacteriophage λ for example, is transferred from one strain of *Escherichia coli* to another strain, its growth efficiency is frequently impaired, with titers on the two strains differing by several orders of magnitude. Phages resulting from the infection of the second strain grow normally upon it, but if they are subsequently propagated on their original host strain, they again grow poorly when transferred back to the second strain. This phenomenon is known as *host-controlled restriction*[1] and was first observed in experiments where bacteriophage λ was transferred between different strains of *E. coli*. It is now known to occur widely throughout different strains and species of bacteria.

Experiments with radioactively labeled bacteriophage preparations showed that restriction of phage growth was accompanied by a rapid degradation of the bacteriophage DNA upon infection of the host cell.[2] However, an infection of the host strain used for propagation of the phage did not lead to a comparable degradation of the phage DNA. If a bacterial cell is to possess a nuclease that will selectively degrade DNA from an invading virus (or other source), it must be able to distinguish this foreign DNA from its own. It does so by a process called *host-controlled modification*. The modification involves a methylation of certain bases within the recognition sequence for the restriction enzyme which protects these sequences from attack by the restriction enzyme.[3] This explains why bacteriophage that survive the infection of a new strain of bacteria subsequently grow normally upon that strain. Their DNA is replicated in the presence of the modifying enzyme and it, like that of the host cell, becomes methylated and hence protected against the restriction system. However, if these phage are subsequently returned to a different strain, the modification is lost on replication and the phage progeny will once again be sensitive to the restriction system when transferred back to the original host cells.

II. RESTRICTION ENZYMES AND THEIR PROPERTIES

Attempts to isolate enzymes responsible for this selective degradation of the phage DNA were based upon the observation of the breakdown of the phage DNA in vivo. For a long time, experiments were unsuccessful as the assays were based upon the conversion of DNA from unmodified phage into acid-soluble oligonucleotides, while DNA from modified phage remained completely insoluble. When it was realized that the primary products resulting from restriction of the phage DNA may be of high molecular weight and an assay designed accordingly, an enzyme was obtained by Messelson and Yuan from *E. coli* strain K which would degrade phage λ DNA provided that the phage had been grown upon a different strain; DNA from phage grown upon *E. coli* K was not sensitive to the enzyme.[4] An analysis of the products of the reaction by sedimentation through sucrose gradients showed that the average size of the DNA fragments was approximately one third that of wild-type λ DNA, thus indicating a very high degree of enzyme specificity. Somewhat similar enzymes were isolated from *E. coli* strain B[5] and *E. coli* cells infected with the phage P1.[4]

The enzymes from *E. coli* strains K and B have unusual properties. They require the cofactors ATP and S-adenosyl-L-methionine (SAM) in addition to magnesium, and degradation of the DNA is accompanied by a very extensive ATP hydrolysis.[6] The enzymes are specific for double-stranded DNA. They contain three different subunits and function as a high molecular weight complex, probably of the form $\alpha \beta_2 \gamma$. The molecular weights of the EcoB subunits are α, 135,000; β, 60,000; γ, 55,000, and the active enzyme, approximately 310,000.[7] Attempts to determine the sequence of nucleotides at which these enzymes function have proved frustrating, and sequences at the positions of DNA breakage have yet to be determined. The enzymes have the unusual property of interacting with DNA molecules at a particular recognition sequence, which may be lost by mutation, or the interaction prevented by modification. They subsequently move along the molecule — probably in either direction — and break it elsewhere.[8,9] The position of breakage does not appear to be definite; however, it is not completely random.[10] Enzymes of this type have been called Type I restriction enzymes.[11]

The properties of Type II restriction enzymes are in marked contrast with those of the Type I, for they recognize a certain target sequence in a double-stranded DNA molecule and break the polynucleotide chains within that sequence to give a characteristic set of DNA fragments. Such enzymes have been prepared from many bacterial sources and are particularly important for biochemical methods of genetic engineering. These endonucleases have been correlated with a specific restriction function in vivo in only a few cases. Nevertheless, they are frequently, if not usually, described as restriction enzymes. They are readily detected and assayed by electrophoresis in agarose gels of digests of a small viral DNA which give characteristic band patterns (Figure 1). Type II enzymes have not been studied extensively as such, since interest has centered upon their use as reagents for the analysis of DNA and genetic manipulations in vitro. Those that have been examined appear to be much smaller than Type I enzymes, contain two (or in some cases four) probably identical subunits, and their only cofactor requirement is magnesium ions.

The first of the Type II enzymes was described and isolated by Smith and Wilcox

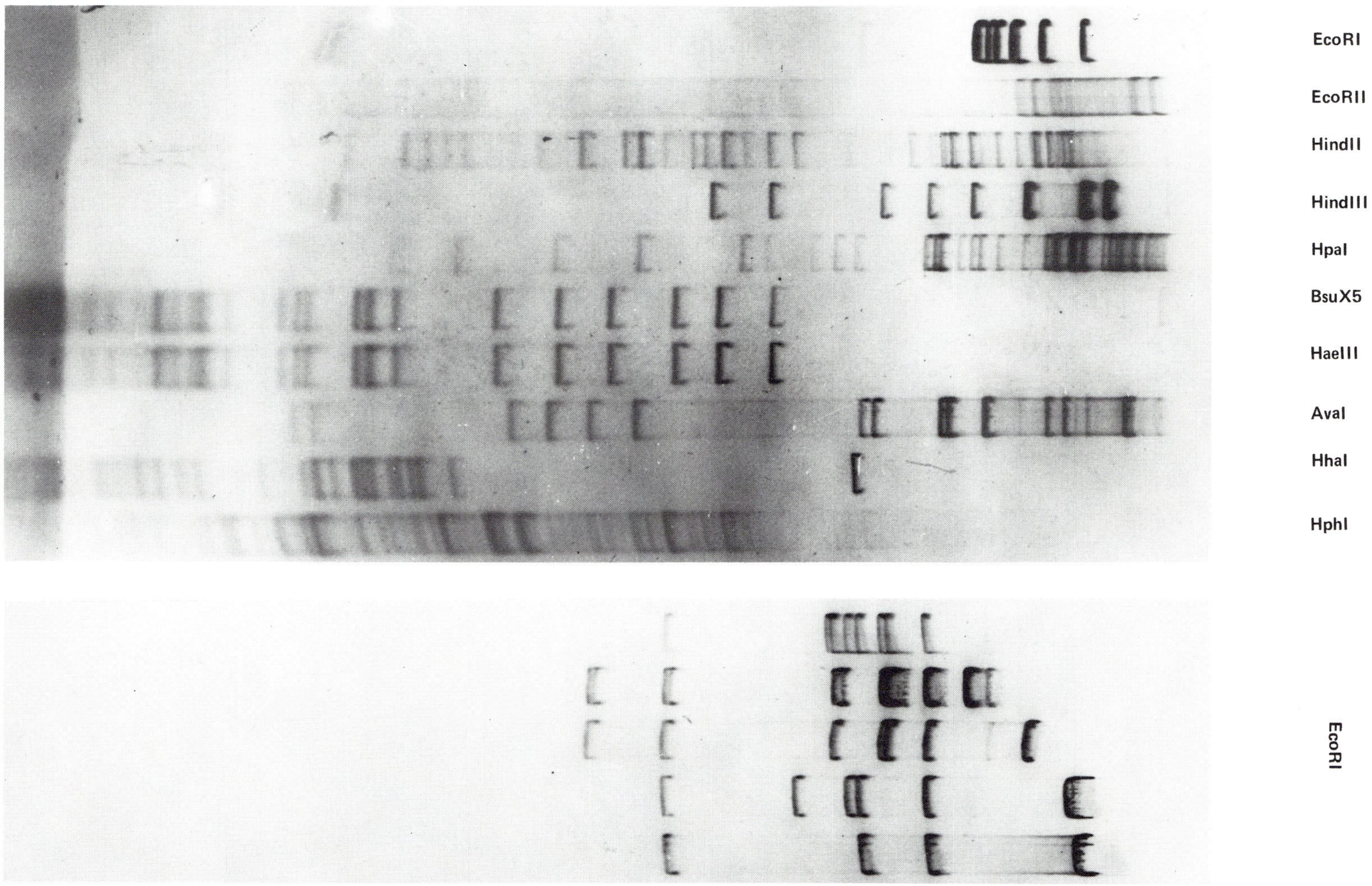

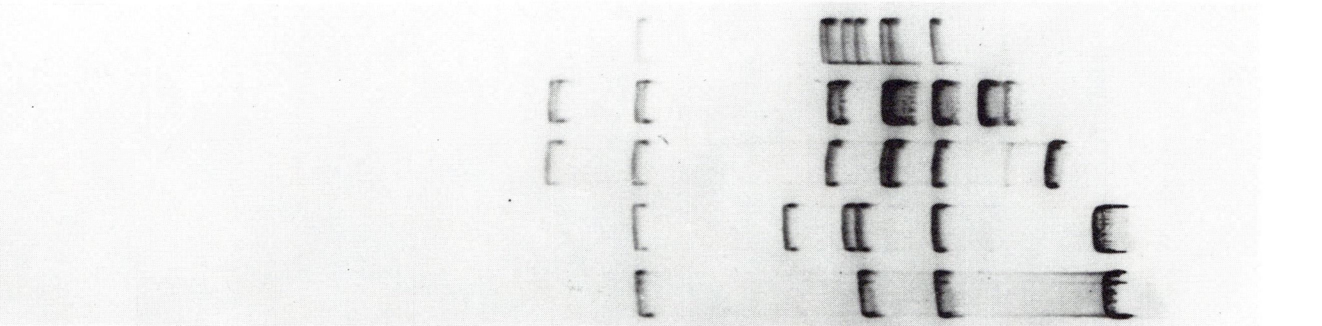

FIGURE 1. Analysis by electrophoresis in agarose gels of bacteriophage DNA digests with restriction endonucleases. The left-hand panel shows digests of the same DNA sample (a hybrid of λ and φ80) with several different enzymes (identified above by their abbreviated names — see Table 5.1). The right-hand panel shows digests of DNA from a number of λ derivatives digested with EcoRI.[15] The gels used for electrophoresis were 1% agarose in 0.04 M trisacetate buffer, pH 8.2, containing ethidium bromide (0.4 µg/ℓ). They were photographed under ultraviolet light on Ilford® FP4 film with a 4× red filter.[34]

from *Haemophilus influenzae.*[12] The nucleotide sequence at the breakage point of T7 DNA by this enzyme was determined, and it was found that the enzyme recognized a sequence of six base pairs.[13] Enzymes with similar properties were subsequently found in other species of *Haemophilus*[14] and in *E. coli* carrying both fi⁺ and fi⁻ plasmids [15-17] The discovery that these enzymes have different nucleotide sequence specificities stimulated a search for enzymes with similar properties in a large number of other organisms and many have now been described (Table 1).

III. RECOGNITION SEQUENCES FOR RESTRICTION ENZYMES

A list of the endonucleases whose recognition sequences are currently known is also given in Table 5.1. A more extensive list of enzymes of this type, with some of their characteristics, is in a recent review by Roberts.[18] This list shows that both the size of the nucleotide sequence recognized by a particular enzyme and the position of the phosphodiester bonds broken* vary. Some of the target sequences consist of a unique hexanucleotide, some are hexanucleotides with a degree of degeneracy at various positions in the sequence, while some recognize pentanucleotide sequences and others, tetranucleotide sequences. In some cases, the specificity of the enzyme may be changed by adjusting the reaction conditions.[19,20]

The size of the target sequence is important, as it determines the size range of fragments produced after digestion of a given DNA preparation. Any tetranucleotide sequence will occur by chance once in 256 base pairs if the base composition of the DNA is uniform. A hexanucleotide sequence will occur, again by chance, once in 4,096 base pairs. Therefore, the size of DNA fragments obtained with these enzymes will usually be of the order of a gene for the average protein and often large enough to contain several genes. If attempts to isolate a DNA fragment or a gene from a restriction enzyme digest are frustrated by the presence, within that fragment, of the recognition sequence for the enzyme, the fragment may be recoverable through the use of another restriction enzyme with a different recognition sequence. Thus, the application of these enzymes for DNA fractionation and isolation of specific sequences is apparent, given the known range of target sequences.

IV. REJOINING OF DNA FRAGMENTS

Type II restriction enzymes sometimes break the phosphodiester bonds between base pairs that are directly opposite to give what are termed "even breaks" or sometimes "flush-ended breaks," whereas others break phosphodiester bonds a few base pairs apart within the target sequence. In the latter case, the fragments have single-stranded projections which may be either 5′ or 3′ projections, and their length varies from one to five nucleotides. The type of restriction endonuclease making these staggered breaks, of which the EcoRI and EcoRII enzymes were the first examples[16,17] and HindIII was a close second,[21,22] is important, since fragments of DNA produced by these enzymes obviously have the ability to reassociate by normal Watson-Crick base pairing between the single-stranded ends of any pair of fragments.** Once fragments have been joined by the cohesion of these single-stranded ends, the covalent bonds are easily made by treatment with polynucleotide ligase.[23,24] The restriction enzymes of this type have therefore rapidly assumed a central role in the construction of new DNA sequences in vitro, and both plasmid and phage vectors have been constructed for use with them.

Other enzymes that generate DNA fragments with cohesive ends have now been

* All of the enzymes are essentially specific for double-stranded DNA.
** A given fragment may also circularize by cohesion between its two ends.

TABLE 1

Nucleotide Sequences Recognized by Restriction Endonucleases

Bacterial source of the enzyme	Abbreviation	Target nucleotide sequence 5′→3′	Ref.
Anabaena subcylindrica	AsuI	G↓-G-N-C-C	35
A. variabilis	AvaI	C-Y-C-G-R-G	36
	AvaII	G↓-A-/T-C-C	36
Arthrobacter luteus	AluI	A-G↓-C-T	37
Bacillus amyloliquefaciens H	BamHI	G↓-G-A-T-C-C	38
B. stearothermophilus 1503-4R	BstI		39
B. brevis S	BbvS	G-C-(A/T)-G-C	40
B. globigii	BglII	A↓-G-A-T-C-T	41
B. subtilis R	BsuR	G-G↓-C-C[a]	42
Brevibacterium luteum	BluII		18,43
Haemophilus aegyptius	HaeIII		21, 44
H. haemoglobinophilus	HhaI		18
Providencia alcalifaciens	PalI		18
Streptococcus faecalis	SfaI		18, 45
Bordetella bronchiseptica	BbrI	A↓-A-G-C-T-T	18
Corynebacterium humiferum	ChuI		18
H. influenzae, d, b	HindIII,HinbIII		21, 22
H. suis	HsuI		18
Brevibacterium albidum	BalI	T-G-G-C↓-C-A	29
B. umbra	BumI	C-A-G-C-T-G	18
Corynebacterium humiferum	ChuII	G-T-Y↓-R-A-C	18
Haemophilus influenzae d, c	HindII,HincII		13, 18
Diplococcus pneumoniae	DpnI, II	↓G-A-T-C	18, 46
Moraxella bovis	MboI		47,48
M. osleonsis	MosI		18
Escherichia coli/RI	EcoRI	G↓-A-A-T-T-C	16
	EcoRI′	R-R-A↓-T-Y-Y	49
	EcoRI[a]	↓A-A-T-T	19
E. coli/R245	EcoRII	↓C-C-(A/T)-G-G	17
E. coli/PI	EcoPI	A-G-A-T-C-T[b]	50
H. aegyptius	HaeI	(A/T)-G-G↓-C-C-(A/T)	18, 44, 51
	HaeII	R-G-C-G-C↓Y	18, 44, 52
H. influenzae H1	Hinh		18, 53
Neisseria gonorrhoea	NgoI		18, 54
H. aphrophilus	HapII	C↓-C-G-G	55
H. parainfluenzae	HpaII		44, 56
Moraxella nonliquefaciens	MnoI		18, 57
H. gallinarum	HgaI	G-A-C-G-C 5/10 base pairs	33
H. haemolyticus	HhaI	G-C-G-C	58
H. influenzae f	Hinf	G-A-N-T-C	18, 44, 59
H. parahaemolyticus	HphI	G-G-T-G-A 8/7 base pairs	31
H. parainfluenzae	HpaI	G-T-T-A-A-C	44, 56
Moraxella bovis	MboII	G-A-A-G-A 8/7 base pairs	18, 32, 47
M. nonliquefaciens	MnlI	C-C-T-C 5/10 base pairs	18
Providencia stuarti 164	PstI	C-T-G-C-A-G	60, 61
Serratia marcescens	SmaI	C-C-C-G-G-G	62
Streptomyces albus	SalI	G-T-C-G-A-C	63
Thermus aquaticus	TaqI	T-C-G-A	64
Xanthomonas badrii	XbaI	T-C-T-A-G-A	65
X. holcicola	XhoI	C-T-C-G-A-G	66
X. papavericola	XpaI		66

TABLE 1 (continued)

Nucleotide Sequences Recognized by Restriction Endonucleases

Bacterial source of the enzyme	Abbreviation	Target nucleotide sequence 5′—3′	Ref.
Xanthomonas badrii	XbaI	T–C–T–A–G–A	65
X. holcicola	XhoI	C–T–C–G–A–G	66
X. papavericola	XpaI		66
X. malvacearum	XmaI	C–C–C–G–G–G	62
	XmaII	C–T–G–C–A–G	62

Note: All of the enzymes break duplex DNA, but the sequence is shown for only one of the strands, and all are written in the direction 5′ to 3′. Where the analysis has been done, the enzymes break the phosphodiester bonds to leave a 5′ phosphate and 3′ hydroxyl group. The arrow shows the position of the bonds broken. Most of the sequences are rotationally symmetrical; this is also true of the position of the bonds broken by the enzymes. Enzymes from the bacteria inset have the same target sequence as that for the preceding entry. The abbreviated description of the enzymes was proposed by Smith and Nathans.[67] Additional properties of these enzymes and a list of several more enzymes for which the target sequences are not yet known are given in the review by Roberts.[18]

[a]   BsuR has recently been shown, like EcoRI, to have a relaxed specificity under certain conditions and cleaves the sequence –N–G↓C–N–.[20]

[b]   This is a sequence methylated by the modification enzyme.

described (Table 1), and although fragments of this sort are the easiest to rejoin, they are not the only ones used for this purpose. DNA fragments with nonoverlapping ends may be joined[25] by the action of a ligase prepared from *E. coli* cells infected with phage T4.[26] Consequently, DNA fragments produced by the action of a number of restriction enzymes may be joined in essentially any combination. The products of these reactions may or may not conserve the recognition sequence for the restriction enzyme used to make the fragments being cloned; therefore, recovery of the cloned fragment may not always be straightforward. This problem could also occur with fragments having cohesive ends, as the sequences in Table 1 show. For example, it could be convenient to insert DNA fragments from digests with BglII or Dpn into a receptor for fragments made with BamI, all of which have the same cohesive ends, but in so doing, the sequence recognized by BglII or BamI would be lost.

Fragments produced by an enzyme generating cohesive ends may be readily joined to fragments with even ends through the use of a suitable adaptor fragment. This can be obtained from a double digest of a viral DNA or a plasmid DNA with an appropriate pair of enzymes, or it could be a synthetic product. Alternatively, the DNA fragment with a cohesive end may first be repaired in reactions with DNA polymerase and nucleoside triphosphates so that its two strands become equal in length. Obviously, the opportunities for manipulation of biochemical reactions on DNA fragments to give various combinations are legion.

An elegant example of some of the sophisticated biochemical operations that can be effected is provided by the cloning of two adjacent fragments of phage λ DNA, which together contain the phage *cI* (i.e., repressor) gene, in the plasmids pCR11 and pMB9.[27] The procedure is summarized in Figure 2. pCR11, which contains single targets for HindIII and EcoR1,[28] was digested with EcoR1 and the resulting 3′ ends repaired with DNA polymerase and deoxynucleoside triphosphates. Digestion with HindIII subsequently gave the required receptor molecule with one flush end and one cohesive end for DNA fragments generated by this enzyme. The desired λ fragments (containing the two parts of the *cI* gene) were isolated from separate digests of the

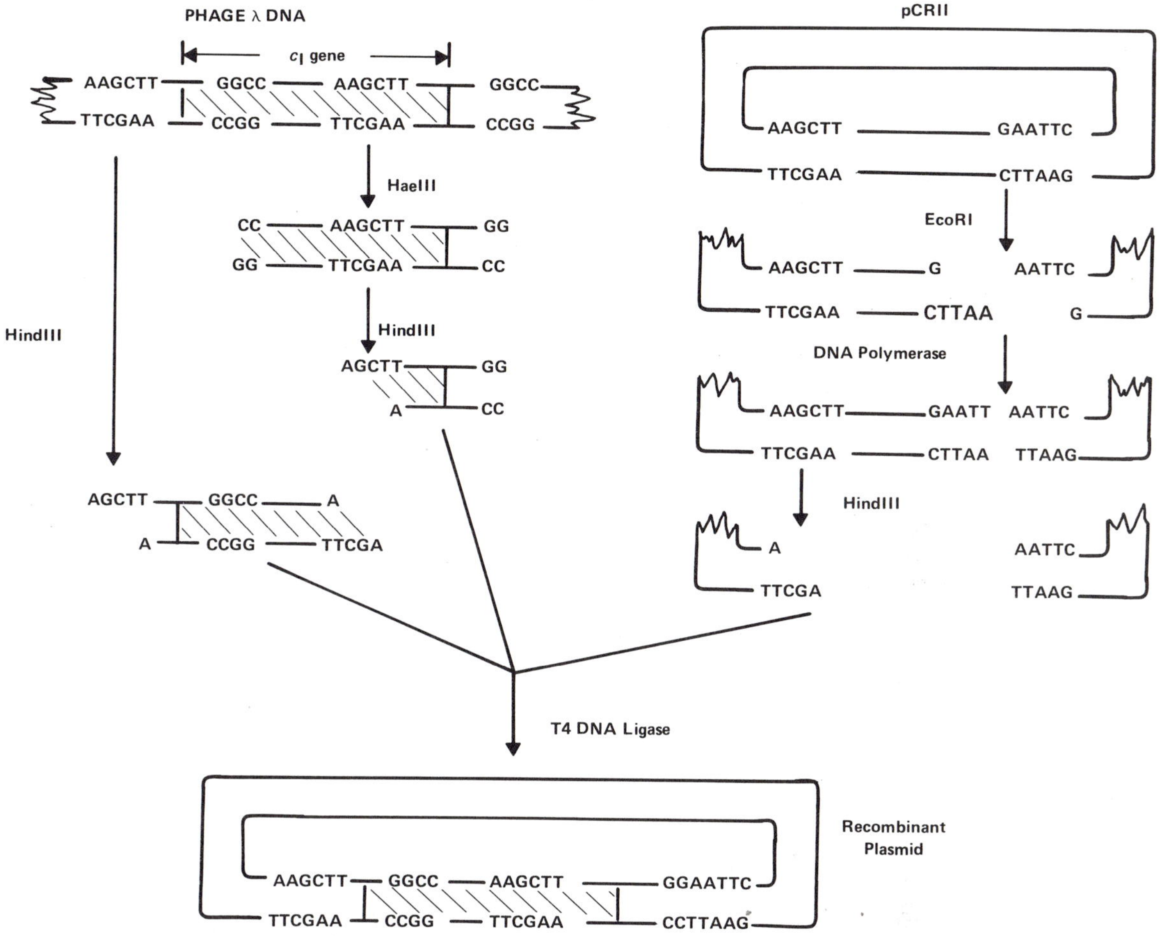

FIGURE 2. Transfer of λ DNA fragments containing the *cI* gene to the plasmid pCR11. The hatched region represents the *cI* gene of λ. The experiment is described in the text, and details are to be found in Reference 27.

phage DNA with HaeIII and HindIII (Figure 2). The fragment isolated from the HaeIII digest, which was 790 base pairs, was further digested with HindIII to give a fragment of 630 base pairs with the HaeIII terminal sequence at one end and the HindIII cohesive end at the other. This DNA fragment and the fragment (520 base pairs) from the HindIII digest of the λ DNA, which contained the other part of the *cI* gene, were then mixed with the recipient plasmid and incubated with polynucleotide ligase from T4-infected *E. coli*; λ-immune clones were isolated from bacteria transformed with the reaction products. The digestion of the recovered recombinant plasmid with HindIII and EcoR1 gave the expected DNA fragments and showed that a functional target sequence for EcoR1 had been regenerated by the joining of the two flush ends, as anticipated. It is not surprising that yields of the desired recombinant were very low (5 λ-immune clones were isolated from bacteria transformed with the reaction products). It seems likely that highly specialized manipulations of this sort may, initially at least, be limited in their application to relatively homogeneous populations of DNA fragments and reverse transcripts of purified mRNA preparations.

Detailed procedures for the use of these various enzymes and the handling and propagation of DNA molecules made by rejoining fragments in vitro are discussed extensively elsewhere in this volume together with a variety of examples of their use. If a given plasmid or phage DNA does not contain a target sequence for an enzyme necessary for the isolation of a desired DNA sequence, it is usually possible to insert the appropriate target sequence into that vector using a DNA fragment from another source. For example, a given plasmid may not contain a target sequence for the enzyme BalI;[29] however, it may contain the target sequence for the enzyme HindIII. An appropriate HindIII fragment from, e.g., bacteriophage λ DNA containing a single target for the BalI enzyme, could be inserted into the plasmid to convert it into a vector for DNA fragments resulting from digestion with BalI. There are many procedural variations, including the use of synthetic oligonucleotides.[30] The point worth emphasizing is that receptors for fragments made with any enzyme that generates cohesive ends or those for fragments made with enzymes making even breaks may be converted into forms for use with the other type of system through the application of some of the now standard biochemical reactions. Yields, however, may often be poor.

More recently, a third type of restriction endonuclease (Type III) has been recognized which resembles the Type 1 enzymes in that the target sequences and points of breakage of the DNA are not coincident, but differs from them in its cofactor requirements (in this it resembles the Type II enzymes) and in breaking the DNA at a specific point that is a measured distance from the recognition or target sequence. Four examples of this have been found,[31-33] three of which appear to recognize pentanucleotide sequences. One of these enzymes breaks DNA molecules about eight base pairs away from this sequence, making staggered breaks one base pair apart.[31] Another recognizes a pentanucleotide sequence, but makes staggered breaks five base pairs apart at a distance of five or ten base pairs (depending upon the strand) from the target sequence.[33] These enzymes would clearly be of much more specialized value in the construction of recombinant DNA molecules. They could, however, be used as reagents recognizing the target pentanucleotide sequence if the ends of the fragments were repaired and subsequently joined to a suitable vector system using the T4 ligase to join nonoverlapping fragments.

V. BIOLOGICAL ROLE OF RESTRICTION ENZYMES

Finally, it is worth reflecting upon the biological role of restriction enzymes. A common view is that they constitute a defence mechanism for bacterial cells against unwanted DNA. However, they are remarkably inefficient and unsuccessful for this purpose, and the whole process appears cumbersome. A more realistic evaluation may be that they provide a barrier to breeding between bacterial species or strains but permit some leakage of heterologous DNA that may be of evolutionary advantage. An alternative view frequently discussed postulates that restriction enzymes may, in fact, be some type of recombination system evolved to provide a very limited and perhaps rather specialized opportunity for recombination. Some recombination is generally regarded as desirable, but the extent of it would, in this case, be limited by the size of the target sequences for the enzymes involved and further constrained by use of the modification reactions. It is an arresting thought that the purposes to which these enzymes are now being widely applied are, in fact, those for which they may have evolved in bacteria.

ACKNOWLEDGMENTS

I am indebted to numerous colleagues and particularly to Dr. R. J. Roberts for permission to include their unpublished results in Table 1. Previously unpublished work from the Department of Molecular Biology, University of Edinburgh was supported in part by the Science Research Council, United Kingdom.

REFERENCES

1. Bertani, G. and Weigle, J. J., *J. Bacteriol.*, 65, 113, 1953.
2. Lederberg, S., *Virology*, 3, 496, 1957; Dussoix, D. and Arber, W., *J. Mol. Biol.*, 5, 37, 1962.
3. Arber, W., *Prog. Nucleic Acid Res. Mol. Biol.*, 14, 1, 1974.
4. Meselson, M. and Yuan, R. T., *Nature*, 217, 1110, 1968.
5. Roulland-Dussoix, D. and Boyer, H. W., Biochim. Biophys. Acta, *195, 219, 1969.*
6. Meselson, M., Yuan, R., and Heywood, J., *Annu. Rev. Biochem.*, 41, 447, 1972.
7. Eskin, B. and Linn, S., *J. Biol. Chem.*, 247, 6183, 1972.
8. Horiuchi, K. and Zinder, N. D., *Proc. Natl. Acad. Sci. U.S.A.*, 69, 3220, 1972.
9. Murray, N. E., Batten, P. L., and Murray, K., *J. Mol. Biol.*, 81, 395, 1973.
10. Bickle, T. A., Brack, C., Reiser, J., and Yuan, R. T., in press.
11. Boyer, H. W., *Annu. Rev. Microbiol.*, 25, 153, 1971.
12. Smith, H. O. and Wilcox, K. W., *J. Mol. Biol.*, 51, 379, 1970.
13. Kelly, T. J. and Smith, H. O., *J. Mol. Biol.*, 51, 393, 1970.
14. Middleton, J. H., Edgell, M. H., and Hutchison, C. A., *J. Virol.*, 10, 42, 1972.
15. Yoshimori, R., Ph.D. Thesis, University California, San Francisco, 1971.
16. Hedgpeth, J., Goodman, H. M., and Boyer, H. W., *Proc. Natl. Acad. Sci. U.S.A.*, 69, 3448, 1972.
17. Bigger, C. H., Murray, K., and Murray, N. E., *Nature New Biol.*, 244, 7, 1973.
18. Roberts, R. J., *CRC Crit. Rev. Biochem.*, 4, 123, 1976.
19. Polisky, B., Greene, P., Garfin, D. E., McCarthy, B. J., Goodman, H. M., and Boyer, H. W., *Proc. Natl. Acad. Sci. U.S.A.*, 72, 3310, 1975.
20. Heininger, K., Horz, W., and Zachau, H. G., *Gene*, in press.
21. Murray, K. and Old, R. W., *Prog. Nucleic Acid Res. Mol. Biol.*, 14, 117, 1974.
22. Old, R. W., Murray, K., and Roizes, G., *J. Mol. Biol.*, 92, 331, 1975.
23. Gellert, M., *Proc. Natl. Acad. Sci. U.S.A.*, 57, 148, 1967.
24. Olivera, B. M. and Lehman, I. R., *Proc. Natl. Acad. Sci. U.S.A.*, 57, 1426, 1967.
25. Sgaramella, V., *Proc. Natl. Acad. Sci. U.S.A.*, 69, 3389, 1972.
26. Weiss, B. and Richardson, C. C., *Proc. Natl. Acad. Sci. U.S.A.*, 57, 1021, 1967.
27. Backman, K., Ptashne, M., and Gilbert, W., *Proc. Natl. Acad. Sci. U.S.A.*, 73, 4174, 1976.
28. Covey, C., Richardson, D., and Carbon, J., *Mol. Gen. Genet.*, 145, 155, 1976.
29. Gelinas, R. E., Myers, P. A., Weiss, G. A., Murray, K., and Roberts, R. J., *J. Mol. Biol.*, 114, 433, 1977.
30. Bahl, C. P., Mariaus, K. J., Wu, R., Stawinsky, J., and Narang, S. A., *Gene*, 1, 81, 1977.
31. Kleid, D., Humayun, Z., Jeffrey, A., and Ptashne, M., *Proc. Natl. Acad. Sci. U.S.A.*, 73, 293, 1976.
32. Brown, N. L., Hutchison, C. A., and Smith, M., unpublished.
33. Brown, N. L. and Smith, M., unpublished.
34. Sharp, P., Sugden, W. B., and Sambrook, J., *Biochemistry*, 12, 3055, 1973.
35. Hughes, S. G., Bruce, T., and Murray, K., unpublished observations.
36. Murray, K., Hughes, S. G., Brown, J. S., and Bruce, S. A., *Biochem. J.*, 159, 317, 1976.
37. Roberts, R. J., Myers, P. A., Morrison, A., and Murray, K., *J. Mol. Biol.*, 102, 157, 1976.
38. Roberts, R. J., Wilson, G. A., and Young, F. E., *Nature*, 265, 82, 1977.
39. Catterall, J. and Welker, N., *J. Bacteriol.*, 129, 1110, 1977.
40. Vanyushin, B. F. and Dobritsa, A. P., *Biochim. Biophys. Acta*, 407, 61, 1975.
41. Pirrotta, V., *Nucleic Acids Res.*, 3, 1747, 1976.
42. Bron, S. and Murray, K., *Mol. Gen. Genet.*, 143, 25, 1975.
43. van Montagu, M., unpublished.
44. Murray, K., Morrison, A., Cook, H., Bruce, S. A., and Roberts, R. J., unpublished.
45. Wu, R., King, C., and Jay, E., unpublished.
46. Lacks, S. and Greenberg, B., *J. Biol. Chem.*, 250, 4060, 1975; Lacks, S. and Greenberg, B., *J. Mol. Biol.*, 114, 153, 1977.
47. Gelinas, R. E., Myers, P. A., and Roberts, R. J., *J. Mol. Biol.*, 114, 169, 1977.
48. Murray, K., Bruce, S. A., and Roberts, R. J., unpublished.
49. Murray, K., Bruce, S. A., and Brown, J. S., unpublished.
50. Brockes, J. P., Brown, P. R., and Murray, K., *J. Mol. Biol.*, 88, 437, 1974.
51. Roberts, R. J., Breitmeyer, J. B., Tabachnik, N. F., and Myers, P. A., *J. Mol. Biol.*, 91, 121, 1975.
52. Barrell, B. G. and Slocombe, P. unpublished.
53. Takanami, M. and Kojo, H., *FEBS Lett.*, 29, 267, 1973.
54. Wilson, G. A. and Young, F. E., unpublished.
55. Sugisaki, H. and Takanami, K., *Nature New Biol.*, 246, 138, 1973.
56. Garfin, D. E. and Goodman, H. M., *Biochem. Biophys. Res. Commun.*, 59, 108, 1974.
57. Baumstark, B. R., Roberts, R. J., and RajBhandary, U. L., unpublished.
58. Roberts, R. J., Myers, P. A., Morrison, A., and Murray, K., *J. Mol. Biol.*, 103, 199, 1976.

59. **Subramanian, K. N., Weissman, S. M., Zain, B. S., and Roberts, R. J.,** *J. Mol. Biol.,* 110, 297, 1977.
60. **Brown, N. L., and Smith, M.,** *FEBS Lett.,* 65, 284, 1976.
61. **Smith, D. L., Blattner, F. R., and Davies, J.,** *Nucleic Acids Res.,* 3, 343, 1976.
62. **Endow, S. and Roberts, R. J.,** *J. Mol. Biol.,* 112, 521, 1977.
63. **Arrand, J. R., Myers, P. A., and Roberts, R. J.,** unpublished.
64. **Sato, S., Hutchison, C. A., and Harris, J. I.,** *Proc. Natl. Acad. Sci. U.S.A.,* 74, 542, 1977.
65. **Zain, B. S. and Roberts, R. J.,** unpublished.
66. **Roberts, R. J., Gingeras, T. R., Olson, J. A., and Myers, P. A.,** unpublished.
67. **Smith, H. O. and Nathans, D.,** *J. Mol. Biol.,* 81, 419, 1973.

Chapter 6

IN VIVO FORMATION OF RECOMBINANT DNA MOLECULES BY IS-ELEMENTS AND TRANSPOSONS

P. Starlinger

TABLE OF CONTENTS

I. INTRODUCTION

The term "genetic engineering" can be broadly used to include all techniques designed to manipulate the genetic constitution of organisms. However, in recent years, the term has been used mainly in a narrower sense, i.e., the enzymatic preparation of recombinant DNA molecules and their subsequent cloning within bacterial cells. These methods have aroused great interest due to the possibility of combining, within one molecule, DNA from various sources and from organisms as unrelated as man and bacteria. Few of these combinations are thought to have existed before in nature.

There is hope that under certain circumstances these new gene combinations can be used for medical or agricultural purposes. However, considerable concern has been expressed about the possibility of unforeseen biohazards. The reason for both the enthusiasm and concern is the widespread belief that new combinations of genes are rarely, if at all, created within living organisms. There are two main reasons for why genes are prevented from entering unusual combinations. First, DNA molecules are confined to cells or nuclei (in higher organisms) and do not mix readily. Specific mechanisms such as fertilization in sexual organisms or conjugation, transduction, and transformation in bacteria are needed to combine DNA molecules, and in most cases, these mechanisms are confined to cells of the same species. Secondly, DNA molecules usually exchange material by genetic recombination. This mechanism, however, does not create new combinations of genes; rather, it leaves the gene arrangements intact and only exchanges allelic gene states. Consequently, the use of the term recombinant DNA molecules for artificially combined DNA molecules is slightly misleading, since a new combination of genes is quite different from the products of the process usually referred to as genetic recombination.

New genetic combinations in higher organisms have been described by cytogeneticists as chromosomal aberrations. These include deletions, duplications, inversions, and translocations. In higher organisms, the chromosome fragments exchanged by these events are often large when compared to the size of single genes. Consequently, effects have been seen at the level of chromosome distribution rather than that of gene function. Also, these events occur within a single species; therefore, the new gene combinations produced are not derived from different species. Recently, it has been found that the new gene combinations involving different species can be created within bacterial cells by natural mechanisms.

1. Plasmids are circular DNA molecules in the molecular weight range from a few million to a few hundred million daltons. They replicate within bacterial cells and carry useful, but frequently dispensable, genetic information. Some are able to promote their own transfer from one cell to another,[1-3] and some of these are not confined to bacteria of the same species. A few have a very broad host range, including many Gram-negative bacteria.[4-7] Therefore, at least the genetic information encoded in the DNA of some plasmids is freely exchanged between very different bacteria.
2. Novel genetic combinations within the same cell can be created with the help of temperate bacteriophage[8,9], episomes,[10,11] and specific DNA elements that only occur within chromosomes, but can be transposed from one DNA site to another. These are called IS-elements (inserted sequence) provided they only exert their influences in position *cis* at the site of integration.[12] Another class, called transposons, carry genes usually coding for resistance to antibiotics.[13,14]
3. Transferring a gene from a bacterial chromosome to a broad host-range plasmid and subsequently transferring the plasmid from one bacterial species to another enables the creation of new gene combinations utilizing the gene pool of Gram-negative bacteria.

Although, the uptake of nonbacterial DNA into bacterial cells in combination with these mechanisms has not been shown to occur, it cannot be excluded that such events occur with a measurable frequency. These in vivo mechanisms must be considered together with biochemical methods of genetic engineering, and their relative usefulness and potential hazards must be evaluated. This chapter will describe some properties of IS-elements, transposons, and bacteriophage mu important in the development of new gene combinations.

IS-elements are unique DNA sequences. Five such sequences are known at present, with lengths of between 800 and 1450 nucleotide pairs. They occur naturally in the chromosome of *Escherichia coli* and some of its plasmids. They can be transposed from these sites to new (more or less specific) locations. Certain effects exerted at these new locations have enabled their detection.

1. A mutation of the gene within which the IS-element is integrated
2. Polar effects on genes located distally to the site of integration, provided the site is within an operon
3. The constitutive expression of a gene that has lost its promoter, provided the IS-element carries its own promoter and is integrated in front of a gene lacking a promoter
4. An increased probability of chromosomal aberrations (i.e., deletions, duplications) adjacent to the site of integration of the IS-element
5. They can be excised from this location so precisely that the original phenotype is restored
6. Two chromosomes, e.g., the *E. coli* chromosome and its sex factor F, can fuse at the site of an IS-element carried by both of them
7. IS-elements have been shown to be part of transposons

IS-elements have not been shown to carry genes coding for proteins with a recognizable function.

Transposons are very similar to IS-elements in all respects mentioned above. However, they are larger and carry recognizable genes in most cases coding for resistance to an antibiotic. Transposition of resistance genes between the different R-plasmids is responsible for the variability of these plasmids causing considerable medical concern. Many transposons carry repeated DNA sequences at their termini in either an inverted or directly repeated form. Some of these repeated sequences have also been detected individually as IS-elements.

Bacteriophage mu is a temperate bacteriophage capable of integrating into any site of the *E. coli* chromosome and its plasmids. In this respect, it has some properties in common with the IS-elements and transposons discussed here.

To unify the nomenclature of these elements and the mutations formed by their integration into genes, rules were suggested by a committee formed at a meeting on DNA insertions in Cold Spring Harbor, New York in 1976.[15] They include designating IS-elements and transposons by an italicized number following the abbreviations of IS and Tn, respectively. A different number is assigned to each independent isolate even if apparently identical to some previous isolate. The abbreviation of the resistance gene(s) coded for by the transposons may be indicated in parenthesis following the name, e.g., Tnlo(Tc). Mutations caused by the integration of these elements are designated by an isolation number followed by a double colon and the name of the transposable element, e.g., *gal*T-N102::IS1.

II. DESCRIPTION OF IS-ELEMENTS AND OF STRUCTURES HAVING SIMILAR PROPERTIES

A. IS-elements
1. Detection
All known mutations caused by the insertion of an IS-element are of spontaneous

origin. For the selection of these mutations, specific techniques have been employed.[16-21] IS-elements were first detected by the polar mutations created at their new sites. Since these mutants revert to wild-type, they are not caused by deletions. The back mutation rate, however, is not increased by mutagens. Since most point mutations can be induced to revert by various mutagens, the mutations discussed here are probably of another type, e.g., inversions, duplications, or DNA insertions. Samples of these polar mutations were subsequently characterized as DNA insertions by biochemical methods and electron microscopy.

Length measurements on the DNA of transducing phage carrying the mutated bacterial gene showed that DNA has been added by the mutation. These measurements were performed directly by sedimentation of the DNA in the ultracentrifuge[22] or indirectly by measuring the buoyant density of the transducing phage in CsCl gradients.[23,24] The increased density of the phage indicated addition of DNA but did not differentiate between duplications and insertions. This distinction is made by hybridization studies. Duplications do not introduce new sequences into the operon, while the insertion of DNA from other regions of the chromosome usually does. RNA transcribed in vitro from a DNA template carrying the mutations is exhaustively hybridized to a DNA devoid of the mutation but otherwise identical. This is followed by a test hybridization of the remaining RNA to the template DNA. These experiments show that the additional DNA is newly introduced and not the result of a duplication of part of the mutated gene.[25,26]

An electron microscopic examination of DNA heteroduplex molecules carrying one mutated and one nonmutated strand shows molecules in which the additional DNA forms a single-stranded loop emerging from the double-stranded molecules. Heteroduplex molecules prepared from DNA carrying independent mutations caused by the insertion of IS-elements allow the detection of complete or partial sequence homologies between these elements and enable the orientation of the IS-elements with respect to the chromosome. If the insertion of the same IS-element results in two independent mutations having the IS-element in inverted order, heteroduplex molecules are formed from the IS-element when identical DNA strands of the phage carrying the mutations are annealed.[26-29]

A heteroduplex analysis was applied to the study of several insertions in the *lac* and *gal* operon of *E. coli* and in bacteriophage λ.[26-29] All insertions studied were found to belong to a few classes of homologous DNA sequences called IS-elements. They have subsequently been discovered in a variety of other DNA molecules.[30-37] The formation of heteroduplex structures prepared from identical DNA strands on a preparative scale, followed by digestion with single-strand-specific endonuclease Sl, enables the isolation of pure IS-DNA by sucrose gradient centrifugation or polyacrylamide-gel electrophoresis.[38,39] This method yields homogeneous DNA preparations. Isolated IS-DNA, or IS-elements integrated at known positions in well-studied DNA molecules like λ DNA have been subjected to the action of a variety of restriction endonucleases.[38,40]

2. Nature of Polar Effects

Most IS-elements cause strong polar effects when integrated into a gene within an operon. This has been exploited in determining the number of transcription units and the gene order within a DNA segment, e.g., the genes coding for ribosomal proteins.[20] The nature of the polar effects is not completely understood. It is believed that the polarity caused by IS-elements has many properties in common with the polarity caused by nonsense codons.[41,42] Both polarities are relieved by the action of the mutation suA, which is known to occur in the gene coding for the transcription termination factor rho.[41,43,44] A hypothesis has been developed to explain both polar effects of nonsense mutations and insertions.[42] It suggests that the transcription factor rho interacts with nascent RNA chains and is translocated along these chains in the direction of transcription until it interacts with RNA polymerase molecules. The complex is subsequently thought to stop transcription at defined sites found both within and at the end of operons.

The interaction of the rho factor with nascent RNA chains is assumed to be impossible when the RNA chains are being translated. Consequently, only rho-sensitive DNA sites at the end of operons are recognized when translation proceeds normally. However, if a nonsense codon causes translation termination, the nascent RNA distal to this codon is free of ribosomes and can therefore interact with the rho factor. If a rho-sensitive site is encountered within the operon, this leads to termination of transcription within the operon. Insertion mutations are believed to carry nonsense codons and may or may not carry rho-sensitive transcription stop signals.[42,45] In vivo studies in which no distal RNA was discovered, suggest that transcription termination may occur for IS1.[46] It is shown that IS2 carries a rho-sensitive stop site for RNA synthesis, while IS1 does not.[45]

There are also differences between the polar effects caused by nonsense mutations and IS-elements. The polarity caused by IS-elements (one orientation for IS2 and IS3 and both orientations for IS1 and IS4) is very strong, leaving less than 1% of the residual synthesis of the enzymes coded by the distal genes. Polar effects observed with nonsense codons are usually much weaker.[47] Therefore, the strong polar effect caused by IS1 insertion into *gal*T must be caused by the IS1-element rather than a rho-sensitive site in the distal part of *gal*T. Since IS1 does not carry rho-sensitive sites in vitro, additional explanations must be sought.[48]

Another polarity difference is the constitutivity of residual synthesis distal to IS-elements, even if the polarity of the IS-elements is relieved considerably by suA mutations. The hypothesis that the IS-element carries a promoter that is unmasked by the removal of a stop signal does not seem tenable mainly because, in the case of IS1-elements within gene *gal*T, deletion of the operator and promoter of that operon abolishes the expression of the distal gene *gal*K.[48] Explanations for these results have been suggested, but more experiments are clearly necessary. IS-elements may thus play a role in the elucidation of mechanisms of transcription termination. A mutation on IS2 which relieves its polarity, has been described.[49]

3. Promoters on IS-elements

IS1 and IS4 are polar in both orientations[27,28,34] IS3 has been found only in one orientation within a bacterial operon;[29] nothing can be said about the other orientation. IS2 has been recurrently isolated; however, if the selection was for polar effects, only one orientation was observed.[27,28,36,37,50]

The suspicion has been raised that the other orientation of IS2 is not polar. It was possible to transfer a mutated *gal* operon containing IS2 in orientation I to an F'-particle carrying an IS2 in orientation II 2.6 kb upstream. A class of constitutive revertants was created when the galactose operon was linked to the IS2-element in orientation II by deletion of the DNA in between, including the IS2 in orientation I. The hypothesis that the constitutive synthesis is due to a promoter carried on the IS-element is supported by the finding that (1) the isolation of *gal*-negative colonies from the constitutive revertants is accompanied by a loss of the IS2-element and (2) the isolation of *gal*-positive revertants from these revertants is accompanied by a reinsertion of IS2 in orientation II.[36] It is possible to isolate RNA molecules that contain covalently bound IS2 RNA and *gal* mRNA.[51] Therefore, IS2 carries a strong promoter and is able, by its integration into one of its orientations in a new position, to activate the expression of a previously silent operon.

4. Formation of Deletions Adjacent to IS-elements

IS1[52] and IS3[53] have the property of inducing deletions adjacent to them. These deletions have one endpoint at a terminus of IS and extend to either side for varying distances. These endpoints are not distributed randomly. The deletion endpoint at the IS-element IS1 does not appear to remove a recognizable part of IS1 since:[54]

1. In a strain already carrying a deletion adjacent to IS1, secondary and even tertiary deletions can be selected for at elevated frequencies.
2. Genetic crosses have been performed between strains carrying deletions adjacent to IS1 isolated from the same parental strain, in which the deletions extended in opposite orientations and the only region of homology, if any, was the IS-element. Recombination to a deletionless type has been observed in these crosses.
3. The IS-element could be visualized directly by the inspection of appropriate heteroduplex molecules in the electron microscope.

The enzyme system responsible for the deletion formation adjacent to IS1 does not seem to be the same as the system for the exact excision of the IS-element (see below). Deletion formation is strongly dependent on temperature and is maximal at 30°C and completely abolished at 37°C. However, no temperature dependence is observed for the exact excision of IS-elements. Mutants have been isolated in which deletion formation is decreased while the exact excision of IS-elements remains uninhibited.[55] IS2 does not induce deletions;[52] aberrations adjacent to IS4 and IS5 have not been reported.

A duplication adjacent to mutation *gal* 3[56-58] has been reported.[59] This mutation is caused by the insertion of an IS-element[30] identified as IS2.[138] The duplication was isolated as an unstable, constitutive revertant to the Gal⁺ phenotype, and its nature as a duplication was deduced from its instability (i.e., segregation of Gal⁻ cells at a frequency of several per cent) and the additional observation that this segregation did not occur in a *rec* A background. It is assumed that the duplication has separated the structural genes of the operon from the IS-element and linked them to a new promoter. Since the IS-element must be inserted very near to the beginning of the structural genes, it is possible that the duplication has one of its termini at the IS-element.

Inversion of IS-elements has not been directly observed. However, some revertants of *gal* OP mutations caused by the integration of *gal* OP are very unstable and segregate Gal⁻ cells at a frequency of 10^{-2} to 10^{-4}.[36] This instability is probably not due to a point mutation. Electron microscopy of one of these mutants did not show a visible alteration of IS2. Thus, small inversions are a possible explanation of this type of mutation.[36,51,60]

5. Mechanism and Specificity of IS-integration

Electron microscopical heteroduplex analysis has indicated that IS-elements integrated at different locations are not circularly permuted.[27,28,34] This is compatible with the assumption that only one integration sequence is present on the IS-elements. The specificity of the integration sites on the recipient chromosome differs for various IS-elements. Nothing is known about IS3 and IS5, since these have each been observed within bacterial or phage genes only once.[29,31] IS3 has also been observed once within the tetracycline gene of the tetracycline transposon Tnlo.[35]

IS1 has been observed at several sites in *gal* T and *lac* Z. The integration does not seem to be random within the genes, since a group of three and another group of two independent IS1 mutations have been found which do not recombine to the wild type within the group and, consequently, must be at identical or very closely linked positions.[34] Also, several copies of IS1 in either orientation have been identified within *gal* OP;[26,28,60] which fail to recombine with each other.

IS2 has been identified independently three times within *gal* OP or possibly between *gal* OP and *gal* E.[27,28,30] All three isolates must be in the same site or in very closely linked positions. IS2 has also been observed in bacteriophage λ and in several plasmids.[33,50,61]

A pronounced specificity is seen for IS4. This IS-element has been isolated at various times in various laboratories at only a single site in *gal* T, since none of these mutants recombine with each other.[18,27,34,62] IS4 has been isolated in this position in both orientations.[34] It has been shown that the integration of bacteriophage λ into *gal* T occurs at the same position as that observed in the mutants described in Reference 62.

This finding of IS4 in both orientations deserves a special discussion. The integration of bacteriophage λ into its major integration site always occurs in the same orientation. If the integration of IS4 does occur in either orientation, this is not easily compatible with the assumption of a specific integration sequence, unless this is symmetrical. Alternatively, the recipient integration site may not be a specific sequence, but rather, a limited region of DNA that allows the interaction of IS4, the recipient DNA, and the proteins involved. The same argument applies to the recurrent isolates of IS1 and IS2 in *gal* OP.

The frequency of occurrence of mutations in the *gal* operon caused by the integration of IS-elements is approximately 10^{-7} per cell plated.[16,17,19,63] The reversion of IS-caused mutations to wild type occurs with a frequency between 10^{-6} and 10^{-8}.[16,17,61,63] If the mutation results from the integration of an IS-element within a structural gene, the reversion to wild type implies that their excision is a very exact event which restores the original DNA sequence at the level of single nucleotides.

Nothing is known at present about the enzymes involved in these processes, since mutations in genes responsible for the excision of IS-elements have not yet been isolated. Both the integration of IS and its excision occur in a *rec* A strain,[20,63] suggesting that the (complete) bacterial recombination system is not necessary. It is not yet known whether the integration of an IS-element into a new position is accompanied by the loss of the same element from another position. The constancy of the number of IS-elements in *E. coli* K12 strains[64] could be used as an argument for an excision and reintegration mechanism, but other explanations are conceivable. Nor is it known whether the transfer intermediate is pure IS-DNA, as in the case of bacteriophage λ, or whether it contains host DNA at both of its ends, as in the case of bacteriophage mu (see Section II.C).

6. Normal Occurrence of IS-elements

In the preceding paragraphs, IS-elements have been described in positions to which they were transposed by a mutational event. The question of the location of IS-elements unassociated with a recent mutation was first investigated by DNA-DNA hybridization.[64] This technique showed that approximately eight copies of IS1 and five copies of IS2-DNA are present in the *E. coli* K12 chromosome.[64] However, these experiments did not show whether these sequences are clustered or dispersed, nor could they show whether all cells carry the IS-DNA in the same position.

In addition to the *E. coli* chromosome, IS-sequences have also been detected on some plasmids of *E. coli*, including the F- and several R-plasmids. Hybridization experiments revealed the presence of one IS2 copy and two IS3 copies in unique positions on the F-plasmid.[11,33,35,61,65] No plasmids consisting exclusively of IS-DNA were found.[64]

DNA sequences present as inverted repeats have been found in *E. coli* DNA isolated under gentle conditions, denatured, and renatured under conditions which primarily allowed the formation of intrastrand double-strands.[66] Inspection of these molecules in the electron microscope and measurement of the length of the DNA double-strands formed revealed lengths corresponding to IS1, IS2 to IS4, the ribosomal genes present in several copies in the *E. coli* chromosome, and to the γ-δ-sequence found on the F-plasmid.[11] Additionally duplex molecules of a 1.0 and 0.5 kb length were also found. It is thus possible that the known IS-sequences (and a few others) are the main sequences present in the *E. coli* chromosome in several copies and in inverted order with respect to each other.

Many of the inverted sequences are separated by unique DNA of the same length, indicating a regular pattern in the insertion of these sequences. The sequences of IS1 length are more frequent than determined by Saedler and Heiss;[64] however, it is not known whether all sequences of this site are IS1 or whether different repeated sequences of this size are present in *E. coli.*[66] As no heteroduplex molecules of the length of IS1 or γ-δ were seen, in a similar study,[67] this question needs further clarification.

Little is known about IS-sequences in other species. IS1 is present in *Salmonella typhimurium* and *Citrobacter freundii,* while IS2 has not been detected in either of these species.[51,139]

7. Role of IS-elements in the Interaction of Plasmid DNAs with Each Other and the Bacterial Chromosome

a. F-plasmid

The integration of the F-plasmid into the *E. coli* chromosome leads to the formation of Hfr-strains. The position of the strains investigated does not seem to be random, as Hfr-strains with a particular origin and direction of transfer have been isolated repeatedly.[68,69]

The excision of the F-plasmid from the chromosome of a Hfr-strain can occur in an irregular fashion, leading to an F′-plasmid that carries some bacterial genes in addition to F-genes. Two types of F′-plasmids can be distinguished by the positions at which the Hfr-DNA is cleaved in the process of F′-formation. In F′-plasmids of Type I, one of these sites is located within the F-DNA, while the other is in the bacterial DNA. The resulting plasmid consequently lacks some F-plasmid DNA and contains some DNA sequences of the bacterium, all of which were located on one side of the integrated plasmid. In Type II F′-plasmids, both cleavage points in the DNA are located within the bacterial DNA on either side of the F-plasmid. The resulting F′-plasmid therefore has the complete F-DNA in addition to two bacterial segments derived from either side of the integrated F-plasmid and joined together at their ends.[10]

Detailed heteroduplex analysis of the DNA of the F- and F′-plasmids[11] has led to the view that, in most instances, the integration of the F-plasmid into the *E. coli* DNA occurs by a recombinational event between a sequence carried on the F-plasmid and a homologous sequence carried by the *E. coli* chromosome. Three such sequences have been identified.[11,33,65] Two are IS2 and IS3, while the third sequence participating in these events is called γ-δ and has a length of 5.2 kb. It may be present in several copies on the *E. coli* chromosome, as judged from the identification of inverted repeats of this length;[66] however, γ-δ has not been identified as a transposable element causing polar mutations.

The integration of the circular F-DNA into the *E. coli* chromosome via recombination between a sequence carried in the F-plasmid and a homologous sequence carried on the chromosome leads to a structure in which the F-DNA is bordered at both ends by the DNA sequence responsible for the recombination. Type II F′-plasmids formed from these structures therefore contain the DNA sequence (i.e., IS-element or γ-δ) responsible for the recombination in one more copy than had been present on the F-particle.

Type I F′-plasmids contain a deletion of F-DNA whose termini must be in a region of the F-DNA that does not carry essential DNA sequences. Investigation of many Type I F′-plasmids has shown that the endpoints of the deleted DNA within the F-chromosome are nonrandom. In many instances, the deletion ends at the terminus of one of the IS-elements or the γ-δ sequence. These nonrandom cleavage events have been called "half-site specific recombination."[11,65]

Due to plasmid incompatibility, two different F'-particles cannot normally coexist within the same bacterial cell. Selection for genes carried by either F'-plasmid leads to the isolation of fused F'-plasmids. In most instances, these have lost considerable parts of one or the other F'-chromosome. The endpoints of these deletions often involve the ends of IS-elements or sequence γ-δ. The mechanism involved may resemble site-specific or half-site-specific recombination.[70-72]

The integration of an F-plasmid into the *E. coli* chromosome may not be mediated by the *E. coli rec* system since F'-plasmids carrying several copies of IS2 or IS3 rarely dissociate, while one carrying two copies of γ-δ frequently dissociates by recombination within or at the γ-δ terminus. This process, however, occurs with equal probability in *rec* $^+$ or *rec* A strains.[11,73] The difference in frequency of recombination at these sequences is not explained by size differences of the homologous sequences. Experiments in which early transfer of the integrated F-plasmid transfer genes in Hfr-crosses was tested yielded measurable frequencies only in strains in which integration is thought to have occurred via a γ-δ sequence.[74] Since *tra* genes are known to be transferred late in Hfr-crosses, these experiments can be interpreted as measuring the excision frequency of a functional F from the Hfr-strain. This suggests that recombinational excision of the IS-elements IS2 and IS3 is rare but occurs frequently at the γ-δ sequence.

While these observations suggest the participation of sequence-specific processes, they fail to explain the apparent lack of homologous genetic recombination. It is not known whether *rec*-dependent recombination is also dependent on specific sequences missing on IS-elements or if IS-sequences are actively prevented from participating in homologous pairing.

P-group plasmids are able to mobilize the bacterial chromosome in *Pseudomonas* strain PAT, but not in PAO.[75] A derivative, R68-45,[76] has been isolated that is capable of chromosome mobilization in PAO. This plasmid has been shown to carry additional DNA of a 2.75 kb length.[77] It is not known whether this DNA is analogous to IS-elements in *E. coli.*

b. R-plasmids

Several *fi* $^+$-type R-plasmids, including R1, R6, and R100, are composed of two segments which segregate from each other under certain circumstances and have the ability to replicate independently. One of these segments, the resistance transfer factor (RTF), carries most of the genes necessary for gene transfer and shares considerable homology with the F-plasmid. The other segment, called resistance determinant (r-determinant) carries most of the genes conferring resistance to various antibiotics.[78-83] An IS1 is located at both junctions between these two segments. Since both IS1s are in the same orientation, it is conceivable that the dissociation of the r-determinant from the RTF is via a recombination between these IS1s.[33,35] This dissociation is a rare event in *E. coli*[79]. However, it is observed frequently after transfer of these plasmids into *Proteus mirabilis* and most pronounced in the late growth phase, if selection is made for antibiotic resistance.

Another structure observed under these conditions is an enlarged plasmid consisting of one RTF and several tandem copies of the r-determinant. Conceivably this event (or at least the first duplication) occurs by recombination within the IS1s bordering the r-determinant.[83]

Independence of these events of a functional bacterial recombination system has not been tested. Since these events are rare in *E. coli* as compared to the late growth phase of *P. mirabilis*, it is possible that a specific recombination system rather than homologous recombination is involved. As in the case of F'-plasmid dissociation, the lack of *rec*-dependent dissociation in *E. coli* is not understood.

A plasmid conferring resistance to tetracycline has been described in *Streptococcus faecalis*.[84,85] If a cell carrying this plasmid is cultured in the presence of tetracycline, amplified plasmid forms occur in which part of the sequence carrying the tetracycline resistance gene(s) is amplified, while another part is present in one copy only.[84,85] The amplified part is bordered at both junctions with the nonamplified sequence by a repeated DNA sequence with a 0.4 kb length.[86] Both sequences have the same orientations with respect to each other. The 0.4 kb sequences are called recombination sequences (RS).

8. IS-elements — Constituents of Transposons

Transposons are mentioned in the introduction to Section II and described in more detail in Section II.B. It must be mentioned here that the inverted repeat bordering the tetracycline transposon Tn10 is IS3[35] and the direct repeat bordering the chloramphenicol transposon Tn9 is IS1.[87]

B. Transposons

There is a rapidly growing number of observations on DNA sequences that can be transposed as defined blocks from one chromosome or plasmid to another and usually carry one or more gene(s) coding for the resistance to an antibiotic. Transposons are often bordered by repeated DNA sequences. The occurrence of IS1 in Tn9 (Cm)[87] and IS3 in Tn10 (Tc) [35] has been mentioned above. The ampicillin transposons Tn2 (Ap) and Tn3 (Ap) carry an inverted repeat of approximately 140 nucleotide pairs length[88,89] while the kanamycin transposon Tn5 (Km) is flanked by inverted DNA repeats of about 1450 nucleotide pairs length.[90] These inverted repeats have not been identified as independently transposable IS-elements. Another transposon for kanamycin resistance (Tn6) does not carry an inverted repeat at its ends[90]; a few other transposons have not been investigated at the molecular level.

1. Detection

The detection of transposons is easier than the detection of IS-elements, because the antibiotic resistance carried by the former can be selected for.

2. Nature of Polar Effects

Tn10 has been integrated into various sites within the histidine operon of *S. typhimurium*.[91] In all cases, the function of the gene(s) distal to the transposon are missing. Insertions of the kanamycin transposon Tn5 into the *lac Z* gene of *E. coli* are polar for the expression of the distal *lacY* gene.[92] Both orientations of Tn10 occur in the former case; [93] in the latter case, the orientation of the Tn5 insertions has not been tested. However, since all mutations investigated were polar, and since it is unlikely that all have been integrated in the same orientation, it can be assumed that polarity is found in both orientations.

The situation is different with the ampicillin transposon, which has been found to have polar effects on the expression of antibiotic genes on plasmid RSF 1010. Apparently, the genes for sulfonamide and streptomycin resistance are organized in one operon, and integration into the former causes polar effects on the latter. In this case, a class of nonpolar integrations in the sulfonamide gene were found and shown, by heteroduplex mapping in the electron microscope, to be integrated in one orientation, while integration in the other orientation showed the polar phenotype.[94] The asymmetry of polarity must mean that the polar signal is not carried by the inverted repeat at the termini of the transposon, but must lie somewhere within the transposon.

3. Promoters on Transposons

There must naturally be promoters on the transposons where the transcription of the resistance genes is initiated. In no case, however, has transcription beyond the border of the transposon into adjacent genes been observed. The large number of independently isolated mutations caused by the integration of Tn10 (Tc) into the histidine operon of *S. typhimurium* and of Tn5 (Km) into the *lac Z* gene of *E. coli* makes it unlikely that in, all of these mutants, the transposon is integrated in the same orientation. If this is so, these transposons cannot carry a promoter capable of transcribing adjacent genes constitutively.

4. Chromosomal Aberrations in the Vicinity of Transposons

Deletion formation has been observed in the vicinity of Tn2 (Ap), Tn3 (Ap), and Tn10 (Tc).[53,93,97] In the case of Tn2 (Ap), internal deletions have been produced enzymatically. Some of these have lost the ability to transpose. It is possible that part of the transposon codes for a transposition function.[94a] Deletions adjacent to Tn2 (Ap) have a limited number of preferential endpoints within phage P22 DNA.[97]

Deletions at the site of a Tn10 (Tc) integrated in the *his* operon of *S. typhimurium* have one endpoint at the site of the transposon. Many are tetracycline-sensitive, which indicates that the gene for tetracycline resistance is (partly) included in the deletion. Deletions, however, often extend into adjacent genes. No preferential endpoints of deletions adjacent to Tn10 have been observed. The loss of tetracycline resistance is more frequent by several orders of magnitude than the exact excision leading to the restoration of the wild-type phenotype of the recipient gene.[91,93]

Mutants caused by the integration of the Tn5 (Km) into *lac Z* of *E. coli* exert polar effects on the distal gene *lacY*. Secondary mutants that are still *lac Z* but are now able to express *lacY* can be readily isolated.[92] When crossed, some of these mutants behave as deletions; others are able to revert. It is thought that only part of the kanamycin transposon, including the resistance gene, has been removed, and another part is left at the site of mutation. It is possible that, similar to the observations made with the ampicillin and tetracycline transposons, one endpoint of the deletion is the end of one of the inverted repeats bordering the transposon.

A large fraction of the tetracycline-sensitive derivatives of tetracycline-resistant cells, in which the tetracycline transposon was integrated into the histidine operon, show alterations within this operon. One quarter carries deletions extending from the point of integration of the transposon in one or the other direction but not in both, as described above. The remaining 75% carry all sites in the *his* operon that have been tested and thus are not deletions. However, they show an anomalous behaviour in crosses and, consequently, may carry other chromosomal rearrangements. Duplications and inversions of DNA segments adjacent to the transposon have been found.[93]

In addition to cause the formation of deletions, Tn3 is a preferred site for the integration of other transposable elements. The exact site for insertion segments seems to be the end of one of the flanking inverted sequences.[96]

5. Specificity of Integration and Excision of Transposons

In cases where inverted or direct repeats of DNA are located at the ends of a transposon, this arrangement of repeats is conserved and no circular permutation has been observed so far. In this respect, transposons are similar to IS-elements. The frequency with which the integration occurs varies. In the case of Tn5 (Km), infection of *E. coli* cells with nonlysogeniziing λ phages carrying the transposon led to approximately 10^{-2} cells which carry an integrated copy of the transposon somewhere in the chromosome.[90] This is considerably more frequent than the integration of one of the known IS-elements into one of the *E. coli* operons, even when the size difference between an operon and the whole chromosome is taken into consideration. The fraction of *lacZ* mutants among the kanamycin resistance cells is approximately 10^{-4}. This is lower by a factor of 10 than the fraction of *lac*DNA over total *E. coli* DNA. Therefore, the probability of integration of Tn5 (Km) into different sites on the *E. coli* chromosome may be variable, and undetected "hot spots" for integration may exist.

Varying frequencies of transposition have been reported for the ampicillin transposon Tn1 (Ap), depending on the donor and recipient chromosome and/or plasmid.[98] The ampicillin and kanamycin transposons Tn2 (Ap) and Tn5 (Km) do not exhibit a strong site-specificity when integrated into plasmid RSF1010 and the *lac* operon of *E. coli*, respectively,[88,92] while for Tn10 (Tc), repeated insertion into the same position within *his*G has been observed. The trimethoprim-streptomycin transposon Tn7 has been found to integrate into only one position on the *E. coli* chromosome between genes *dna*A and *ilv*. However, there are at least eight different sites for this transposon on the much smaller DNA of plasmid RP4.[99]

The exact excision of Tn5 (Km) from *lacZ* of *E. coli* and of Tn10 (Tc) from the histidine operon of *S. typhimurium* has been described.[90-93] Most mutants caused by the integration of these transposons revert to the full wild-type, indicating that the original mutation had not led to the deletion of part of the recipient chromosome. Exact excision, however, is in both cases rarer by orders of magnitude than inexact excision, which leads either to the production of deletions or removal of only part of the transposon, as described in the preceding section.

The chloramphenicol transposon Tn9 (Cm) is rather unstable on the λ chromosome (100). It is lost at high frequency during the multiplication of λ, and heteroduplex analysis indicates that one single copy of IS1 is left behind when most of the transposon, including the resistance gene, has been lost.[87] This indicates possible homologous genetic recombination between the IS1 copies. However, Section II.A reported that an analogous loss of the r-determinant from F-like R-factors does not occur in *E. coli*. It is possible that the *red* genes of bacteriophage λ are more active in this process than the *rec* system of *E. coli*. However, even in the absence of both the functional *red* and *rec* system, the loss still occurs at a considerable frequency, indicating the possible participation of a still unknown enzyme coded by a bacteriophage λ gene.

In the case of transposons, the concomitant loss from one site and the integration into another can be easily selected. In the case of the Tn5 (Km), nearly all of the *lacZ*[+] revertants are kanamycin-sensitive, indicating that direct transposition is a rare event.[92] The same is true for most of the partial revertants, in which the polar effect on the distal *lacY* gene is relieved, while *lacZ* gene still remains in the mutated state. On the other hand, 0.1% to 1% of Lac[+] Sm[r] recombinants from crosses between Hfr lac[+] Sm[s] and F[-] *lac*::Tn5 (Km) Sm[r] carrying two different *lacZ* mutants caused by the integration of Tn5 (Km) into this gene remains kanamycin-resistant. It is not known whether (1) the additional Tn5 had been transposed from the site in lacZ in the form of a copy or (2) the two transposons had been integrated independently.

It is also possible that the transposition of the kanamycin transposon occurs in an analogous manner to bacteriophage mu (see Section II.C). In this phage, viable pro-mu progeny is not viable by itself.[102] If the same were true in the case of the kana-DNA molecule carrying some bacterial DNA at both ends of the bacteriophage chromosome.[101] Exact excision is possible only in certain mutants, and in these cases, the geny are produced only when the replication of the integrated prophage produces a mycin transposon, the exact excision would not lead to reintegration. The finding of kanamycin-resistant recombinants between two Tn5 (Km) insertions in *lacZ* is not in contradiction with this hypothesis if one considers the possibility that the transpositions of Tn5 (Km) from a previous site might occur from a chromosome arm that has already been replicated. In this case, its loss might not be detected.[92]

The transposition of the ampicillin transposon Tn1 may be as frequent as 10^{-2}. With this frequency, one would expect that the Tn1 (Ap) is highly mutagenic. This has not been shown to be the case. The failure to detect high mutagenicity may be related to the finding that a plasmid already carrying an Tn1 (Ap) has a severely reduced probability of acquiring a second transposon of the same type.[103]

In all cases studied, the integration and excision of transposons has been found to occur in *rec*A mutants.[14]. In summary, most of the effects observed with IS-elements are also observed in one or another transposon.

6. Normal Occurrence of Transposons

All transposons characterized so far have been first detected on a variety of plasmids. While they can be transposed to the chromosome of *E. coli*, they are not found here normally. The organism in which the genes for antibiotic resistance have originated is not known. It has been speculated that as the Streptomycetes produce most of the antibiotics, they are also the species in which the resistance genes originated. Although the kinetic constants of certain of the antibiotic inactivating enzymes isolated from Streptomycetes and from Gram-negative Enterobacteriaceae carrying R-plasmids are similar,[104] their chemical properties and immunological behaviour are different in one case which has been examined in detail. The variant of chloramphenicol acetyltransferase purified from *Streptomyces acrimycini*[105] differs in many important respects from the three types of enzymes found in association with R-plasmids in *E. coli* and the four variants specified by plasmids in Staphylococci.[106] As might be expected, sequence analysis[140] of the chloramphenicol acetyltransferase variants reveals marked differences which preclude the hypothesis that their plasmid-linked structural genes are recent acquisitions from Streptomycetes. The length of time the IS3 copies flanking Tn10 or the IS1 copies flanking Tn9 have been associated with the respective resistance genes is also unknown.

C. Bacteriophage Mu

Bacteriophage mu is a temperate bacteriophage whose properties have been reviewed recently.[9,107,107a] Unlike bacteriophage λ, mu has no, or at most a very weak, integration-site specificity.[108] Integration into genes abolishes gene function. Consequently, mu is a potent mutagen[109] and shows many analogies to IS-elements and transposons. These properties will be discussed briefly.

The presence of bacteriophage mu within a cell is detected by its immunity to superinfection. If mu-DNA is integrated into a particular gene, gene function is lost.[109] If the gene is part of an operon, the function of the genes distal to the integration of mu is also lost.[23,110,111] This is true for all of the many mu insertions in *lac* Z and, therefore, most probably for both orientations of integration. This suggests that mu DNA does not carry a promoter capable of transcribing adjacent genes. Otherwise, mutations would have been detected in which the synthesis of the genes distal to the mu DNA was constitutive. No such cases have been reported.

Integration of mu always occurs without alteration of its gene order.[112-114] Therefore, neither random circular permutation (as in the case of the phages with a head-filling maturation mechanism) nor an ordered circular permutation upon transition from the vegetative to the prophage state (as in the case of bacteriophage λ) occurs with this phage. Consequently, it has been concluded that bacteriophage mu always uses the same DNA sequence for integration. No, or at most a very limited, specificity of integration is found on the recipient DNA. Many mu-induced mutations in gene *lac* Z or *E. coli* have been investigated and found to map at different sites.[108] However, it is not certain that there is no preferred site of integration on the *E. coli* chromosome. As is the case with several transposons, the fraction of auxotrophic or otherwise identifiable mutations among the lysogens is only 1 to 3%. Since the fraction of all genes whose mutation leads to auxotrophy must be larger than this, an explanation must be sought for this apparent discrepancy. One or a few preferred sites of integration outside the genes leading to auxotrophy and a large number of less preferred integration sites distributed at random over the rest of the chromosome would be compatible with the experimental findings.

Mutations caused by ordinary phage mu do not or, at most, very rarely ($< 10^{-10}$) revert.[23,108] However, mutations (termed muX) have been isolated in which the excision process is altered in such a way that the original host sequence can be restored, resulting in a reversion to the wild-type phenotype. However, the excised bacteriophage DNA is unable to replicate without a helper phage.[102]

It is interesting to note that the muX mutations are caused by the integration of IS1 into a certain gene of mu.[116] It has been shown that during replication, the DNA of prophage mu is not excised from the host DNA.[115] About 15% of the mutants caused by the integration of bacteriophage mu have been found in genetic crosses to contain deletions.[117,118] As these investigations were performed with wild-type mu, and since it is not easy to isolate reversions of these mutations, it is not clear whether the deletions were produced on integration of the bacteriophage or if they originated afterwards by processes similar to those described for IS1, IS3, and several transposons. Other chromosomal aberrations in the vicinity of a mu prophage have not yet been described.

III. USE OF TRANSPOSONS AND IS-ELEMENTS IN GENETIC ENGINEERING

It can be seen from the preceding sections that the study of IS-elements, transposons, and bacteriophage mu is still in a state of development where new findings are frequently made and many important aspects of the biology of these elements are not yet fully understood. Therefore, any discussion of the possible use of these elements in the construction of new combinations of DNA will be highly speculative and most likely overlook many important possibilities that will be brought out by future research. However, bearing this in mind, several possibilities will be discussed together with a few experiments involving bacteriophage mu, by means of which new combinations of DNA sequences have already been obtained.

The integration of IS-elements, transposons, or bacteriophages (i.e., mu) into a virtually unlimited number of sites on bacterial chromosomes shows that new combinations of DNA can be produced in vivo. In the case of transposons, this process leads to a new combination of recognizable phenotypes. If it becomes feasible, by appropriate selection procedures, to associate genes other than resistance determinants with two inverted repeats (e.g., of IS3), it may prove possible to obtain transposons for many other gene functions. Whether these combinations are produced in vitro by methods of nucleic acid enzymology or in the cell by natural processes will probably depend on the relative convenience of the procedures.

The in vivo construction of transposons from IS-elements will be difficult in the near future due to the lack of a selection method for the presence of the IS-elements. Also, the rate of integration of an IS-element within a given operon or gene is small and in the order of 10^{-7} cell plated. If the formation of a transposable element containing a specific gene necessitates the integration, in an inverted order, of two IS-elements in the vicinity of this gene, it is not likely to occur with more than the square of the single probabilities. If an attempt was made, without any further selection, to look for cells in which this structure has been transposed to a plasmid, one could easily end up with probabilities of 10^{-17} which are of no use in bacterial genetics.

This may, however, be greatly facilitated if transposons are used instead of single

IS-elements, since the presence of the former can be selected. Tn10 (Tc) has been used in ingenious ways to place mutations at given sites on chromosomes and use these in subsequent genetic manipulations.[119] A method for mobilizing bacterial genes with the help of bacteriophage mu and integrating the mobilized gene into a plasmid will be described below.

The formation of new combinations of genes within a single cell, however, does not seem to be a very important goal at the moment, unless very special questions are to be answered. A transfer between species is more exciting and promises to be more useful. This may be attained by first transposing genes onto a plasmid and then transferring this plasmid into other bacterial species. Plasmids with a wide host range like RP4 will soon make it possible to freely mix the DNA content of most or all Gram-negative bacteria. Whether this can be extended to other bacterial species will depend on the success of transformation experiments or the discovery of plasmids with an even wider host range.

A general method for the mobilization of bacterial genes by bacteriophage mu has been devised. This method makes use of the fact that mu can integrate into both the host chromosome and a plasmid and is followed by recombination events in the mu DNA and the emergence of plasmids carrying the gene of interest sandwiched between two mu prophages. These structures are unstable because of recombination between the two mu copies. However, a series of spontaneously occurring inversions and deletions, eliminates part of the mu chromosome, thus stabilizing the mu arrangements on the plasmids.[120,121] If the plasmid is RP4, it can transfer freely between many bacterial species. The bacterial gene, however, remains attached to the plasmid under these circumstances and is not transposed to the new chromosome.

The discovery of transposons and IS-elements, however, has opened another possibility that may be of greater importance in the future. The existance of IS-elements and transposons in bacteria has shown that enzyme systems exist that are able to recognize very specifically the terminal sequences of larger DNA segments and integrate them into many sites in DNA. The role played by the large central portion of these elements is not yet known. However, the size of IS-elements and many transposons makes it very unlikely that more than small terminal parts are actually involved in the integration processes. The identification, isolation, and successful ligation of these terminal sequences to other DNA segments should make these segments capable of integration into host DNA. This may prove useful in cases where it is desirable to introduce several genes, e.g., for a whole metabolic pathway, into a new host and would be difficult to have each on a separate, compatible plasmid.

It would be of very great interest to discover similar systems in higher organisms. Observations on a variety of eukaryotic organisms indicate that transposable elements may exist in these organisms.[123-129] If it would be possible to identify and isolate these, a natural way might be found to integrate foreign DNA into the DNA of eukaryotes. A final goal of genetic engineering is the introduction of foreign DNA sequences into higher organisms. At the moment, the combination of these sequences with virus DNA is discussed; however, it is by no means certain that the introduction of a virus able to integrate into host DNA can or will be regarded as safe enough to be applied in, e.g., human medicine. Should it prove possible to detect a transposon system in higher cells, it might also be possible to construct safe vectors consisting of only a few integration sequences in addition to the material to be introduced. Such DNA molecules may be safer vectors than any that are presently discussed. In the long range, the safety may be increased if it would be possible to find systems with only a limited number of integration sites in the host genome, such as IS4 or Tn7 in bacteria, since the danger of impairing a vital gene function would be consequently diminished.

IV. DOES THE DISCOVERY OF IS- AND OTHER TRANSPOSABLE ELEMENTS CONTRIBUTE ARGUMENTS TO THE DEBATE ABOUT POSSIBLE HEALTH HAZARDS OF GENETIC ENGINEERING?

The debate about the health hazards of genetic engineering is mainly centered around the assumption that combinations of DNA sequences will be created that have never before existed in nature. The crossing of the barrier between prokaryotes and eukaryotes has been considered especially dangerous.[130] In all these arguments, it is assumed that such exchanges do not occur naturally. However, the transposition of genes within the same bacterial cell is now widely known. The second condition for the exchange of DNA between organisms is a mechanism which transfers the DNA from a cell of one species to a cell of another species. There is very little known in this field; however, the experiments on the Ti-plasmid of *Agrobacter tumefaciens* at least indicate the possibility of total or partial plasmid transfer from a bacterium to a plant cell.[131-135]

Additionally, the uptake of DNA from the surrounding medium into higher cells has been widely discussed, and the uptake of DNA into bacterial cells (especially those of the Gram-positive species) has been amply documented. Up to now, it was not considered very likely that these DNAs could be incorporated into the genetic material of the recipient cells. The discovery of IS-elements and transposons and of phage mu, however, makes it very likely that eukaryotic DNA at least occasionally is integrated into a bacterial DNA. If this is conceded, it is not unlikely that such processes will have occurred sufficiently often during the long periods of biological evolution to exert their influences. Should it be that these possibilities do exist and are realized in nature, the danger of genetic engineering may be smaller than is presently assumed.

Though it would be unwise to advocate a decrease in the stringency of safety conditions as laid down in the National Institutes of Health Guidelines[136] or the Williams Report[137] current research about natural transposition mechanisms may, in the long run, contribute significantly to the evaluation of possible risks involved in the application of genetic engineering.

Acknowledgments

I thank Drs. D. Berg, D. Botstein, H. Boyer, A. Bukhari, L. Chow, D. Clewell, S. Cohen, M. Green, N. Grindley, R. Hedges, N. Kleckner, L. MacHattie, M. van Montagu, B. Rak, H. Saedler, J. Schell, and R. Shaw for communicating unpublished information and sending manuscripts prior to publication and Drs. J. Besemer, H. Chadwell, and R. Ehring for critical readings of the manuscript. Work in the author's laboratory was supported by the Deutsche Forschungsgemeinschaft through SFB 74.

REFERENCES

1. **Meynell, G. G.,** *Bacterial plasmids,* MacMillan Press, London and Basingstoke, 1972.
2. **Falkow, S.,** *Infectious multiple drug resistance,* Pion, London, 1975,
3. **Helinski, D. R.,** Plasmids, *Fed. Proc. Fed. Am. Soc. Exp. Biol.,* 35, 2024, 1976.
4. **Datta, N., Hedges, R. W., Shaw, E. J., Sykes, R. P., and Richmond, M. H.,** Properties of an R factor from *Pseudomonas aeruginosa, J. Bacteriol.,* 108, 1244, 1971.
5. **Olsen, R. H. and Shipley, P.,** Host range and properties of the *Pseudomonas aeruginosa* R factor 1922, *J. Bacteriol.,* 113, 772, 1973.
6. **Beringer, J.,** R factor transfer in *Rhizobium leguminosarum, J. Gen. Microbiol.,* 84, 188, 1974.

7. **Hedges, R. W., Jacob, A. E., and Smith, J. T.,** Properties of an R factor from *Bordetella bronchiseptica, J. Gen. Microbiol.,* 84, 199, 1975.
8. **Franklin, N.,** Illegitimate recombinations, in *The Bacteriophage Lambda,* Hershey, A. D., Ed., Cold Spring Harbor Laboratory, New York, 1971, 175.
9. **Bukhari, A. I.,** Bacteriophage mu as a transposition element, *Annu. Rev. Genet.,* 10, 389, 1976.
10. **Scaife, J.,** Episomes, *Annu. Rev. Microbiol.,* 21, 601, 1967.
11. **Davidson, N., Deonier, R. C., Hu, S., and Ohtsubo, E.,** The DNA sequence organization of F and F-primes and the sequences involved in Hfr formation, *Microbiology-1974,* Schlessinger, D., Ed., 1975, 56.
12. **Starlinger, P. and Saedler, H.,** IS-elements in microorganisms, *Curr. Top. Microbiol. Immunol.,* 75, 111, 1976.
13. **Cohen, S. N. and Kopecko, D. J.,** Structural evolution of bacterial plasmids: role of translocating genetic elements and DNA sequence insertions, *Fed. Proc.,* 35, 2031, 1976.
14. **Cohen, S. N.,** Transposable genetic elements and plasmid evolution, *Nature,* 263, 731, 1976.
15. **Campbell, A., Berg, D., Botstein, D., Novick, R., and Starlinger, P.,** Nomenclature for transposable elements of DNA, in *DNA insertions, Plasmids and Episomes,* Bukhari, A., Shapiro, J., and Adhya, S., Eds., Cold Spring Harbor Laboratory New York, 1977 in press.
16. **Malamy, M. H.,** Frameshift mutations in the lactose operon of *E. coli, Cold Spring Harbor Symp. Quant. Biol.,* 31, 189, 1966.
17. **Shapiro, J.,** The Structure of the Galactose Operon in *E. coli* K-12, Ph.D. thesis, University of Cambridge 1967.
18. **Adhya, S. L. and Shapiro, J. A.,** The galactose operon of *E. coli* K-12. I. Structural and pleiotropic mutations of the operon, *Genetics,* 62, 231, 1969.
19. **Saedler, H. and Starlinger, P.,** 0°-mutations in the galactose operon in *E. coli.* I. Genetic characterization, *Molec. Gen. Genetic.,* 100, 178, 1967.
20. **Jaskunas, S. R., Lindahl, L., and Nomura, M.,** Isolation of polar insertion mutants and the correction of transcription of ribosomal protein genes in *E. coli, Nature,* 256, 183, 1975.
21. **Ketner, G. and Campbell, A.,** Operator and promoter mutations affecting diverging transcription in the bio cluster of *Escherichia coli, J. Mol. Biol.,* 96, 13, 1975.
22. **Malamy, M. H.,** Some properties of insertion mutations in the lac operon, in *The Lactose Operon,* Beckwith, J. R. and Zipser, D., Eds., Cold Spring Harbor Laboratory New York, 1970, 359.
23. **Jordan, E., Saedler, H., and Starlinger, P.,** 0°- and strong polar mutations in the gal operon are insertions, *Mol. Gen. Genet.,* 102, 353, 1968.
24. **Shapiro, J. A.,** Mutations caused by the isertion of genetic material into the galactose operon of *Escherichia coli, J. Mol. Biol.,* 40, 93, 1969.
25. **Michaelis, G., Saedler, H., Venkov, P., and Starlinger, P.,** Two insertions in the galactose operon having different sizes but homologous DNA sequences, *Mol. Gen. Genet.,* 104, 731, 1969.
26. **Hirsch, H. J., Saedler, H., and Starlinger, P.,** Insertion mutations in the control region of the galactose operon of *E. coli.* II. Physical characterization of the mutations, *Mol. Gen. Genet.,* 115, 266, 1972.
27. **Fiandt, M., Szybalski, W., and Malamy, M. H.,** Polar mutations in lac, gal and phage λ consist of a few DNA sequences inserted with either orientation, *Mol. Gen. Genet.,* 119, 223, 1972.
28. **Hirsch, H. J., Starlinger, P., and Brachet, P.,** Two kinds of insertions in bacterial genes, *Mol. Gen. Genet.,* 119, 191, 1972.
29. **Malamy, M. H., Fiandt, M., and Szybalski, W.,** Electron microscopy of polar insertions in the lac operon of *Escherichia coli, Mol. Gen. Genet.,* 119, 207, 1972.
30. **Ahmed, A., and Scraba, D.,** The nature of the gal3 mutation of *Escherichia coli, Mol. Gen. Genet.,* 136, 233, 1975; Fiandt, M., Ahmed, A., and Szybalski, W., personal communication, 1976.
31. **Blattner, F. R., Fiandt, M., Hass, K. K., Twose, P. A., and Szybalski, W.,** Deletions and insertions in the immunity region of coliphage λ: revised measurement of the promoter-startpoint distance, *Virology,* 62, 458, 1974.
32. **Bukhari, A. and Froshauer, S.,** Some properties of the TnC element, in *DNA Insertions, Plasmids and Episomes,* Bukhari, A., Shapiro, J., and Adhya, S., Eds., Cold Spring Harbor Laboratory, New York, 1977, in press.
33. **Hu, S., Ohtsubo, E. Davidson, N., and Saedler, H.,** Electron microscope heteroduplex studies of sequence relations among bacterial plasmids. XII. Identification and mapping of the insertion sequences IS1 and IS2 in F and R plasmids, *J. Bacteriol.,* 122, 764, 1975.
34. **Pfeifer, D., Kubai-Maroni, D., and Habermann, P.,** Specific sites for integration of IS-elements within the transferase gene of the galactose operon of *E. coli* K-12, in *DNA Insertions, Plasmids and Episomes,* Bukhari, A., Shapiro, J., and Adhya, S., Eds., Cold Spring Harbor Laboratory, New York, 1977, in press.
35. **Ptashne, K. and Cohen, S. N.,** Occurrence of insertion sequence (IS) regions on plasmid deoxyribonucleic acid as direct and inverted nucleotide sequence duplications, *J. Bacteriol.,* 122, 776, 1975.

36. **Saedler, H., Reif, H. J., Hu, S., and Davidson, N.,** IS2, a genetic element for turn-off and turn-on of gene activity in *E. coli, Mol. Gen. Genet.,* 132, 265, 1974.

37. **Saedler, H., Kubai, D., Nomura, M., and Jaskunas, S. R.,** IS1 and IS2 mutations in the ribosomal protein genes of *E. coli* K-12, *Mol. Gen. Genet.,* 141, 85, 1975.

38. **Schmidt, F., Besemer, J., and Starlinger, P.,** The isolation of IS1 and IS2 DNA, *Mol. Gen. Genet.,* 145, 145, 1976.

39. **Ohtsubo, H. and Ohtsubo, E.,** Isolation of inverted repeat sequences including IS1, IS2 and IS3 in *Escherichia coli* plasmids, *Proc. Natl. Acad. Sci. U.S.A.,* 73, 2316, 1976.

40. **Grindley, N. D. F.,** Physical mapping of IS1 by restriction endonucleases, in *DNA Insertions, Plasmids and Episomes,* Bukhari, A., Shapiro, J., and Adhya, S., Eds., Cold Spring Harbor Laboratory, New York, 1977, in press.

41. **Richardson, J. P., Grimley, C., and Lowery, C.,** Transcription termination factor rho activity is altered in *Escherichia coli* with suA gene mutations, *Proc. Natl. Acad. Sci. U.S.A.,* 72, 1725, 1975.

42. **Adhya, S., Gottesman, M., de Crombrugghe, B., and Court, D.,** Transcription termination regulates gene expression, in *RNA Polymerase,* Losick, R. and Chamberline, H., Ed., Cold Spring Harbor Laboratory, New York, 1976, 719.

43. **Ratner, D.,** Evidence that mutations in the suA polarity suppressing gene directly affect termination factor rho, *Nature,* 259, 151, 1976.

44. **Das, A., Court, D., and Adhya, S.,** Isolation and characterization of conditional lethal mutants of *Escherichia coli* defective in transcription termination factor rho, *Proc. Natl. Acad. Sci. U.S.A.,* 73, 1959, 1976.

45. **de Crombrugghe, B., Adhya, S., Gottesman, M., and Pastan, I.,** Effect of rho on transcription of bacterial operons, *Nature New Biol.,* 241, 260, 1973.

46. **Starlinger, P., Saedler, H., Rak, B., Tillmann, E., Venkov, P., and Waltschewa, L.,** mRNA distal to polar nonsense and insertion mutations in the gal operon of *E. coli, Mol. Gen. Genet.,* 122, 279, 1973.

47. **Martin, R. G.,** Control of gene expression, *Annu. Rev. Genet.,* 3, 181, 1969.

48. **Besemer, J. and Herpers, M.,** Suppression of polarity of insertion mutations within the gal operon of *E. coli, Mol. Gen. Genet.,* 157, 259, 1977.

49. **Tomich, K. and Friedman, D. I.,** Mutations in an insertion sequence that relieve polarity, in *DNA Insertions, Plasmids and Episomes,* Bukhari, A., Shapiro, J., and Adhya, S., Eds., Cold Spring Harbor Laboratory, New York, 1977, in press.

50. **Mosharrafa, E., Pilacinski, W., Zissler, J., Fiandt, M., and Szybalski, W.,** Insertion sequence IS2 near the gene for prophage λ excision, *Mol. Gen. Genet.,* 147, 103, 1976.

51. **Rak, B.,** Gal mRNA initiated within IS2. I. Hybridization studies, *Mol. Gen. Genet.,* 149, 135, 1976.

52. **Reif, H. J. and Saedler, H.,** IS1 is involved in deletion formation in the gal region of *E. coli* K-12, *Mol. Gen. Genet.,* 137, 17, 1975.

53. **Foster, T. J.,** R factor mediated tetracycline resistance in *Escherichia coli* K-12 dominance of some tetracycline sensitive mutants and relief of dominance by deletion, *Mol. Gen. Genet.,* 143, 339, 1976.

54. **Reif, H. J. and Saedler, H.,** Chromosomal rearrangements in the gal region of *E. coli* K-12 after integration of IS1, in *DNA Insertions, Plasmids and Episomes,* Bukhari, A., Shapiro, J., and Adhya, S., Eds., Cold Spring Harbor Laboratory, New York, 1977, in press.

55. **Nevers, P., Reif. H. J., and Saedler, H.,** A mutant of *Escherichia coli* defective in IS1-mediated deletion formation, in *DNA Insertions, Plasmids and Episomes,* Bukhari, A., Shapiro, J., and Adhya, S., Eds., Cold Spring Harbor Laboratory, New York, 1977, in press.

56. **Lederberg, E. M.,** Genetic and functional aspects of galactose metabolism in *E. coli, Microbial Genetics,* 10th Symp. Soc. Gen. Microbiol., Hayes, W. and Clowes, R. D., Eds., University Press, Cambrridge, England, 1960, 115.

57. **Hill, C. W. and Echols, H.,** Properties of a mutant blocked in inducibility of messenger RNA for the galactose operon of *Escherichia coli, J. Mol. Biol.,* 19, 38, 1966.

58. **Morse, M. L. and Pollock, B. F.,** Reversion instability in the galactose operon of *Escherichia coli, J. Bacteriol.,* 99, 567, 1969.

59. **Ahmed, A.,** Mechanism of reversion of the gal-3 mutation of *Escherichia coli, Mol. Gen. Genet.,* 136, 243, 1975.

60. **Saedler, H., Besemer, J., Kemper, B., Rosenwirth, B., and Starlinger, P.,** Insertion mutations in the control region of the gal operon of *E. coli.* I. Biological characterization of the mutations, *Mol. Gen. Genet.,* 115, 258, 1972.

61. **Hu, S., Ohtsubo, E., and Davidson, N.,** Electron microscope heteroduplex studies of sequence relations among plasmids of *Escherichia coli:* structure of F13 and related F- primes, *J. Bacteriol.,* 122, 749, 1975.

62. **Shimada, K., Weisberg, R. A., and Gottesman, M. E.,** Prophage λ at unusual chromosomal locations. II. Mutations induced by bacteriophage λ in *Escherichia coli* K-12, *J. Mol. Biol.,* 80, 297, 1973.

63. Jordan, E., Saedler, H., and Starlinger, P., Strong polar mutations in the transferase gene of the galactose operon in *E. coli, Mol. Gen. Genet.,* 100, 296, 1967.

64. Saedler, H. and Heiss, B., Multiple copies of the insertion-DNA sequences IS1 and IS2 in the chromosome of *E. coli* K-12, *Mol. Gen. Genet.,* 122, 267, 1973.

65. Hu, S., Ptashne, K., Cohen, S. N., and Davidson, N., The $\alpha\beta$-sequence of F is IS3 *J. Bacteriol.,* 123, 687, 1975.

66. Chow, L. T., The organization of insertion sequences on the *E. coli* chromosome, in *DNA Insertions, Plasmids and Episomes,* Bukhari, A., Shapiro, J., and Adhya, S., Eds., Cold Spring Harbor Laboratory, New York, 1977, in press.

67. Deonier, C. R. and Hadley, R. G., Distribution of inverted IS-length sequences in the *E. coli* K-12 genome, *Nature,* 264, 191, 1976.

68. Broda, P., The formation of Hfr strains of *Escherichia coli* K-12, *Genet. Res.,* 9, 35, 1967.

69. Curtiss, R., III and Stallions D. R., Probability of F integration and frequency of stable Hfr donors in F⁺-populations of *Escherichia coli* K-12, *Genetics,* 63, 27, 1969.

70. Press, R., Glansdorff, N., Miner, P., de Vries, J., Kadner, R., and Maas, W. K., Isolation of transducing particles of ϕ80 bacteriophage that carry different regions of the *Escherichia coli* genome, *Proc. Natl. Acad. Sci. U.S.A.,* 68, 795, 1971.

71. Palchaudhuri, S., Maas, K., and Ohtsubo, E., Fusion of two F-prime factors in *Escherichia coli* studied by electron microscope heteroduplex analysis, *Mol. Gen. Genet.,* 146, 215, 1976.

72. Besemer, J. and Kubai, D., Isolation and characterization of ϕ80dgal transducing phages that carry gal operator promoter insertion mutations, *Mol. Gen. Genet.,* 148, 79, 1976.

73. Ohtsubo, E., Deonier, R. C., Lee, H. J., and Davidson, N., Electron microscope heteroduplex studies of sequence relations among plasmids of *Escherichia coli.* IV. The F sequence in F14, *J. Mol. Biol.,* 89, 565, 1974.

74. Broda, P., Meacock, P., and Achtman, M., Early transfer of genes determining transfer functions by some Hfr strains in *Escherichia coli* K-12, *Mol. Gen. Genet.,* 116, 336, 1972.

75. Stanisich, V. A. and Holloway, B. W., Chromosome transfer in *Pseudomonas aeruginosa* mediated by R factors, *Genet. Res.,* 17, 169, 1971.

76. Haas, D. and Holloway, W., R factor variants with enhanced sex factor activity in *Pseudomonas aeruginosa, Mol. Gen. Genet.,* 144, 243, 1976.

77. Jacob, A. E., Cresswell, J. M., and Hedges, R. W., Molecular characterization of the P group plasmid R68 and variants with enhanced chromosome mobilizing ability, *FEMS Microbiology Letters,* in press.

78. Nisioka, T., Mitani, M., and Clowes, R., Composite circular forms of R. factor deoxyribonucleic acid molecules, *J. Bacteriol.,* 97, 376, 1969.

79. Cohen, S. N. and Miller, C. A., Non-chromosomal antibiotic resistance in bacteria. III. Isolation of the discrete transfer unit of the R factor R1, *Proc. Natl. Acad. Sci. U.S.A.,* 67, 510, 1970.

80. Silver, R. P. and Falkow, S., Studies on resistance transfer factor deoxyribonucleic acid in *Escherichia coli, J. Bacteriol.,* 104, 340, 1970.

81. Rownd, R. and Mickel, S., Dissociation of RTF and r-determinants of the R factor NR 1 in *Proteus mirabilis, Nature New Biol.,* 234, 40, 1971.

82. Sharp, P. A., Cohen, S. N., and Davidson, N., Electron microscope heteroduplex studies of sequence relations among plasmids of *Escherichia coli.* II. Structure of drug resistance (R) factors and F factors, *J. Mol. Biol.,* 75, 235, 1973.

83. Rownd, R. H., Perlman, D., and Goto, N., Structure and replication of R. factor DNA in *Proteus mirabilis,* in *Microbiology 1974,* Schlessinger, D., Ed., American Society for Microbiology, Washington D. C. 1975, 76.

84. Clewell, D. B., Yagi, Y., and Bauer, B., Plasmid-determined tetracycline resistance in *Streptococcus faecalis:* evidence for gene amplification during growth in presence of tetracycline, *Proc. Natl. Acad. Sci. U.S.A.,* 72, 1720, 1975.

85. Yagi, Y. and Clewell, D. B., Plasmid-determined tetracycline resistance in *Streoptococcus faecalis:* tandemly repeated resistance determinants in amplified forms of pAMα DNA, *J. Mol. Biol.,* 102, 583, 1976.

86. Yagi, Y. and Clewell, D. B., Identification and characterization of a small sequence located at two sites on the amplifiable tetracycline resistance plasmid pAMαl in *Streptococcus faecalis, J. Bacteriol.,* 429, 400, 1977.

87. MacHattie, L. A. and Jackowski, J. B., Structure and excision properties of an IS1 bracketed drug resistance determinant carried by phage λ, in *DNA Insertions, Plasmids and Episomes,* Bukhari, A., Shapiro, J., and Adhya, S., Eds., Cold Spring Harbor Laboratory, New York, 1977, in press.

88. Heffron, R., Rubens, C., and Falkow, S., Translocation of a plasmid DNA sequence which mediates ampicillin resistance: molecular nature and specificity of insertion, *Proc. Natl. Acad. Sci. U.S.A.,* 72, 3623, 1975.

89. **Kopecko, D. J. and Cohen, S. N.**, Site-specific recA-independent recombination between bacterial plasmids: involvement of palindromes at the recombinational loci, *Proc. Natl. Acad. Sci. U.S.A.*, 72, 1373, 1975.

90. **Berg, D. E., Davies, J. Allet, B., and Rochaix, J.-D.**, Transposition of R factor genes to bacteriophage λ, *Proc. Natl. Acad. Sci. U.S.A.*, 72, 3628, 1975.

91. **Kleckner, N., Chan, R. K., Tye, B. K., and Botstein, D.**, Mutagenesis by insertion of a drug-resistance element carrying an inverted repetition, *J. Mol. Biol.*, 97, 344, 1975.

92. **Berg, D. E.**, Insertion and excision of the transposable kanamycin resistance determinant TnK(1), in *DNA Insertions, Plasmids and Episomes*, Bukhari, A., Shapiro, J., and Adhya, S., Eds., Cold Spring Harbor Laboratory, New York, 1977, in press.

93. **Botstein, D. and Kleckner, N.**, Translocation and illegitimate recombination by the tetracycline resistance element Tn10, in *DNA Insertions, Plasmids and Episomes*, Bukhari, A., Shapiro, J., and Adhya, S., Eds., Cold Spring Harbor Laboratory, New York, 1976.

94. **Heffron, F., Rubens, C., and Falkow, S.**, Transposition of a plasmid DNA sequence which mediates ampicillin resistance: general description and epidemiological considerations, in *DNA Insertions, Plasmids and Episomes*, Bukhari, A., Shapiro, J., and Adhya, S., Eds., Cold Spring Harbor Laboratory, New York, 1977, in press.

94a. **Heffron, F. A., Bedinger, P., Champoux, J., and Falkow, S.**, Deletions affecting transposition of an antibiotic resistance gene, in *DNA Insertions, Plasmids and Episomes*, Bukhari, A., Shapiro, J., and Adhya, S., Eds., Cold Spring Harbor Laboratory, New York, 1977, in press.

95. **Chan, R. K. and Botstein, D.**, Genetics of bacteriophage P22. I. Isolation of prophage deletions which affect immunity to superinfection, *Virology*, 49, 257, 1972.

96. **Brevet, J., Nisen, P., Kopecko, J., and Cohen, S. N.**, Promotion of insertions and deletions by translocating segments of DNA carrying antibiotic resistance genes, in *DNA Insertions, Plasmids and Episomes*, Bukhari, A., Shapiro, J., and Adhya, S., Eds., Cold Spring Harbor Laboratory, New York, 1977, in press.

97. **Weinstock, G. and Botstein, D.**, Specialized transduction of ampicillin resistance by phage P22, in *DNA Insertions, Plasmids and Episomes*, Bukhari, A., Shapiro, J., and Adhya, S., Eds., Cold Spring Harbor Laboratory, New York, 1977, in press.

98. **Bennet, P. M. and Richmond, M. H.**, The translocation of a discrete piece of DNA carrying an amp gene between replicons in *Escherichia coli*, *J. Bacteriol.*, 126, 1, 1976.

99. **Barth, P. T., Datta, N., Hedges, R. W., and Grinter, N. J.**, Transposition of a DNA sequence encoding trimethoprim and streptomycin resistances from R483 to other replicons, *J. Bacteriol.*, 125, 800, 1976.

100. **Gottesman, M. M. and Rosner, J. L.**, Acquisition of a determinant for chloramphenicol resistance by coliphage λ, *Proc. Natl. Acad. Sci. U.S.A.*, 74, 5041, 1975.

101. **Bukhari, A. I.**, Reversal of mutator phage mu integration, *J. Mol. Biol.*, 96, 87, 1975.

102. **Bukhari, A. F. and Taylor, A. L.**, Influence of insertions on packaging of host sequences covalently linked to bacteriophage mu DNA, *Proc. Natl. Acad. Sci. U.S.A.*, 72, 4399, 1975.

103. **Robinson, M. K., Bennet, P. M., and Richmond, M. H.**, The inhibition of TnA translocation by TnA, *J. Bacteriol.*, 429, 407, 1977.

104. **Davies, J. E. and Rownd, R.**, Transmissible multiple drug resistance in enterobacteriaecae, *Science*, 176, 758, 1972.

105. **Zaidenzaig, Y. and Shaw, W. V.**, FEBS Letters, in press Affinity and Hydrophobic Chromatography of Three Variants of Chloramphenicol Transacetylase Specified by R Factors in *Escherichia coli*, *FEBS Lett.*, 62, 266, 1976.

106. **Shaw, W. V. and Hopwood, I. A.**, *J. Gen. Microbiol.*, *in press*.

107. **Howe, M. M. and Bade, E. G.**, Molecular biology of bacteriophage mu, *Science*, 190, 624, 1975.

107a. **Coutourier, M.**, The integration and excision of bacteriophage mu, *Cell*, 7, 155, 1976.

108. **Bukhari, A. I., and Zipser, D.**, Random insertion of mu-1 DNA with a single gene, *Nature New Biol.*, 236, 240, 1972.

109. **Taylor, A. L.**, Bacteriophage-induced mutation in *Escherichia coli*, *Proc. Natl. Acad. Sci. U.S.A.*, 50, 1043, 1963.

110. **Toussaint, A.**, Insertion of phage mu-1 within prophage λ: a new approach for studying the control of the late functions in bacteriophage λ, *Mol. Gen. Genet.*, 106, 89, 1969.

111. **Daniell, E., Roberts, R., and Abelson, J.**, Mutations in the lactose operon caused by bacteriophage mu, *J. Mol. Biol.*, 69, 1, 1972.

112. **Howe, M. M.**, Transduction by bacteriophage mu-1, *Virology*, 55, 103, 1973.

113. **Wijffelman, C. A., Westmaas, G. C., and van de Putte, P.**, Vegetative recombination of bacteriophage mu-1 in *Escherichia coli*, *Mol. Gen. Genet.*, 116, 40, 1972.

114. **Coutourier, M. and van Vliet, F.**, Vegetative recombination in bacteriophage mu-1, *Virology*, 60, 1, 1974.

115. **Ljungquist, E. and Bukhari, A.** quoted in Bukhari, A. I., Bacteriophage mu as a transposition element, *Annu. Rev. Genet.,* 10, 389, 1976.
116. **Bukhari, A., Froshauer, S., Chow, L., Ljungquist, E., Khatoon, H., and de Bruijn, F.,** Integration and excision of bacteriophage mu, in *DNA Insertions, Plasmids and Episomes,* Bukhari, A., Shapiro, J., and Adhya, S., Eds., Cold Spring Harbor Laboratory, New York, 1977, in press.
117. **Howe, M. M. and Zipser, D.,** Host deletions caused by the integration of bacteriophage mu-1, *American Soc. Microbiol. Abstr.,* No. 208, 235, 1974.
118. **Cabezon, T., Faelen, M., de Wilde, M., Bollen, A., and Thomas, R.,** Expression of ribosomal protein genes in *Escherichia coli, Mol. Gen. Genet.,* 137, 125, 1975.
119. **Kleckner, N., Roth, J., and Botstein, D.,** Genetic engineering in vivo using translocatable drug-resistance elements: a short manual of new methods in bacterial genetics, in *DNA Insertions, Plasmids and Episomes,* Bukhari, A., Shapiro, J., and Adhya, S., Eds., Cold Spring Harbor Laboratory, New York, 1977, in press.
120. **Faelen, M. and Toussaint, A.,** Bacteriophage mu-1: a tool to transpose and to localize bacterial genes, *J. Mol. Biol.,* 104, 525, 1976.
121. **Faelen, M., Toussaint, A., van Montagu, M., van den Elsacker, S., Engler, G., and Schell, J.,** In vivo genetic engineering: the mu mediated transpositions of chromosomal DNA segments onto transmissible plasmids, in *DNA Insertions, Plasmids and Episomes,* Bukhari, A., Shapiro, J., and Adhya, S., Eds., Cold Spring Harbor Laboratory, New York, 1977, in press.
123. **McClintock, B.,** Controlling elements and the gene, *Cold Spring Harbor Symp. Quant. Biol.,* 21, 197, 1956.
124. **McClintock, B.,** The control of gene action in maize, *Brookhaven Symp. Biol.,* 18, 162, 1965.
125. **McClintock, B.,** II, The role of the nucleus. Genetic systems regulating gene expression during development, *Dev. Biol.,* Suppl. 1, 84, 1967.
127. **Peterson, P. A.,** The position hypothesis for controlling elements in maize, in *DNA Insertions, Plasmids and Episomes,* Bukhari, A., Shapiro, J., and Adhya, S., Eds., Cold Spring Harbor Laboratory, New York, 1977, in press.
128. **Green, M. M.,** The genetics of a mutable gene at the white locus of *Drosophila melanogaster, Genetics,* 56, 467, 1967.
129. **Green, M.,** The case for DNA insertion mutation in *Drosophila,* in *DNA Insertions, Plasmids and Episomes,* Bukhari, A., Shapiro, J., and Adhya, S., Eds., Cold Spring Harbor Laboratory, New York, 1977, in press.
130. **Sinsheimer, R. L.,** The hazards of recombinant DNA, *Trends in Biochemical Sciences,* 4, N171, 1976.
131. **Zaenen, I., van Larebeke, N., Teuchy, H., van Montagu, M., and Schell, J.,** Supercoiled circular DNA in crown gall-inducing *Agrobacterium* strains, *J. Mol. Biol.,* 86, 109, 1974.
132. **van Larebeke, N., Engler, G., Holsters, M., van den Essacker, S., Zaenen, I., Schilderoort, R. A., and Schell, J.,** Large plasmids in *Agrobacterium tumefaciens* essential for crown gall inducing ability, *Nature,* 252, 169, 1974.
133. **Bomhoff, G., Klapwijk, P. M., Kester, H. C. M., Schilderoort, R. A., Hernalsteens, J. P., and Schell, J.,** Octopine and nopaline synthesis and breakdown genetically controlled by a plasmid of *Agrobacterium tumefaciens, Mol. Gen. Genet.,* 145, 177, 1976.
133a. **Currier, T. C. and Nester, E. W.,** Evidence for diverse types of large plasmids in tumor-inducing strains of *Agrobacterium, J. Bacteriol.,* 126, 157, 1976.
134. **van Larebeke, N., Genetello, C., Schell, J., Schilderoort, R. A., Hermans, A. K., Hernalsteens, J. P., and van Montagu, M.,** Acquisition of tumor-inducing ability by non-onogenic *Agrobacteria* as a result of plasmid transfer, *Nature,* 255, 742, 1975.
135. **van Montagu, M., Genetello, C., Engler, G., Harnasteens, J.-P., Zaenen, I., van Larebeke, N., de Picker, A., and Schell, J.,** Inserts and gene transposition in a large plasmid of *Agrobacterium tumefaciens,* in *DNA Insertions, Plasmids and Episomes,* Bukhari, A., Shapiro, J., and Adhya, S. Eds., Cold Spring Harbor Laboratory, New York, 1977, in press.
136. **National Institutes of Health,** Guidelines for research involving recombinant DNA molecules, U.S. Department of Health, Education, and Welfare, Public Health Service, 1976.
137. *Report of the Working Party on the Practice of Genetic Manipulation,* London, Her Majesty's Stationary Office, 1976.
138. **Fiandt, M., Ahmed, A., and Szybalski, W.,** personal communication.
139. **Saedler, H.,** personal communication.
140. **Shaw, W.,** personal communication.

Chapter 7

DEVELOPMENT OF *BACILLUS SUBTILIS* AS A CLONING SYSTEM

F. E. Young and G. A. Wilson

TABLE OF CONTENTS

I. INTRODUCTION

Recombinant molecule technology emanated from years of fundamental studies in microbial genetics. A seminal discovery was DNA-mediated transformation of bacteria. Through this procedure, DNA could be manipulated in a test tube and subsequently reintroduced into the bacterial cell. Without this method of genetic exchange, the elegant techniques of splicing and rejoining of DNA would be merely analytical tools. Another major factor was the elucidation of the chromosomal architecture and function of a few model organisms. Thus, years of basic research on *Escherichia coli* genetics could be applied to the cloning of *E. coli* genes and eukaryotic genes in this organism. The advent of DNA-mediated transformation of *E. coli* has led some researchers to suggest that this organism should be selected as the sole microbe for genetic engineering studies. In fact, in the formulation of the original National Institutes of Health (NIH) guidelines, little emphasis was given to the development of alternative systems, and no intellectual or economic stimulation was provided for the implementation of alternative approaches. However, others have recommended that additional systems should be explored and, if appropriate, developed. This review is aimed at (1) discussing some of the advantages and difficulties of the *Bacillus subtilis* system, (2) summarizing the current state of cloning in this system, and (3) outlining the possible avenues that might be followed in the refinement of *B. subtilis* for recombinant molecule technology.

II. ADVANTAGES OF THE *BACILLUS SUBTILIS* SYSTEM

If this model is to transcend the parochial pedantic pedagogical prejudice of its proponents and warrant public support, some unique advantages must be present. At least three can be raised for consideration. First, the organism is one of the few well-studied model systems in which a morphogenetic event occurs. For years, sporulation has been investigated as a prokaryotic example of differentiation. Although genes regulating sporulation could be mapped through an analysis of mutants,[1,2] few direct examples of the effect of gene dosage could be developed using discrete fragments of DNA. One exception was the discovery of merodiploids by Anagnostopolous and co-workers.[3] The extensive chromosomal duplication and rearrangement that accompanies the formation of merodiploids precluded a precise determination of single genes.[4] The recent cloning of the *pheA-spoB* segment of the *B. subtilis* chromosome in *E. coli* by Szulmajster[5] and Ito[6] represents a new approach to the purification and analysis of spore functions.

The importance of this group of organisms as a source of exoenzymes provides still another impetus for the development of cloning in bacilli. In a recent review, Priest emphasized that all of the 48 species of this genus described in *Bergey's Manual* secrete soluble extracellular enzymes.[7] Enzymes for the degradation of starch and protein constitute the majority of these proteins. It is interesting to note that two major types of amylases have been described and isolated: those that liquify (e.g., from *B. amyloliquefaciens*) and those that saccharify (e.g., from *B. amylosacchariticus*). While these enzymes have been characterized extensively and thoroughly mapped on the chromosome of *B. subtilis*,[8-10] it has not been possible to develop strains with multiple copies of these enzymes. Instead, heterospecific transformation has resulted in the replacement of one allele by another. Although approximately 40 other enzymes have been identified,[7] none currently assumes the commercial importance of the amylases and proteases.[11,12] The application of recombinant DNA techniques may permit a more detailed study of the various factors that regulate amylase production as well as elucidate the control mechanisms involved in enzyme secretion. Of particular commercial importance is the development of strains that synthesize a variety of amylases (e.g., α- and β-amylase) with varying substrate specificity and pH optima. Finally, it will be essential to develop organisms that can secrete other proteins through fusion of structural genes with genes regulating secretion of exoenzymes. Thus, recombinant DNA technology has a major impact on the fermentation industry.

A third and highly significant attribute is the nonpathogenic nature of this organism. It should be recalled that this soil microorganism does not infect healthy human beings. In fact, it has been eaten for at least 1000 years in the orient in the form of a vegetable cheese produced by *B. natto,* an organism closely related to *B. subtilis.* Although spores will survive degradation in the gastrointestinal (G.I.) tract, bacilli do not. Skin is also an inhospitable environment, as this organism is rapidly inactivated by the short-chain fatty acids present on the skin.[13] Additionally, serum contains B-lysins that rapidly lyse bacilli.[14] Nevertheless, under rare conditions, *B. subtilis* can produce infections in man. Such infections invariably occur in a host that is compromised by a traumatic injury, chemotherapy, or a major disease. Three recent review articles provide a clear clinical appreciation for infection with bacilli.[15-17] Farrar[15] divides infections with nonpathogenic bacilli into three classes: (1) local infections of damaged organs such as the eye, (2) mixed infections with organisms of recognized pathogenicity, (3) and disseminated infections in patients with underlying disorders. According to Idhe and Armstrong[17] these infections may also occur in immunologically compromised hosts. Due to the infrequent occurrence of infections with bacilli, such cases are occasionally described as case reports in the medical literature. Unfortunately, the organisms are rarely speciated and merely designated *B. subtilis* by exclusion. One striking exception is the case report of Reller,[18] in which the diagnosis of a bacteremia due to *B. subtilis* was established by Dr. R. Gordon, an expert in taxonomy of bacilli. The predisposing factor may have been the talc filler in 40 *meperidine* tablets that the patient injected intravenously during the week immediately prior to his hospital admission. Repeated introduction of dirt and talc into the blood stream can and will produce bacteremia or septicemia. Thus, in the absence of major disturbances to hemostasis, *B. subtilis* is not a pathogen.

III. EXTENT OF GENETIC EXCHANGE

Prior to the discovery of transformation of *B. subtilis* by Spizizen,[19] there were no significant genetic studies in bacilli. The subsequent observation of transduction of bacteriophage SP10[20] and PBS1[21] provided additional methods of genetic exchange. Although *B. subtilis* can both excrete DNA[22] and incorporate large fragments from gently lysed protoplasts,[23] no convincing evidence for conjugation has been presented. A number of bacilli can donate genetic information by DNA-mediated transformation (Figure 1); however, only three are transformable — *B. subtilis*,[19] *B. licheniformis*,[24] and *B. amyloliquefaciens*.[25] The low frequency of transformation of *B. amyloliquefaciens* has inhibited the exploitation of this system. Nevertheless, numerous bacilli can be used in transformation studies of both *B. subtilis* and *B. licheniformis*, as first demonstrated by Marmur and co-workers.[26] However, there is a considerable variation in the efficiency of heterologous transformation with foreign DNA. The region between the *pur*A locus and the *rif* locus is most highly conserved,[27] as is a segment located near the *leu* genes. The most efficient markers for heterospecific transformation are the antibiotic resistant markers that affect ribosomal subunits or RNA polymerase. Once the foreign gene has been integrated to form an intergenotic or hybrid region, the frequency of transformation is greatly increased,[28] as shown in Table 1 and illustrated in Figure 2.

These heterologous and intergenotic transfection studies have clearly established that the primary barrier to interspecific transformation is the extent of homology surrounding the integration site of the donor DNA.[28,29] In fact, the sequential introduction of foreign DNA can result in a gradual replacement of homologous regions by foreign DNA or heterologous alelles.[30] Restriction of heterologous DNA does not present an effective barrier to heterospecific transformation.[31,32] Therefore, if DNA could be inserted into plasmids or bacteriophages that can propagate in *B. subtilis*, it should function in a fashion analogous to the *E. coli* model system. Alternatively, if foreign DNA such as pMB9 could be integrated into the chromosome of *B. subtilis*, this foreign region could be used as the site of recombination for DNA fragments cloned in this vector. The current status of each of these procedures will be discussed, and future directions will be indicated in subsequent sections.

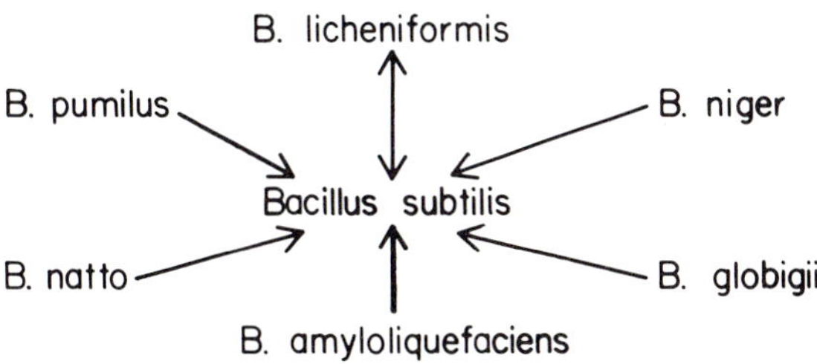

FIGURE 1.   Range of genetic exchange among bacilli. The direction of the arrow indicates the usual direction of exchange from donor to recipient.

TABLE 1

Interspecific Transformation of *Bacillus subtilis* 168

Marker	Heterologous	Source of DNA intergenotic	Homologous
His⁺	22	77,400	111,000
Rif^R	76,000	1,670,000	650,000
φ105	409	2,267	5,060

Note: This table shows the results of transformation of 10^8 cells of *B. subtilis* 168 by DNA from different sources. Heterologous refers to the use of *B. amyloliquefaciens* DNA as donor; homologous, to the use of *B. subtilis* DNA as donor; and intergenotic, to the use of DNA from a transformed strain of *B. subtilis* as donor (i.e., DNA from one of the colonies isolated during the heterologous transformation experiment).

Data from Wilson, G. A. and Young, F. E., *J. Bacteriol.*, 111, 705, 1972; Wilson, G. A. and Young, F. E., in *Bacterial Transformation,* Archer, L. J., Ed., Academic Press, New York, 1973, 269; and Wilson, G. A. and Young F. E., unpublished.

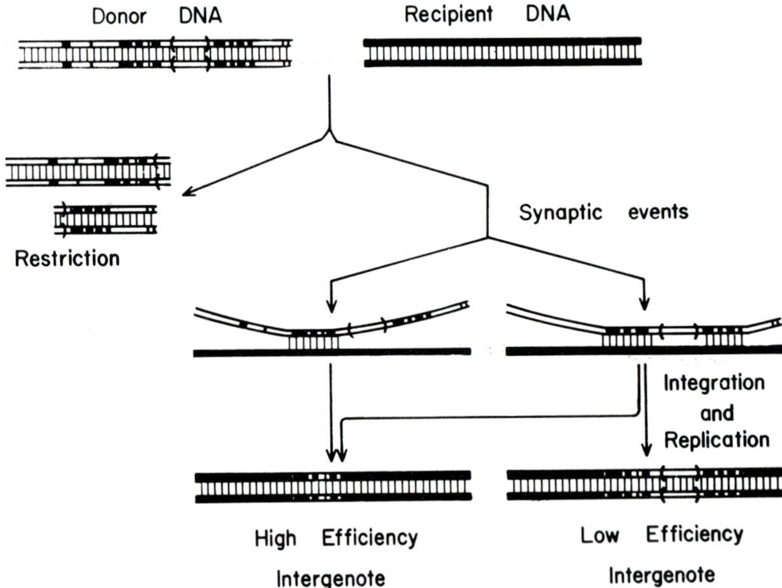

FIGURE 2.   Integration of foreign DNA during the process of heterospecific transformation. Double-stranded DNA is depicted as two heavy lines. Regions of identical or complementary base sequences are solid black. Light vertical lines indicate proper base pairing between the two strands of DNA. The area enclosed by parentheses is the nucleotide sequence recognized by a restriction endonuclease. In this figure, scission of the DNA occurs at the recognition site. The low-efficiency intergenote is probably not distinguishable from the original heterologous donor by transformation. (From Wilson, G. A. and Young, F. E., *J. Bacteriol.*, 3, 713, 1972. Reprinted with permission of American Society of Microbiology.)

IV. DEVELOPMENT OF PLASMID VECTORS

Unlike the *Enterobacteriaceae,* the genus *Bacillus* does not have a rich supply of plasmids encoding antibiotic resistance. In fact, plasmids were only identified in *B. pumilus* by Lovett in 1973.[33] Subsequent analyses in related bacilli by Lovett and co-workers[34,35] and others [36,37] have established the existence of plasmids in *B. subtilis* and *B. natto.* Unfortunately, these plasmids do not contain selectable markers. Using the Kill[+] trait encoded by the plasmid pL10, Lovett was able to detect the transfer of this *B. pumilus* plasmid to *B. subtilis* by a circuitous route. For instance, the plasmid can be transferred among *B. pumilus* strains by PBSl-, PBPl-, and PMBl-mediated transduction[38,39] with direct scoring of Kill[+] colonies, whereas this trait is not readily scored in *B. subtilis.* The transfer to *B. subtilis* 168 could only be accomplished by congression with DNA from *B. subtilis* ATCC 7003.[40] Transformation of *B. subtilis* 168 strain JH86 from Phe[-] Trp[-] Spo[-] to Trp[+] Phe[-] Spo[-] resulted in transfer of Kill[+] in 7/200 Trp[+] clones. Analysis of two of these revealed the presence of pL10. Furthermore, this plasmid, which was derived originally from *B. pumilus,* was maintained by *B. subtilis* 168 through the spore state and during 100 successeive transfers in rich medium. Therefore, such plasmids are replicated and exist in sufficient numbers to be used as cloning vectors in *B. subtilis.* In fact, some strains can harbor more than one plasmid.[41] However, the lack of convenient selection techniques mitigates against their use. Although other *Bacillus* plasmids may be detected that encode antibiotic resistance, none have been reported to date. In view of the amplifiable plasmids encoding tetracycline resistance in *Streptococcus faecalis,*[42] it may be useful to screen for plasmids after the induction with either tetracycline or erythromycin. Since the development of an all *Bacillus* system could minimize the hyopthetical objections to "mixed cloning systems," it is imperative to undertake extensive screening experiments to identify plasmids that would afford direct selection either due to their antibiotic-resistant or heavy metal-resistant traits.

An alternative approach involves the introduction of plasmids from other species into *B. subtilis.* The first successful report involved the transfer of RPl plasmid from *E. coli* to *B. subtilis.*[43] Regrettably, this preliminary observation has not been confirmed. More recently, Ehrlich[44] demonstrated that one plasmid encoding tetracycline resistance (pT127) and four plasmids encoding chloramphenicol resistance (pC194, pC221, pC223, and pUB112) could be introduced into *B. subtilis* by DNA-mediated transformation. Transformation to kanamycin and neomycin resistance could not be detected with plasmids pS117Sn[r] and pK545Km[r]Nm[r]. Although the frequency of the original transformation was low (100 colonies per microgram DNA), the transformation efficiency with ccc DNA isolated from the transformants was increased as much as 50-fold. It is important to note that treatment of the DNA by restriction endonuclease markedly reduced the transformation efficiency to less than 0.5% of the original activity. Treatment of the cleaved preparation by ligase resulted in an increase to 30 to 50% of the original activity. The acquisition of plasmids from Gram-positive bacilli is not restricted to *S. aureus.* Clewell[45] demonstrated the transformation of *B. subtilis* to Tet[R] by plasmid DNA from *S. faecalis.* Unlike the transformation with plasmids of *S. aureus* in Ehrlich's study, the plasmids from *S. faecalis* are reduced in molecular weight following uptake by *B. subtilis.*

The plasmids from *S. aureus* have been characterized extensively by Dubnau and co-workers[46] and Lovett and co-workers.[47] These plasmids are both maintained and readily amplified by *B. subtilis.* Furthermore, it has been possible to clone the gene(s) in the tryptophan biosynthetic pathway of *B. licheniformis, B. pumilus,* and *B. subtilis* in pUB110 and complement the genetic defect in the *trpC2* locus of *B. subtilis* 168.[48] Since there has been extensive characterization of the plasmids from *S. aureus,* these should provide important cloning vectors. The exact copy number and the proportion of plasmids that are in the extrachromosomal state vs. integrated into the chromosome of *B. subtilis* are not known. Further studies are essential to explore these exciting observations in detail and to develop these cloning vehicles.

An alternate approach is the development of a mixed cloning system. As first demonstrated by Chang and Cohen,[49] site-specific endonuclease digests of two plasmids can be joined to form a chimeric plasmid. This chimeric plasmid, which is derived from pSC101, and the gene encoding penicillin resistance from the *S. aureus* plasmid pI258 could function in *E. coli* as well as be transferred among *E. coli* strains by conjugation. Analogously, we recommend that recombinant chimeric plasmids might be developed using *B. subtilis* bacteriophage genes instead of fragments derived from plasmids. Since the bacteriophage φ3T has a gene encoding thymidylate synthetase and converts Thy⁻ *B. subtilis* to Thy⁺,[50] it was suggested that this bacteriophage and its fragments might provide ideal cloning segments.[51] Furthermore, it can be digested with a number of site-specific endonucleases and still retain the Thy⁺ transforming activity.[52] The development of chimeric plasmids composed of the gene encoding thymidylate synthetase (thyP3) in pSC101 and pMB9 was accomplished simultaneously and independently by Ehrlich and co-workers[53,54] and Duncan and co-workers.[55,56] Since chimeric plasmid DNA isolated from *E. coli* readily transformed *B. subtilis* from Thy⁻ to Thy⁺, it is apparent that DNA propagated in *E. coli* is not restricted by *B. subtilis*. Similar results were obtained by Mahler and Halvorson with cloned chromosomal genes.[57] While the *thyP3* gene and the entire bacteriophage genome from bacteriophage φ3T is integrated between the two chromosomal markers encoding thymidine biosynthesis in *B. subtilis*,[58] the *thyP3* gene cloned in pMB9 is integrated near or in the bacterial locus encoding thymidylate synthetase, *thyA*.[59] The failure of the chimeric plasmid to exist either as a self-replicating plasmid in *B. subtilis* or integrated within the chromosome precluded the further use of this vector. Accordingly, we selected a derivative of pCD1, pCD2 as the cloning vector. As shown in Figure 3, the gene encoding thymidylate synthetase (*thyP3*) is bordered by two *Bgl*II sites. This gene can be excised from pCD1 and transferred to pMB9 to form pCD3. This plasmid also transforms Thy *E. coli* to Thy⁺, demonstrating that the segment bounded by the *Bgl*II sites is the *thyP3* gene. The plasmid formed after removal of the *thyP3* gene, pCD2, carries a 0.51 Mdalton region of bacteriophage φ3T that hybridizes to the chromosome of *B. subtilis* but does not transform Thy⁻ to Thy⁺. This plasmid, pCD2, has been used to clone foreign genes in *E. coli* and *B. subtilis* through the generation of pCD4 and pCD6. The foreign gene was obtained from bacteriophage β22 since the gene encoding thymidylate synthetase in this virulent bacteriophage has no homology with the *B. subtilis* chromosome.[60] Transformation studies with pCD4 and pCD6 have established that cccDNA, but not linear DNA, can transform *B. subtilis* from Thy⁻ to Thy⁺. Furthermore, analyses of EcoRI digests of these transformed clones by the Southern hybridization using pCD5 as the "hot probe" have established that the entire chimeric plasmid is incorporated by a Campbell-like model into the chromosome of *B. subtilis*. Apparently, the site of recombination is within the small segment derived from bacteriophage φ3T. This model of recombination is not limited to chimeric plasmids with extensive chromosomal homology. An *E. coli* plasmid (pBR322) containing the *E. coli* gene encoding thymidylate synthetase will also transform *B. subtilis* by a Campbell-like mode. Thus, there are two pathways for DNA-mediated transformation: (1) the integration of linear, single-stranded fragments from linear, double-stranded fragments and (2) the integration of covalently closed, circular DNA. The latter results in the incorporation of extensive regions of foreign DNA that can subsequently serve as recombination sites. Therefore, it is predicted than any fragment of DNA cloned in pMB9 or pBR322 would recombine with the chromosome of *B. subtilis* clones that have integrated these vectors.

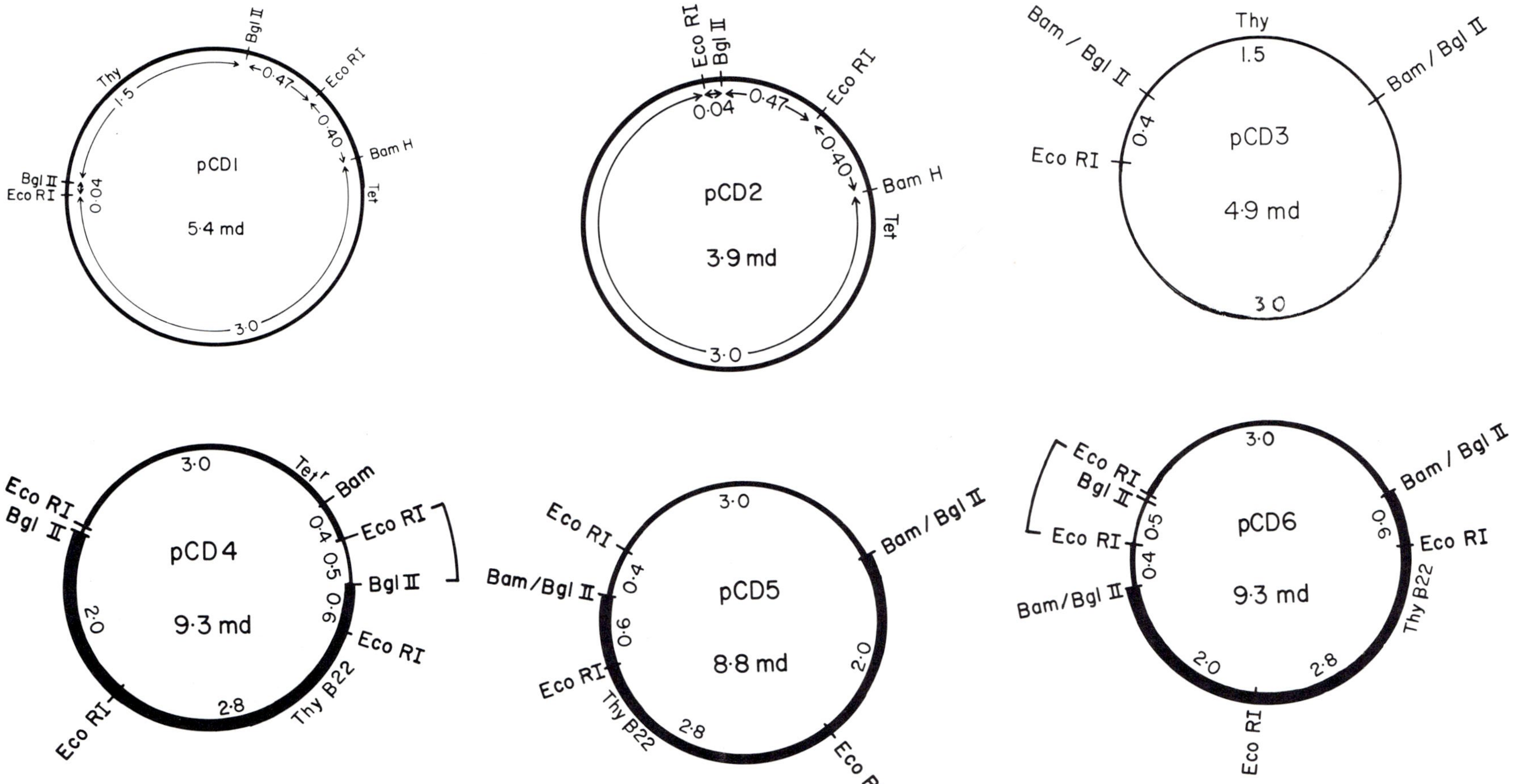

FIGURE 3. Physical map of plasmids. Each was derived from an analysis of the fragments generated by site-specific endonucleases. The location of cleavage by each endonuclease and the molecular weights of the fragments are included. For pCD4, 5, and 6, the fine lines defined by brackets donote the DNA segment(s) from bacteriophage ϕ3T; the moderately heavy line, DNA from pMB9; and the heavy line, the DNA from bacteriophage β22.

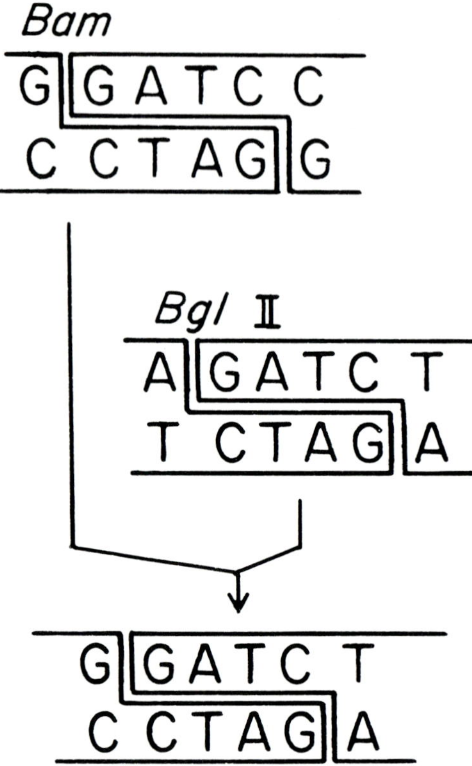

FIGURE 4.   *Bam/Bgl* fusion. Incubation of DNA with site-specific endonucleases *Bgl*II and *Bam*HI in the presence of ligase produces a unique site.

An additional technical procedure should be considered in the development of recombinant plasmids. As first discussed by R. Roberts, it is possible to select only recombinant plasmids in vitro. For example, the genus *Bacillus* contains numerous site-specific endonucleases. Two discovered in our laboratory[61,62] generate a common internal sequence GATC: BamHI G↓GATCC[63] and BglII A↓GATCT.[64] Following the addition of these two enzymes plus ligase to a mixture of digested DNA, only the recombinant plasmid survives (Figure 4). This method has been used extensively to develop the plasmids shown in Figure 3.

V. DEVELOPMENT OF BACTERIOPHAGE VECTORS

The development of bacteriophage vectors is primitive as compared to the *E. coli* model system. Three major lysogenic bacteriophages have been explored: φ3T, φ105, and SP02. The bacteriophage φ3T is a useful vector since it already has a selectable trait (Thy⁺) and can be readily mutated to produce a collection of deletion mutations as shown by Ito.[65] In addition, it is amendable to physical mapping by the combination of nuclease and transformation analyses.[66] Since φ3T cannot propagate above 40°C and *sus* mutations have not been isolated, it has been difficult to establish a genetic map of this virus. Nevertheless, it remains an attractive phage vehicle. Conceivably, bacterial genes could be directly fused to the *thy*P3 locus, introduced by congression to Thy⁻-competent cells, and subsequently mobilized by Thy⁻ variants of bacteriophage φ3T. These spontaneous variants occur at a surprisingly high frequency (approximately 3%) as noted by Tucker in his original publication.[50]

TABLE 2

Potential φ Vectors under Development

Characteristics	Bacteriophages			
	φ3T	φ105	SPO2	SP50
Host range	*Bacillus subtilis*168 *B. globigii*	*B. subtilis* 168 *B. globigii* *B. amyloliquefaciens*[a]	*B. subtilis* 168 *B. globigii* *B. amyloliquefaciens*[b]	*B. subtilis* *B. amyloliquefaciens*[c]
Temperate φ	+	+	+	0
φ Mutations	Deletion	sus ts[d]	sus ts[d]	ts[d]
Single enzyme sites	None known	*Bg*lII *Sa*lG1	*Bg*lII *Sa*lG1	*Bg*lI *Bg*lII
Selectable traits	Thy⁺	0	0	0
Transfection	+	+	+	+

[a] Strains H and F.
[b] Strains F, H, K, N, and T.
[c] Strains F, H, N, and T.
[d] Temperature sensitive.

Bacteriophage φ105 and its related derivatives offer some significant advantages Rutberg and co-workers established the attachment site of this virus[66] and have enabled the construction of a genetic map.[67-69] Recently, Dean et al.[71] have elucidated the physical map and established that bacteriophage φ105 is similar to bacteriophage ϱ14 isolated originally by Dr. J. Hoch. Due to a single Sa1G1 site located in the nonessential immunity region, the former should be convenient cloning vehicle.

SPO2, a temperate bacteriophage that integrates between *ery* and *lin* on the chromosome of *B. subtilis*,[72,73] has also been partially characterized. Unlike many *Bacillus* bacteriophage, stable *ts* and *sus* mutants are isolated readily.[72,74] We have characterized the physical map of this virus using EcoRI, Bg1I, Bg1II, and SalG1. The availability of single Bg1II and SalG1 sites could provide regions for cloning, while the isolation of deletion mutants should aid in establishing the nonessential genes that may be replaced by foreign DNA. Detailed electron microscopic analyses by Chow et al.[75-76] indicated that both SPO2 and φ105 have cohesive ends. Recent studies using restriction endonucleases have established this contention for bacteriophage φ105[77] and support it for bacteriophage SPO2.[78] The essential properties of these viruses and one virulent bacteriophage are shown in Table 2. A more detailed review of *Bacillus* bacteriophages is provided by Hemphill and Whiteley.

VI. HOST PROPERTIES THAT INFLUENCE THE DEVELOPMENT OF CLONING VECTORS

At present, no disabled strains of *B. subtilis* have been developed. Nevertheless, a few properties already known about *B. subtilis* may be relevant to the development of cloning vehicles. First, it should be possible to mutate the organism to preclude sporulation. Piggott and Coote[2] have summarized the previously mapped sporulation markers and estimated that at least 30 and probably between 40 to 50 sporulation loci exist. It is important to note that these are widely scattered throughout the chromosomal map. Some Spo⁻ mutants, particularly *spo*0 mutants, are still transformable and therefore can be readily used as recipients. The introduction of a Spo0 mutation has an additional advantage. The most active autolysin, *N*-acyl-muramyl L-alanine amidase, is resistant to protease action and decreases during stationary phase[80] whereas the hexaminidase is highly sensitive to protease activity and usually inactivated by the protease that is released during the early stage of sporulation. The lack of protease activity in Spo0 mutants results in the flash lysis of *B. subtilis* at the end of logarithmic phase of growth;[81] this significantly reduces survival. An extremely important nutrient for survival is D-alanine. Not only is D-ala an essential component of the peptidoglycan of the cell wall but it is also required for the maintenance of active transport.[82] Adenine transport is particularly sensitive.[83] Because *B. subtilis* undergoes thymineless death and mutants defective in thymine biosynthesis can be used as recipients for chimeric plasmids or phages carrying genes encoding thymidylate synthetase, this auxotrophic trait should also be added. Therefore, the initial phenotype of the vehicle might be Spo0⁻, Dal⁻, Pur⁻, Thy⁻. Experiments are in progress to evaluate the stability of this strain in nature as compared to the wild-type *B. subtilis* 168.

At present, there is limited information concerning the exchange of genes among bacilli in nature. In view of the abundance of bacteriophages in soil,[84] it is likely that transduction occurs in the environment. Accordingly, it may be important to introduce mutations in genes that are known to inhibit transduction such as *mot* in the case of the PBS1[85] and defects in glucosylation of teichoic acid.[86,87] The latter blocks adsorption of a large number of bacteriophages that infect *B. subtilis,* including the transducing virus SP10. An additional problem is the excretion of DNA by *B. subtilis* strain 168. Studies involving germinating spores and vegetative cells have clearly established that *B. subtilis,* like other transformable organisms, excretes DNA during growth even in the absence of cell lysis.[22] Therefore, it may be necessary to isolate mutants that are still transformable but do not excrete DNA. Obviously, many experiments will be required to develop a safe vehicle for cloning experiments, and these suggestions should be merely used as a guide to such studies.

VII. CONCLUSION

B. subtilis should prove to be an extremely useful system for cloning genes. Sufficient background information about the genetic map, temperate bacteriophages, and plasmids is available to enable its development. The recent identification of plasmids that can be transferred to *B. subtilis* from *S. aureus* and the production of chimeric plasmids should provide important directions to the development of cloning systems. The usual obstacles that plague research and the intractable nature of biological systems will continue to present sufficient challenges to make such studies interesting and exciting. The safety of the system, the high degree of transformation, and the importance of the organism to the fermentation industry should greatly stimulate its exploration. Therefore, vigorous programs directed towards the development of both cloning vectors and vehicles should be encouraged.

Acknowledgments

The authors are indebted to Scott Graham, Marshall Williams, Craig Duncan, and Edward Rubin for discussions of various aspects of this work.

REFERENCES

1. **Young, F. E. and Wilson, G. A.,** Genetics of *Bacillus subtilis* and other gram positive sporulating bacilli, in *Spores,* 5th ed., Halvorson, H. O., Hanson, R., and Campbell, L. L., Eds., American Society for Microbiology, Washington, D. C., 1972, 77.
2. **Piggot, P. J. and Coote, J. G.,** Genetic aspects of bacterial endospore formation, *Bacteriol. Rev.,* 40. 908, 1976.

3. **Karmazyn, C., Anagnostopoulos, C., and Schaeffer, P.** Dominance of spoOA mutations in spo⁺/spo merodiploid strains of *Bacillus subtilis*, in *Spores*, 5th ed., Halvorson, H. O., Hanson, R., and Campbell, L. L., Eds., American Society for Microbiology, Washington, D.C., 1972, 126.

4. **Anagnostopoulos, C.,** Genetic analysis of *Bacillus subtilis* strains carrying chromosomal rearrangements, in *Modern Trends in Bacterial Transformation and Transfection,* Portoles, A., Lopez, R., and Espinosa, M., Eds., North-Holland, Amsterdam, 1977, 211.

5. **Szulmajster, K.,** personal communication.

6. **Ito, J.,** personal communication.

7. **Priest, F. G.,** Extracellular enzyme synthesis in the genus *Bacillus, Bacteriol. Rev.,* 41, 711, 1977.

8. **Yamaguchi, K., Nagata, Y., and Maruo, B.,** Genetic control of the rate of α-amylase synthesis in *Bacillus subtilis, J. Bacteriol.,* 119, 410, 1974.

9. **Yamaguchi, K., Nagata, Y., and Maruo, B.** Isolation of mutants defective in α-amylase from *Bacillus subtilis*: genetic analyses, *J. Bacteriol.,* 119, 416, 1974.

10. **Yoneda, Y., Yamane, K., Yamaguchi, K., Nagata, Y., and Maruo, B.,** Transformation of *Bacillus subtilis* in α-amylase productivity by deoxyribonucleic acid form *B. subtilis* var. *amylosacchariticus, J. Bacteriol.,* 120, 1144, 1974.

11. **Erickson, R. J.** Industrial applications of the Bacilli: a review and prospectus, in *Microbiology-1976,* Schlessinger, D., Ed., American Society for Microbiology, Washington, D. C., 1976, 406.

12. **Ingle, M. B. and Boyer, E. W.,** Production of industrial enzymes by *Bacillus* species, in *Microbiology-1976,* Schlessinger, D., ed., American Society for Microbiology, Washington, D. C., 1976, 420.

13. **Sheu, C. W., Salomon, D., Simmons, J. L., Sreevalsan, T., and Freese, E.,** Inhibitory effects of lipophilic acids and related compounds on bacteria and mammalian cells, *Antimicrob. Agents Chemother.,* 7, 349, 1975.

14. **Bornside, G. H., Merritt, C. B., and Weil, A. L.,** Reversal by ferric iron of serum inhibition of respiration and growth of *Bacillus subtilis, J. Bacteriol.,* 87, 1443, 1964.

15. **Farrar, W. E.,** Serious infections due to "non-pathogenic" organisms of the genus *Bacillus, Am. J. Med.,* 34, 134, 1963.

16. **Pearson, H. E.,** Human infections caused by organisms of the *Bacillus* species, *Am. J. Clin. Pathol.,* 53, 506, 1970.

17. **Ihde, D. C. and Armstrong, D.,** Clinical spectrum of infection due to *Bacillus* species, *Am. J. Med.,* 55, 839, 1973.

18. **Reller, L. B.,** Endocarditis caused by *Bacillus subtilis, Am. J. Clin. Pathol.,* 60, 714, 1973.

19. **Spizizen, J.,** Transformation of biochemically deficient strains of *Bacillus subtilis* in deoxyribonuclease, *Proc. Natl. Acad. Sci. U.S.A.,* 44, 1072, 1958.

20. **Thorne, C. B.,** Transduction in *Bacillus subtilis, J. Bacteriol.,* 83, 106, 1962.

21. **Takahashi, I.,** Genetic transduction in *Bacillus subtilis, Biochem. Biophys. Res. Commun.,* 5, 171, 1961.

22. **Streips, U. N. and F. E. Young,** Transformation in *Bacillus subtilis* using excreted DNA, *Mol. Gen. Genet.,* 133, 47, 1974.

23. **Bettinger, G. E. and Young, F. E.,** Transformation of *Bacillus subtilis*: transforming ability of deoxyribonucleic acid in lysates of L-forms or protoplasts, *J. Bacteriol.,* 122, 987, 1975.

24. **Gwinn, D. D. and Thorne, C. B.,** Transformation of *Bacillus licheniformis, J. Bacteriol.,* 87, 519, 1964.

25. **Coukoulis, H. and Campbell, L. L.,** Transformation in *Bacillus amyloliquefaciens, J. Bacteriol.,* 105, 319, 1971.

26. **Marmur, J., Falkow, S., and Mandel, M.,** New approaches to bacterial taxonomy, *Ann. Rev. Microbiol.,* 17, 329, 1963.

27. **Chilton, M. D. and McCarthy, B. J.,** Genetic and base sequence homologies in *Bacillus, Genetics,* 62, 697, 1969.

28. **Wilson, G. A. and Young, F. E.,** Intergenotic transformation of the *Bacillus subtilis* genospecies, *J. Bacteriol.,* 111, 705, 1972.

29. **Wilson, G. A. and Young, F. E.,** Intergenotic and heterospecific transformation: the mechanism of restriction of genetic exchange in *Bacillus subtilis,* in *Bacterial Transformation,* Archer, L. J., Ed., Academic Press, New York, 1973, 269.

30. **Erickson, R. J. and Young, F. E.,** Mechanisms of interspecific genetic exchange in the Bacilli and their potential application in the industrial laboratory, in *Developments in Industrial Microbiology,* Vol. 19, 1978, 245.

31. **Young, F. E. and Wilson, G. A.,** Transformation of bacteria by heterologous DNA, in *Modern Trends in Bacterial Transformation and Transfection,* Portoles, A., Lopez, R., and Espinosa, M., Eds., North-Holland, Amsterdam, 1977, 193.

32. **Trautner, T. A., Pawlek, B., Bron, S. and Anagnostopoulos, C.,** Restriction and modification in *B. subtilis*: biologic aspects, *Mol. Gen. Genet.,* 131, 181, 1974.

33. **Lovett, P. S.,** Plasmid in *Bacillus pumilus* and the enhanced sporulation of plasmid-negative variants, *J. Bacteriol,* 115, 291, 1973.

34. **Lovett, P. S. and Bramucci, M. G.,** Biochemical studies of two *Bacillus pumilus* plasmids, *J. Bacteriol.,* 120, 488, 1974.

35. **Lovett, P. S. and Bramucci, M. G.,** Plasmid deoxyribonucleic acid in *Bacillus subtilis* and *Bacillus pumilus, J. Bacteriol.,* 124, 484, 1975.

36. **Tanaka, T., Kuroda, M., and Sakaguchi, K.,** Isolation and characterization of four plasmids from *Bacillus subtilis, J. Bacteriol.,* 129, 1487, 1977.

37. **Tanaka, T. and Koshikawa, T.,** Isolation and characterization of four types of plasmids from *Bacillus subtilis (natto), J. Bacteriol.,* 131, 699, 1977.

38. **Bramucci, M. G. and Lovett, P. S.,** Low-frequency PBS1-mediated plasmid transuction in *Bacillus pumilus, J. Bacteriol.,* 127, 829, 1976.

39. **Bramucci, M. G. and Lovett, P. S.,** Selective plasmid transduction in *Bacillus pumilus, J. Bacteriol.,* 131, 1029, 1977.

40. **Lovett, P. S., Duvall, E. J., and Keggins, K. M.,** *Bacillus pumilus* plasmid pPL10: properties and insertion into *Bacillus subtilis* 168 by transformation, *J. Bacteriol.,* 127, 817, 1976.

41. **Lovett, P. S. and Bramucci, M. G.,** Plasmid DNA in Bacilli, in *Microbiology-1976,* Schlessinger, D., Ed., American Society for Microbiology, Washington, D. C., 1976, 388.

42. **Clewell, D. B., Yoshihiko, Y., and Bauer, B.,** Plasmid-determined tetracycline resistance in *Streptococcus faecalis*: evidence for gene amplification during growth in presence of tetracycline, *Proc. Natl. Acad. Sci. U.S.A.,* 72, 1720, 1975.

43. **Domardskii, I. V., Levadnaya, T. B., Sitnikov, B. S., Rassadin, A. S., and Denisova, T. S.,** Transformation of *Bacillus subtilis* by isolated DNA of R plasmids, *Dokl. Akad. Nauk SSSR,* 226, 143, 1976.

44. **Ehrlich, S. D.,** Replication and expression of plasmids from *Staphylococcus aureus* in *Bacillus subtilis, Proc. Natl. Acad. Sci. U.S.A.,* 74, 1680, 1977.

45. **Clewell, D.,** personal communication.

46. **Gryczan, T. J. and Dubnau, D.,** Construction and properties of chimeric plasmids in *Bacillus subtilis, Proc. Natl. Acad. Sci. U.S.A.,* 75, 1428, 1978.

47. **Lovett, P.,** personal communication.

48. **Keggins, K. M., Lovett, P. S., and Duvall, E. J.,** Molecular cloning of genetically active fragments of *Bacillus* DNA in *Bacillus subtilis* and properties of the vector plasmid pUB110, *Proc. Natl. Acad. Sci. U.S.A.,* 75, 1423, 1978.

49. **Chang, A. C. Y. and Cohen, S. N.,** Genome construction between bacterial species *in vitro*: Replication and expression of *Staphylococcus* plasmid genes in *Escherichia coli, Proc. Natl. Acad. Sci. U.S.A.,* 71, 1030, 1974.

50. **Tucker, R. G.,** Acquisition of thymidylate synthetase activity by the thymine-requiring mutant of *Bacillus subtilis* following infection by the temperate phage $\phi3$, *J. Gen. Virol.,* 4, 489, 1969.

51. **Young, F. E.,** *Appendix A,* Fed. Regist., 41, 27922, 1976.

52. **Graham, R. S., Young, F. E., and Wilson, G. A.,** Effect of site-specific endonuclease digestion on the *thy*P3 gene of bacteriophage $\phi3T$ and the ThyP11 gene of bacteriophage p11, *Gene,* 1, 169, 1977.

53. **Ehrlich, S. D., Bursztyne, H., Stroynowski, I., and Lederberg, J.,** Cloning of the thymidylate synthetase gene of the phage phi-3-T, in *Recombinant Molecules: Impact on Science and Society,* Beers R. F. and Bassett, E. G., Eds., Raven Press, New York, 1977, 69.

54. **Ehrlich, S. D., Bursztyn-Pettegrew, H., Stroynowski, I., and Lederberg, J.,** Expression of the thymidylate synthetase gene of the *Bacillus subtilis* bacteriophage phi-3-T in *Escherichia coli, Proc. Natl. Acad. Sci. U.S.A.,* 73, 4145, 1976.

55. **Young, F. E., Duncan, C., and Wilson, G. A.,** Development of the *Bacillus subtilis* model system for recombinant molecule technology, in Recombinant Molecules: Impact on Science and Society, Beers, R. F., and Bassett, E. G., Eds., Raven Press, New York, 1977, 33.

56. **Duncan, C. H., Wilson, G. A., and Young, F. E.,** Transformation of *Bacillus subtilis* and *Escherichia coli* by a hybrid plasmid pCD1, *Gene,* 1, 153, 1977.

57. **Mahler, I. and Halvorson, H. O.,** Transformation of *Escherichia coli* and *Bacillus subtilis* with a hybrid plasmid molecule, *J. Bacteriol.,* 131, 374, 1977.

58. **Williams, M. T. and Young, F. E.,** Temperate *Bacillus subtilis* bacteriophage $\phi3T$: chromosomal attachment site and comparison with temperate bacteriophages $\phi105$ and SPO2, *J. Virol.,* 21, 522, 1977.

59. **Williams, M. T. and Young, F. E.,** Analysis of thymidylate synthetase encoded by the genome of *Bacillus subtilis* phages $\phi3T$ and $\varrho11$, in *Abstr. Annu. Meet. Amr. Soc. Microbiol.,* 325, 1977.

60. **Duncan, C. H., Wilson, G. A., and Young, F. E.,** Mechanism of integrating foreign DNA during transformation of *Bacillus subtilis, Proc. Natl. Acad. Sci. U.S.A.,* 75, 3664, 1978.

61. **Wilson, G. A. and Young, F. E.,** Isolation of a sequence-specific endonuclease (*Bam*I) from *Bacillus amyloliquefaciens* H., *J. Mol. Biol.,* 97, 123, 1975.

62. **Wilson, G. A. and Young, F. E.**, Restriction and modification in the *Bacillus subtilis* genospecies, in *Microbiology-1976,* Schlessinger, D., Ed., American Society for Microbiology, Washington, D.C., 1976, 350.

63. **Roberts, R. J., Wilson, G. A., and Young, F. E.**, Recognition sequence of specific endonuclease *Bam*HI from *Bacillus amyloliquefaciens* H., *Nature,* 265, 82, 1977.

64. **Pirrotta, V.**, Two restriction endonucleases from *Bacillus globigii, Nucleic Acids Res.,* 3, 1747, 1976.

65. **Ito, J.** personal communication.

66. **Thomas, E. K., Hailparn, E. M., Wilson, G. A., and Young, F. E.**, Physical mapping of bacteriophage DNA by exonuclease III and endodeoxyribonuclease *Bam*HI, in *Molecular Mechanisms in the control of Gene Expression,* Nierlich, D. P., Rutter, W. J., and Fox, D. F., Eds., Academic Press, New York, 1976, 605.

67. **Peterson, A. M. and Rutberg, L.**, Linked transformation of bacterial and prophage markers in *Bacillus subtilis* 168 lysogenic for bacteriophage ϕ105, *J. Bacteriol.,* 98, 874, 1969.

68. **Rutberg, L.**, Mapping of a temperate bacteriophage active on *Bacillus subtilis., J. Virol.,* 3, 38, 1969.

69. **Rutberg, L. and Armentrout, R. W.**, Low frequency rescue of a genetic marker in deoxyribonucleic acid from *Bacillus* bacteriophage ϕ105 by superinfecting bacteriophage, *J. Virol.,* 6, 768, 1970.

70. **Armentrout, R. W. and Rutberg, L.**, Mapping of prophage and mature deoxyribonucleic acid from temperature *Bacillus* bacteriophage ϕ105 by marker rescue, *J. Virol.,* 6, 760, 1970.

71. **Dean, D. H., Perkins, J. B., and Zarley, C. D.**, A potential temperate phage molecular vehicle for *B. subtilis* 168, in *Spores VII,* American Society for Microbiology, Washington, D.C., in press.

72. **Inselberg, J. W., Eremenko-Volpe, T., Greenwald, L., Meadow, W. L., and Marmur, J.**, Physical and genetic mapping of the SPO2 prophage on the chromosome of *Bacillus subtilis* 168, *J. Virol.,* 3, 627, 1969.

73. **Young, F. E. and Wilson, G. A.**, Chromosomal map of *Bacillus subtilis,* in *Spores VI,* Gerhardt, P., Costilow, R. N., and Sadoff, H. I., Eds., American Society for Microbiology, Washington, D.C., 1975, 596.

74. **Yasunaka, K., Tsukamato, H., Okubo, S., and Horiuchi, T.**, Isolation and properties of suppressor-sensitive mutants of *Bacillus subtilis* bacteriophage SPO2, *J. Virol.,* 5, 819, 1970.

75. **Chow, L. T., Boice, L., and Davidson, N.** Map of the partial sequence homology between DNA molecules of *Bacillus subtilis* bacteriophage SPO2 and ϕ105, *J. Mol. Biol.,* 68, 391, 1972.

76. **Chow, L. T. and Davidson, N.**, Electron microscope study of the structures of the *Bacillus subtilis* prophages SPO2 and ϕ105, *J. Mol. Biol.,* 75, 257, 1973.

77. **Scher, B. M., Dean, D. H., and Garro, A. J.**, Fragmentation of *Bacillus* bacteriophage ϕ105 DNA by restriction endonuclease *Eco*RI: evidence for complementary single-stranded DNA in the cohesive ends of the molecule, *J. Virol.,* 23, 377, 1977.

78. **Yoneda, Y. and Young, F. E.**, unpublished observations.

79. **Hemphill, H. E. and Whiteley, H. R.**, Bacteriophages of *Bacillus subtilis, Bacteriol. Rev.,* 39, 257, 1975.

80. **Young, F. E.**, Autolytic enzyme associated with cell walls of *Bacillus subtilis, J. Biol. Chem.,* 241, 3462, 1966.

81. **Brown, W. C. and Young, F. E.**, Dynamic interactions between cell wall polymers, extracellular proteases and autolytic enzymes, *Biochem. Biophys. Res. Commun.,* 38, 564, 1970.

82. **Clark, V. L. and Young, F. E.**, D-alanine incorporation into macromolecules and effects of D-alanine deprivation on active transport in *Bacillus subtilis* 168, *J. Bacteriol.,* 133, 1339, 1978.

83. **Clark, V. L. and Young, F. E.**, unpublished observations.

84. **Reanney, D. C.**, Extrachromosomal elements as possible agents of adaptation and development, *Bacteriol. Rev.,* 40, 552, 1976.

85. **Joys, T. M.**, Correlation between susceptibility to bacteriophage PBS1 and motility in *Bacillus subtilis, J. Bacteriol.,* 90, 1575, 1965.

86. **Young, F. E.**, Requirement of glycosylated teichoic acid for absorption of phage in *Bacillus subtilis* 168, *Proc. Natl. Acad. Sci. U.S.A.,* 58, 2377, 1967.

87. **Yasbin, R. E., Maino, V. C., and Young, F. E.**, Bacteriophage resistance in *Bacillus subtilis* 168, W23 and interstrain transformants, *J. Bacteriol.,* 125, 1120, 1976.

Chapter 8

POTENTIAL APPLICATIONS OF MOLECULAR CLONING TO GENETIC THERAPY*

R. O. Roblin

TABLE OF CONTENTS

I. INTRODUCTION

Recently developed molecular cloning or recombinant DNA molecule techniques[1,2] will uniquely facilitate the isolation, amplification, and characterization of human DNA sequences coding for specific proteins — essential primary steps in possible further attempts at gene therapy.[3] This chapter will concentrate upon some recent developments in molecular cloning and illustrate how they *could* be applied to the isolation of specific human genes and the development of new approaches to molecular therapy of human genetic diseases. Whether molecular cloning techniques *should* be used in this way raises ethical, political, and social issues, some of which the author, among many others, has grappled with elsewhere.[4-6] This chapter emphasizes the technical aspects; however, this focus on technical possibilities should not be equated with the current advocacy of further attempts at human genetic therapy.[3]

II. ISOLATION OF HUMAN DNA SEQUENCES

Recombinant DNA molecule techniques have already made possible the cloning of a variety of specific eukaryotic DNA segments in *Escherichia coli*. Thus, cloning of DNA sequences coding for sea urchin histones,[7] yeast tRNAs,[8] chicken ribosomal RNA,[9] and ribosomal RNA of *Euglena gracilis* chloroplasts[10] have recently been reported. In principle, cloning of restriction endonuclease fragments of human DNA should be easily accomplished, based upon this experience with cloning DNA segments from other organisms. However, the current National Institutes of Health (NIH) Guidelines for Recombinant DNA Research require P3/P4-EK-2 containment conditions for the cloning of human (and other primate) DNA segments,[11] and only a few certified P3 and P4 containment facilities currently exist in the U.S. Since these facilities are in demand for a variety of experiments, there are practical barriers to cloning random human DNA sequences by means of recombinant DNA techniques.

*   This chapter was supported by the National Cancer Institute, National Institutes of Health, Bethesda, Maryland, under Contract No. N01-C0-25423 with Litton Bionetics, Inc., Kensington, Maryland.

One alternative to the random cloning approach for isolating and cloning specific human DNA sequences is to first purify a specific messenger RNA (mRNA) species and then prepare a double-stranded complementary DNA (cDNA) molecule by reverse transcription of the mRNA in vitro.[12,13] After modification of the termini of the cDNA molecule, it can be joined to an appropriate vector and cloned by transfection of *E. coli*. This approach has been used by several groups of investigators to clone DNA sequences from rabbit[14-17] and human[18] globin genes as well as sequences from the human placental lactogen[19] and rat insulin genes.[20] In several cases,[14-16,20] the cloned DNA is known to contain only part of the protein coding region. However, in the case of the rabbit β-globin cDNA sequences synthesized by Efstratiadis et al.[12] and cloned by Maniatis et al.,[13] the entire β-globin coding region appears to have been cloned in the PMB9 plasmid. In addition, this recombinant plasmid DNA apparently also includes approximately 40 and 110 base pair rabbit DNA sequences, which correspond to untranslated regions adjacent to the 5' and 3' ends of the messenger RNA, respectively. Thus, application of synthetic and linkage techniques used by these investigators may make it possible to clone a full-length copy of any human gene whose mRNA can be isolated in reasonably pure form. This appears to be currently feasible for human α- and β-globin mRNA,[18] as well as for human placental lactogen mRNA.[19]

III. CHARACTERIZATION OF CLONED HUMAN DNA SEQUENCES

Isolation of cloned, specific human DNA segments would only be the first step in an approach to human genetic therapy. Structural and functional characterization of the cloned human DNA would be a necessary prerequisite to further steps. The ways in which restriction endonuclease cleavage can be used to characterize cloned eukaryotic DNA sequences and determine their orientation vis-a-vis vector DNA are illustrated in recent publications.[17,21] Recently developed DNA sequencing techniques[22,23] can be used to characterize the nucleotide sequence of the cloned DNA fragments.[24] In cases where the amino acid sequence of the protein coded by the cloned human gene is known, a comparison of the amino acid and nucleotide sequences provides unambiguous identification of the cloned gene.

Functional characterization of the cloned human DNA implies in vitro demonstration that it contains the required DNA sequence signals for correct initiation and termination of transcription and translation. Recent experiments have shown that various purified DNAs, including that of a Col E1 plasmid carrying a segment of *Drosophila melanogaster* DNA that codes for the histone proteins, are efficiently transcribed upon microinjection into *Xenopus laevis* oocytes.[25] Preliminary results suggesting translation of mRNA made from microinjected Simian virus 40 (SV40) DNA have been described in the same report.[25] This and other in vitro systems for transcription and translation should make possible functional characterization of cloned human DNA. This is necessary to ensure that the cloned DNA segments retain not just the DNA sequence coding for the desired protein, but also the appropriate adjacent DNA sequences, which regulate the processes of transcription and translation.

Selection for functional gene expression by complementation during initial isolation of recombinants may facilitate the cloning of functionally active human DNA sequences. Apparent complementation of *E. coli* mutations by yeast DNA segments cloned in a phage λ vector[26] and the Col E1 plasmid vector[27] has been reported. In those cases where human and bacterial cells share common biochemical pathways, it may be possible to clone functional human genes by virtue of their ability to complement bacterial mutations, which inactivate the analogous bacterial metabolic pathway. However, differences in the molecular details of transcription, translation, and posttranslational modification mechanisms between human and *E. coli* cells may limit the applicability of this approach.

Intermediate steps of amplification of cloned human genes and their separation from the initial cloning vector DNA appear to be simple extensions of already developed techniques. Cloning human genes with either bacteriophage or plasmid vectors will readily provide microgram quantities of the desired human DNA segments. Cleavage of the recombinant DNA with the restriction enzyme initially used to fragment the human DNA regenerates the human DNA fragment with cohesive ends. The human DNA fragment could then be separated from the DNA of the initial cloning vector by, for example, electrophoresis on agarose gels. Isolation, amplification, and characterization of specific human DNA fragments are thus technically quite feasible using currently available molecular cloning techniques.

IV. REINTRODUCTION OF CLONED HUMAN DNA INTO HUMAN CELLS

The final molecular step in genetic therapy would be reintroduction of the cloned, functional, specific human DNA into the cells of a patient with a genetic disease. For a small group of genetic diseases, this might involve an intermediate step in which bone marrow cells from a patient were removed, the desired gene inserted in the cells in vitro, and the cells subsequently transplanted back into the patient.[5] Reintroduction of the desired human DNA has been pictured (e.g., Reference 3) as involving physical integration of the functional human DNA, carried on a suitable vector, into the chromosomal DNA of the patient's cells. An alternative possibility, incorporation of the desired gene into a plasmid-like replicon capable of replication in the nuclei of human cells, might also be considered.

Vectors possibly capable of integrating cloned human DNA sequences into the chromosomal DNA of human cells have recently been described.[28-30] The vector described by Goff and Berg,[29] constructed as a general transducing virus for animal cells, exhibits some of the general characteristics required of a vector for possible human genetic therapy. This vector (SVGT-1) was derived from the DNA of SV40 virus by specific restriction endonuclease cleavages, which removed most of the "late" (virus structural protein) genes from SV40 DNA. Hamer et al.[30] derived the same vector and also pointed out that since this fragment of SV40 DNA contains all the genetic information required for transformation of mammalian cells,[31] it should prove an effective vector for integration of exogenous DNA into chromosome DNA of cells from a variety of species.

Goff and Berg showed that a fragment of bacteriophage λ DNA could be linked by means of poly dA:dT joints to SVGT-1 vector DNA in vitro and that this recombinant DNA could be replicated and encapsidated upon infection of permissive monkey cells in the presence of a temperature-sensitive mutant SV40 helper virus.[29] SVGT-1 can accept foreign DNA sequences up to about 2000 base pairs in length; thus, it could serve as a vector for the cloned DNA segments encoding the α and β chains of hemoglobin.[14,15,24] However, in experiments already reported,[29] Goff and Berg were unable to detect the synthesis of more than a trace amount of RNA complementary to the cloned λ DNA fragment in monkey cells infected with λ-SVGT-1 virus in the presence of helper tsA58 SV40 virus. This was not due to a general failure of transcription, since SV40-specific RNA was detected in the infected cells. The orientation of the λ-SVGT-1 was also not responsible, since only traces of λ-specific RNA were observed in cells infected with λ-SVGT-1 recombinants carrying the λ DNA insert in both possible orientations with respect to the SV40 DNA. In addition, the inserted λ DNA segment carried several DNA sites where λ-specific mRNA transcription is known to be initiated in bacterial cells.

Goff and Berg considered several possible explanations for the very low level of synthesis of λ-specific RNA sequences following λ-SVGT-1 infection.

... first, transcription of the λ DNA segment did not occur because the poly(dA:dT) segment proximal to the λ DNA sequence interrupted transcription at that point; second, some not yet understood feature of the λ DNA sequence precludes its transcription by the mammalian RNA polymerase; third, the λ DNA segment is transcribed, but the lifetime of the transcript is too brief to permit its detection; or fourth, the major late SV40 16S RNA transcript is initiated from a SV40 DNA sequence that is missing from the SVGT-1 vector.[29]

In contrast, Hamer et al.[30] were able to detect transcription of the *E. coli* DNA segment inserted in their vector upon infection of monkey cells. Since the bacterial DNA segment inserted in their experiments coded for the production of an *E. coli* suppressor transfer RNA (tRNA) species, they also looked for the production of functional suppressor tRNA in monkey cells.[30] Their failure to detect the formation of this molecule suggests that the mechanisms of processing precursor RNA species into mature RNA molecules may differ in prokaryotic and eukaryotic cells. Goff and Berg's difficulty in obtaining detectable transcription from the λ-specific DNA segment of λ-SVGT-1 suggests that more specific knowledge regarding mechanisms of mRNA synthesis and processing,[32] capping,[33] and methylation[34] may be required before controlled transcription of cloned human DNA sequences in vectors like SVGT-1 can be achieved. Molecular cloning has already made a contribution in this area by facilitating the determination of the nucleotide sequence of the 5′ untranslated portion of rabbit β-globin mRNA,[24] although other techniques[35] have also provided the same information.

SVGT-1 and the vector described by Hamer et al.[30] have at least one possible drawback as a vector for possible human genetic therapy, as a consequence of the fact that these vector DNA molecules retain the genetic capacity to code for the synthesis of the SV40 A cistron product, the "T antigen." The transformation of cells by mutant SV40 viruses, which synthesize temperature-sensitive A cistron proteins, produces cells that exhibit temperature-dependent malignancy-associated phenotypic characteristics.[36-40] This suggests that the SV40 A cistron product may be responsible for the ability of SV40 virus to cause tumors upon injection into newborn hamsters.[41] However, several SV40-transformed human cells, known to express the A cistron product (T antigen), fail to form tumors when transplanted into athymic nude mice.[42] Since many, but not all human tumor cells form tumors in nude mice,[42-44] it is not yet clear whether expression of the SV40 A cistron product in human cells is sufficient to make them malignant. Further selective cleavage of the SV40 DNA by restriction enzymes to produce potential vectors lacking the A cistron gene region should be possible. These vectors, however, would then probably be dependent upon coinfection with wild-type SV40 virus for integration, since the SV40 A cistron product appears to be required for integration of SV40 DNA into the DNA of the infected cell.[36,38] This would probably eliminate their usefulness as potential vectors for human genetic therapy.

V. CONCLUSIONS

Molecular cloning via recombinant DNA molecule techniques has made the isolation, amplification, and characterization of human DNA sequences coding for specific proteins technically feasible. The NIH Guidelines for Recombinant DNA Research currently restrict execution of experiments involving human DNA because of the shortage of certified P3 and P4 level containment laboratories.[11] However, as increasing numbers of P3 facilities are constructed and certified, the importance of this limitation will diminish.

The construction of vectors derived from SV40 DNA, which may be capable of integrating cloned human DNA sequences back into the chromosomal DNA of human cells, has recently been reported.[28-30] The first experiments[29,30] with such vectors have suggested that more must be learned regarding the molecular details of mRNA synthesis and processing in mammalian cells before the functional capacity of cloned DNA sequences in human cells can be accurately predicted.

The technical feasibility of all the molecular operations required of a potential genetic therapy has not yet been demonstrated. However, molecular cloning techniques have dramatically increased the rate of acquisition of knowledge regarding the organization and functioning of genes in mammalian cells, an essential prerequisite for realistic approaches to possible human genetic therapy. Thus, it may be time to initiate a more detailed assessment of the issues raised by the developing technical potential for human genetic therapy. This assessment should help distinguish cases of somatic gene therapy, which raise primarily questions of safety and efficacy of a possible treatment for an existing individual, from more difficult cases involving possible genetic treatment of those unborn. The author believes that wide-ranging discussions of the issues raised by the developing potential for human genetic therapy should precede further[3] actual attempts at genetic therapy in human patients.

REFERENCES

1. **Cohen, S. N.,** The manipulation of genes, *Sci. Am.,* 233, 24, 1975.
2. **Abelson, J.,** Recombinant DNA: examples of present-day research, *Science,* 196, 159, 1977.
3. **Rogers, S.,** Reflections on issues posed by recombinant DNA molecule technology. II., *Ann. N.Y. Acad. Sci.,* 265, 66, 1976.
4. **Friedmann, T. and Roblin, R.,** Gene therapy for human genetic disease?, *Science,* 175, 949, 1972.
5. **Roblin, R.,** Reflections on issues posed by recombinant DNA molecule technology. I., *Ann. N.Y. Acad. Sci.,* 265, 59, 1976.
6. **Roblin, R.,** Human genetic therapy: outlook and apprehensions, in *Health Handbook,* Chacko, G. K., Ed., North-Holland, Amsterdam, 1978, pp. 103—114.
7. **Kedes, L. H., Chang, A. C. Y., Houseman, D., and Cohen, S. N.,** Isolation of histone genes from unfractionated sea urchin DNA by subculture cloning in *E. coli, Nature,* 255, 533, 1975.
8. **Beckmann, J. F., Johnson, P. F., and Abelson, J.,** Cloning of yeast transfer RNA genes in *Escherichia coli, Science,* 196, 205, 1977.
9. **McClements, W. and Skalka, A. M.,** Analysis of chicken ribosomal RNA genes and construction of lambda hybrids containing gene fragments, *Science,* 196, 195, 1977.
10. **Lomax, M. I., Helling, R. B., Hecker, L. I., Schwartzbach, S. D., and Barnett, W. E.,** Cloned ribosomal RNA genes from chloroplasts of *Euglena gracilis, Science,* 196, 195, 1977.
11. **National Institutes of Health,** Guidelines for Recombinant DNA Research, *Fed. Regist.,* July 7, 1976, 27902.
12. **Efstratiadis, A., Kafatos, F. C., Maxam, A. M., and Maniatis, T.,** Enzymatic *in vitro* synthesis of globin genes, *Cell,* 7, 279, 1976.
13. **Rougeon, F. and Mach, B.,** Stepwise biosynthesis *in vitro* of globin genes from globin mRNA by DNA polymerase of avian myeloblastosis virus, *Proc. Natl. Acad. Sci. U.S.A.,* 73, 3418, 1976.
14. **Rougeon, F., Kourilsky, P., and Mach, B.,** Insertion of a rabbit β-globin gene sequence into an *E. coli* plasmid, *Nucleic Acids Res.,* 2, 2365, 1975.
15. **Rabbits, T. H.,** Bacterial cloning of plasmids carrying copies of rabbit globin messenger RNA, *Nature,* 260, 221, 1976.
16. **Higuchi, R., Paddock, G. V., Wall, R., and Salser, W.,** A general method for cloning eukaryotic structural gene sequences, *Proc. Natl. Acad. Sci. U.S.A.,* 73, 3146, 1976.
17. **Maniatis, T., Kee, S. G., Efstratiadis, A., and Kafatos, F. C.,** Amplification and characterization of a β-globin gene synthesized *in vitro, Cell,* 8, 163, 1976.
18. **Wilson, J. T., Forget, B. G., Wilson, L. B., and Weissman, S. M.,** Human globin messenger RNA: importance of cloning for structural analysis, *Science,* 196, 200, 1977.

19. **Shine, J., Seeburg, P., Baxter, J., and Goodman, H. M.,** personal communication, 1977.

20. **Ullrich, A., Schine, J., Chirgwin, J., Pictet, R., Tischer, E., Rutter, W. J., and Goodman, H. M.,** Rat insulin genes: construction of plasmids containing the coding sequences, *Science,* 196, 1313, 1977.

21. **Lin, A. L., Paddock, G. V., Heindell, H. C., and Salser, W.,** Nucleotide sequences from a rabbit alpha globin gene inserted in a chimeric plasmid, *Science,* 196, 192, 1977.

22. **Maxam, A. M. and Gilbert, W.,** A new method for sequencing DNA, *Proc. Natl. Acad. Sci. U.S.A.,* 74, 560, 1977.

23. **Sanger, F. and Coulson, A. R.,** A rapid method for determining sequences in DNA by primed synthesis with DNA polymerase, *J. Mol. Biol.,* 94, 441, 1975.

24. **Efstratiadis, A., Kafatos, F. C., and Maniatis, T.,** The primary structure of rabbit β-globin mRNA as determined from cloned DNA, *Cell,* 10, 571, 1977.

25. **Mertz, J. E. and Gurdon, J. B.,** Purified DNAs are transcribed after microinjection into xenopus oocytes, *Proc. Natl. Acad. Sci. U.S.A.,* 74, 1502, 1977.

26. **Struhl, K., Cameron, J. R., and Davis. R. W.,** Functional genetic expression of eukaryotic DNA in *Escherichia coli, Proc. Natl. Acad. Sci. U.S.A.,* 73, 1471, 1976.

27. **Ratzkin, B. and Carbon, J.,** Functional expression of cloned yeast DNA in *Escherichia coli, Proc. Natl. Acad. Sci. U.S.A.,* 74, 487, 1977.

28. **Ganem, D., Nussbaum, A. L., Davoli, D., and Fareed, G. C.,** Propagation of a segment of bacteriophage λ-DNA in monkey cells after covalent linkage to a defective simian virus 40 genome, *Cell,* 7, 349, 1976; Nussbaum, A. L., Davoli, D., Ganem, D., and Fareed, G. C., Construction and propagation of a defective simian virus 40 genome bearing an operator from bacteriophage λ, *Proc. Natl. Acad. Sci. U.S.A.,* 73, 1068, 1976.

29. **Goff, S. P. and Berg, P.,** Construction of hybrid viruses containing SV40 and λ phage DNA segments and their propagation in cultured monkey cells, *Cell,* 9, 695, 1976.

30. **Hamer, D. H., Davoli, D., Thomas, C. A., Jr., and Fareed, G. C.,** Simian virus 40 containing an *Escherichia coli* suppressor gene, *J. Mol. Biol.,* 112, 155, 1977.

31. **Graham, F. L., Abrahams, P. J., Mulder, C., Heijneker, H. L., Warnaar, S. O., deVries, F. A. J., Fiers, W., and Van der Eb, A. J.,** Studies on in vitro transformation by DNA and DNA segments of human adenoviruses and simian virus 40, *Cold Spring Harbor Symp. Quant. Biol.,* 39, 637, 1974.

32. **Brawerman, G.,** Eukaryotic messenger RNA, *Ann. Rev. Biochem.,* 43, 621, 1974; Lewin, B., Units of transcription and translation: the relationship between heterogeneous nuclear RNA and messenger RNA, *Cell,* 4, 11, 1975; Lewin, B., Units of transcription and translation: sequence components of heterogeneous nuclear RNA and messenger RNA, *Cell,* 4, 77, 1975.

33. **Ensinger, M. J. and Moss, B.,** Modification of the 5′ terminus of mRNA by an RNA (guanine-7-)-methyl transferase from HeLa cells, *J. Biol. Chem.,* 251, 5283, 1976.

34. **Furuichi, Y., Morgan, M., Shatkin, A. J., Jelinek, W., Salditt-Georgieff, M., and Darnell, J. E.,** Methylated, blocked 5′ termini in HeLa cell mRNA, *Proc. Natl. Acad. Sci.,* 72, 1904, 1975.

35. **Baralle, F. E.,** Structure-function relationship of 5′ non-coding sequence of rabbit α- and β-globin mRNA, *Nature,* 267, 279, 1977.

36. **Tegtmeyer, P.,** Function of simian virus 40 gene A in transforming infection, *J. Virol.,* 15, 613, 1975.

37. **Osborn, M. and Weber, K.,** Simian virus 40 gene A function and maintenance of transformation, *J. Virol.,* 15, 636, 1975.

38. **Kimura, G. and Itagaki, A.,** Initiation and maintenance of cell transformation by simian virus 40: a viral genetic property, *Proc. Natl. Acad. Sci.,* 72, 673, 1975.

39. **Brugge, J. S. and Butel, J. S.,** Role of simian virus 40 gene A function in maintenance of transformation, *J. Virol.,* 15, 619, 1975.

40. **Tenen, D. G., Martin, R. G., Anderson, J., and Livingston, D. M.,** Biological and biochemical studies of cells transformed by simian virus 40 temperature-sensitive gene A mutants and A mutant revertants, *J. Virol.,* 22, 210, 1977.

41. **Butel, J. S., Tevethia, S. S., and Melnick, J. L.,** Oncogenicity and cell transformation in papovavirus SV-40: the role of the viral genome, in *Advances in Cancer Research,* Vol. 15, Klein, G. and Weinhouse, S., Eds., Academic Press, New York, 1972, 1.

42. **Stiles, C. D., Desmond, W., Jr., Sato, G., and Saier, M. H., Jr.,** Failure of human cells transformed by simian virus 40 to form tumors in athymic nude mice, *Proc. Natl. Acad. Sci.,* 72, 4991, 1975.

43. **Schmidt, M. and Good, R. A.,** Transplantation of human cancers to nude mice and effects of thymus grafts, *J. Natl. Cancer Inst.,* 55, 81, 1975.

44. **Shimosato, Y., Kameya, T., Nagai, K., Hirohashi, S., Koide, T., Hayashi, H., and Nomura, T.,** Transplantation of human tumors in nude mice, *J. Natl. Cancer Inst.,* 56, 1251, 1976.

Chapter 9

SOCIAL AND POLITICAL ISSUES IN GENETIC ENGINEERING

E. Loechler, T. McLellan, R. Park, D. Shore, S. Thacher, and P. Youderian*

TABLE OF CONTENTS

I. INTRODUCTION

Recently developed techniques in molecular genetics have become the subject of debate among scientists and a matter of public concern. Recombinant DNA technology allows molecular biologists to combine the genetic material of higher organisms, the eukaryotes, with that of a bacterium, and to copy it at will by inserting these hybrid DNAs into a bacterial host. There is no direct evidence that this type of recombination between eukaryotes and prokaryotes occurs in nature, and the debate over this type

---

* The Recombinant DNA Group of Science for the People was formed early in the summer of 1976 from the Genetics and Social Policy Group of Science for the People in response to the growing concern expressed by the Cambridge, Massachusetts community surrounding intentions by both Harvard University and the Massachusetts Institute of Technology to conduct recombinant DNA research. The Genetics and Social Policy Group had been concerned with the health hazards of recombinant DNA research since the beginning of the controversy. The primary activity of the group to date has been the preparation of articles, position papers, press releases, criticisms, and suggestions in order to better inform the public and its decision-making bodies of the recognized need for the public regulation of science policy in this area.

of research has focused on the risk that such hybrids could escape from the laboratory, establish themselves in nature, and cause unforeseen health hazards. Researchers in the field claim that recombinant DNA technology holds enormous potential for the detailed examination of gene function in eukaryotes, as well as the potential for the commercial production of large volumes of specific proteins and other practical industrial, agricultural, and health-related applications. "It is only a matter of time before the race begins to exploit these developments commercially. A new industry with untold potential is about to appear . . ."[1]

It is the public at large that will have to live with the potential benefits and suffer from the harmful consequences resulting from the industrial exploitation of genetic manipulation of bacteria and eukaryotes. The right of scientists to regulate themselves without informed public consent has become a serious issue, bringing the debate concerning recombinant DNA techniques under close public scrutiny. Gene manipulation techniques may result in a more detailed understanding of the basis of life and may be the first major step toward the ability to control our own genetics.

A. Development of the Issue

At a virus workshop in Cambridge, Massachusetts during the summer of 1971, a student of one of the originators of the recombinant DNA technique described a planned experiment involving the joining of the DNA from the monkey virus SV40 with that of bacteriophage lambda and the subsequent insertion of this virus into *Escherichia coli*. SV40 is known to cause tumors in some animals, but does not appear to cause tumors in humans. An instructor in the workshop raised the possibility that such a hybrid might escape from the laboratory, survive, and enable SV40 to infect humans. The experiment was abandoned.[2]

Paul Berg and several others considered the potential for the new gene-splicing techniques. During July 1974, they addressed a letter to the scientific community,[3] asking that there be a self-imposed moratorium on certain recombinant experiments until the risks involved had been assessed and necessary safeguards had been proposed. The National Institutes of Health (NIH) established a committee charged with the responsibility of assessing the hazards of the research and deciding what lines of research, if any, should be followed; the question of whether or not research should be conducted at all was never addressed. From its inception, the committee interpreted their charge to mean the establishment of guidelines for carrying out recombinant DNA research safely, foregoing the conclusion that the research should be done at all.

In February 1975, the NIH sponsored a large conference of experts in the field of molecular genetics at Asilomar, California. At the meeting, basic classifications of risk were established and a need for the development of safe vectors and hosts for use in these experiments was expressed. There was some discussion of what organisms could be used safely as hosts. *E. coli* had the advantage of being well characterized and, at the time, it was the most convenient organism to employ because of the immense amount of research that had previously been done on its genetics. Other hosts, less well known genetically, were considered because they might not normally reside in or infect humans. A consensus emerged from the Asilomar meeting that it might be possible to manipulate strains of *E. coli* in the laboratory so that they would have a decreased probability of survival outside the laboratory;[4] subsequently, work was begun on the construction of weakened strains. The hazards of using *E. coli*, (whose pathogenicity is both extensive and poorly characterized) vs. using micoorganisms of more limited habitat and pathogenicity were considered to be outweighed by both the convenience of using genetically well-characterized *E. coli* and the expense and time it might take to characterize the genetics of an alternative.

The first draft of the Guidelines (called "Current Guidelines for Research on Recombinant DNA Molecules'') was prepared at Woods Hole, Massachusetts in July 1975.[5] The earliest criticism of the Guidelines was expressed at the Cold Spring Harbor Phage Meetings Cold Spring Harbor, New York in August 1975.[6] They were in response to the concern that the Woods Hole draft substantially lowered the safety standards set at Asilomar. A letter signed by 50 biologists urged that the most hazardous experiments be curtailed, that mammalian DNA be cloned under more stringent conditions than those specified in the Woods Hole Guidelines, and that the representation on the guidelines committee be broadened to include representation from scientists not involved in the cloning experimentation and from the general public.[7]

By December 1975, compromises on many aspects of the guidelines had been reached. Some kinds of recombinant DNA research were banned outright, such as experiments with the DNA of high risk pathogens. However, the cloning of viruses known to cause cancer in animals, for which a moratorium had been asked earlier,[3] were permitted under very restrictive conditions. The NIH held a public hearing on the guidelines in February 1976. A number of people from outside the molecular biology community came to voice opposition to the proposed safety provisions and were left with the feeling that such a voice had been denied. Donald Frederickson, the Director of NIH, hoped that the guidelines for the research had found ''A reasonable balance . . . between concerns to 'go slow' and those to progress rapidly.''[8] Senator Edward Kennedy (Democrat, Massachusetts), a leading congressional supporter of research in the health sciences, criticized the scientists' attempt at self regulation ''because scientists alone decided to impose a moratorium, and scientists alone decided to lift it.''[9]

The members of the Recombinant DNA Molecule Program Advisory Committee which drew up the Guidelines were primarily from the field of molecular genetics, and the majority were personally interested in the conduct of recombinant DNA experimentation.[10] As a result of pressure to have greater public input into the Guidelines, a few outsiders were added to the committee subsequent to the essential decisions concerning safety provisions; however, they appeared to have had very little input at the meetings. Experts in the fields directly concerned with the hazards of recombinant DNA research, such as epidemiologists and microbial ecologists, were consulted during the formulation of the Guidelines but were not involved directly in the decisions affecting safety measures. The final form of the Guidelines was released on June 23, 1976. On the same day, public opposition had found a forum in the chambers of the City Council of Cambridge, Massachusetts.

Public involvement in the recombinant DNA debate had been growing as the final version of the Guidelines was in preparation. In Ann Arbor, Michigan during the Spring of 1976, the construction of a P3 physical containment facility (see below) for research at the University of Michigan was debated before the Regents. It brought experts from across the country to debate the merits of the provisional Guidelines and weakened hosts and vectors. Concern about the research was voiced by faculty, students, and members of the Ann Arbor Community. After some delay, the Regents voted to allow the construction to proceed.

A similar hearing was held by the Committee on Research Policy at Harvard University, where the construction of a containment facility in the Harvard Biological Laboratories was planned. Brought to the attention of Cambridge Mayor Alfred E. Vellucci by publicity surrounding this hearing, recombinant DNA research became an issue before the city government. After two crowded public hearings, which exposed the deep rift in the molecular biology community over the wisdom of pursuing recombinant DNA research in the city, the Council voted for a moratorium on those experiments judged to be most hazardous and established a citizen review board to examine the issue and recommend appropriate action. The Cambridge Experimentation Review

Board, composed of eight city residents chosen by the City Councilors, were neither involved in the debate nor were they molecular biologists. They recommended that the research be allowed to proceed within the city, but that much more stringent safety procedures should be followed than those provided by the NIH guildelines. The Cambridge citizens' panel requested that the biological containment procedures and laboratory monitoring techniques for P3 research be strengthened.

There were two other unique requests. The first was that a Cambridge Biohazards Committee be set up to oversee all recombinant DNA research in the city. The second was to petition the U.S. Congress to establish uniform standards for recombinant work, both profit and nonprofit, and to establish a registry of all workers participating in the research for future epidemiological studies. Commenting on the role of scientists in the self-regulation of their research, the panel concluded:

Decisions regarding the appropriate course between the risks and benefits of potentially dangerous scientific inquiry must not be adjudicated within the inner circles of the scientific establishment. Moreover, the public's awareness of scientific results that have an important impact on society should not depend on crisis situations.[11]

The health hazards of the research are also under consideration by the Environmental Protection Division of the New York State Attorney General's Office and by the City of San Diego, California, both of which have sponsored public hearings on the matter.

B. Protection of the Environment

As an afterthought to the preparation of the Guidelines, the NIH prepared their Draft Environmental Impact Statement (EIS) on recombinant DNA research guidelines.[12] The National Environmental Policy Act (NEPA) requires that all major federal actions be evaluated for their effects on the environment to study, develop, and describe appropriate alternatives and to recognize the world-wide and long-range character of environmental problems.[13] Since its enactment in 1970, the NEPA has had a significant effect on the planning and execution of many federal and private projects. The preparation of a draft impact statement, solicitation of comments from concerned parties, and consideration of alternatives have led many projects to reduce their adverse environmental effects. For this procedure to work effectively, the law requires that the predictive impact statement and consideration of alternatives be done prior to undertaking action. In the case of recombinant DNA research, the EIS was released well after the NIH had both condoned the research and been directly and actively supporting it in numerous laboratories.

The statement tends to rationalize the safety precautions provided rather than examine them critically and suggest alternatives to the research. It provides little or no evidence that the NIH has undertaken, or plans to undertake, the experimental evaluation of either the hazards of recombinant DNA research or those of the ecology and pathogenicity of *E. coli* and its plasmid vectors. The EIS was prepared long after the decisions had been made to establish the level of hazard of various experiments, employ *E. coli* as a host in this research, and support this research through direct NIH funding. Only the Cambridge Experimentation Review Board has requested an experimental evaluation of the hazards. They have recommended that the City of Cambridge request funds from NIH for the health monitoring of laboratory workers engaged in the research.

Notwithstanding the amount of work that has gone into the preparation of the Guidelines, they are limited in scope and do not address the general problem of the hazards of genetic manipulation. There are many ways of manipulating genetic material other than those involving the specific gene-splicing techniques considered in the

Guidelines. Moreover, the Guidelines are restricted in application, as they only apply to research funded by the NIH and have no jurisdiction over industrial research. The latter may be conducted on a much larger scale and could pose greater hazards. The only enforcement mechanism for NIH sponsored projects is the ex post facto withdrawal of funding; as such, the Guidelines are only suggestions for the safe conduct of research. Since many of its provisions are also suggestions or recommendations, research can actually be conducted according to the guildeines with far fewer safety precautions than envisioned by the authors. The broader problems, both of hazards other than those anticipated with the cloning of viral or toxin genes and of other available techniques that would produce the same resulting hybrids, have not been seriously considered.

II. HAZARDS

Several basic concerns have been expressed concerning the hazards of recombinant DNA research. The hybrid, laboratory-created organisms may never have occurred before in nature and may therefore be uniquely pathogenic to humans or other organisms. The risks of this type of experimentation are presently nonquantifiable and cannot be predicted because there is no way of knowing in advance the properties of these hybrid organisms. The release and survival of these organisms outside the laboratory may be irreversible because they are self-replicating. Serious doubts have been raised that the organisms can be contained by the presently proposed modes of containment. This is a special concern because this type of research is extensively planned in industry and academia. Some of the more important objections that have been raised to the containment plans for these novel microorganisms are presented below.

The particular strain of intestinal bacteria, *E. coli* K12, which has been chosen as a host for recombinant DNA molecules has been severely criticized as particularly dangerous. Related strains of *E. coli* are normal residents of the human intestine and pharynx (in about 15% of all humans) and are human pathogens of the skin, urinary tract, and G.I. tract.[14,15] *E. coli* is one of several agents that can cause epidemic diarrhea of the newborn, a condition described as "a leading cause of infant morbidity and mortality in all parts of the world."[16] *E. coli* can cause anything from mild traveler's diarrhea to a cholera-like illness and is responsible for one third of the cases of meningitis in newborn children. Also, it and other Gram-negative bacteria have been well documented as causing increasing sickness and death in hospital patients due to septicemia (bacteria in the blood) in recent years in a Boston hospital.[17] A similar trend has been observed across the nation for hospital-acquired infections.[16]

There is no evidence that *E. coli* K12 can cause disease in the same way as other strains of *E. coli*. The former has been maintained by serial laboratory culture since it was first isolated from human feces 55 years ago.[18] The transfer of a recombinant plasmid to healthier strains of *E. coli* or other Gram-negative bacteria living side by side with *E. coli* poses a major threat for creating a bacterium with new or increased disease-causing potential.

This is made more likely by the fact that *E. coli* K12 may be able to survive and multiply outside the laboratory. Studies where K12 was fed to a very small number of human subjects[18,19] demonstrated no colonization by K12 of the intestine. It persisted in the feces for 3 to 6 days, and in two of eight subjects,[18] K12 appeared to have multiplied during its passage through the gut. The sample size in both studies was insufficient to allow the conclusion that *E. coli* K12 does not colonize in humans. It is likely that human populations under a variety of unexamined conditions might be considerably more susceptible than the subjects used in these experiments. It is not known what factors lead to colonization of the gut; however, sheep starved for 2 days have been demonstrated to support K12 multiplication and plasmid transfer.[20]

Following what some have termed the inevitable escape from the laboratory, it is possible that *E. coli* carrying foreign DNA could cause disease or pose a serious danger by itself. In addition, many ways exist in nature by which the foreign DNA carried by K12 could be transmitted to healthy *E. coli* or other Gram-negative bacteria which might subsequently pose a health hazard. Foreign DNA is incorporated directly into an autonomously replicating piece of DNA in *E. coli*, a plasmid. It could be introduced into another *E. coli* or Gram-negative bacterium by, for example, an infection of *E. coli* K12 by a viral or conjugational factor which could incorporate the foreign DNA and facilitate its transmission to another cell. Recombinant DNA from dead *E. coli* could transfect other bacteria, allowing survival of the foreign genes. Although the transformation of *E. coli* by *E. coli* DNA has not been well characterized in vivo and its frequency cannot be as yet quantitated, it is known that the DNA of virulent *Pneumococcus* bacteria injected into a mouse can transform benign *Pneumococcus* to a virulent state.[21] The fact that most recombinant plasmids used at the present contain one or more drug-resistance markers increases the possibility of escape by providing a potential for the selective enrichment of these plasmids. The widespread use of antibiotics in our society makes this likelihood particularly serious.

Should a Gram-negative bacterium carrying recombinant DNA find itself in the outside world, it would have rapid access to a whole range of ecological niches. *E. coli*, for example, is found in warm-blooded animals, some insects, soil, water near sewage outfalls, and any other place contaminated by human and animal feces.[15] Other experiments imply that *E. coli* or its plasmids (on which the foreign DNA is carried) can be transferred easily between domestic animals and humans. For example, a high correlation exists between the types of antibiotic-resistent *E. coli* found in animals and humans living in close proximity.[22] Furthermore, one cannot say that selective pressure will act to remove recombinant DNA molecules from the environment before they are established in a viable organism. It has been asserted that *E. coli* carrying nonessential DNA on a plasmid would be lost from the population after many generations due to its inferior rate of reproduction.[23] However, many of the plasmid vectors in current use, such as the bacteriophage λ,[24] may actually increase the cell's rate of reproduction. One mathematical model of a situation where part of a bacterial population carries a nonselected plasmid and thus grows less efficiently shows that these plasmids will become successfully established in the bacterial population.[25] Thus, it is likely that some recombinant DNA-containing plasmids will survive in the environment for a substantial period of time. In this light, it seems unwise to employ an organism (i.e., *E. coli* K12) that has so extensive a host range, including humans, for this research. Given our ignorance of the habitat and ecology of *E. coli*, it is impossible to foresee how organisms carrying recombinant DNA will behave in the future.

A. Physical Containment

Will physical containment prevent the release of *E. coli* K12? The NIH Guidelines describe four levels of physical containment to be used in these experiments, ranging from careful microbiological technique (P1 and P2) to special hoods, airtight rooms, and other paraphenalia used for biological warfare agents (P4).[26] Perhaps the most comprehensive study of the efficacy of physical containment is Wedum's report on the Army's biological warfare laboratory at Fort Detrick, Maryland.[27] It was commissioned by NIH when the debate on how to contain the organisms created by recombinant DNA experiments began. No one expects containment to be perfect, and indeed, there is no question that even workers protected by the highest levels of containment could receive infectious doses of organisms against which they had been vaccinated.

Under P4 conditions at Fort Detrick with 45 personnel at risk daily (some of whom administered aerosol exposures of highly infectious agents to live animals), six work-related infections were reported between 1952 and 1959 in one building, and one infection for the years 1959 to 1969 in another.[27]

In a P3-type facility where *Tularemia* (a plague-like organism) was studied for 6 years, 14 cases occurred among workers who had been vaccinated with a dead form of tularemia. The introduction of a live vaccine in 1960 prevented the appearance of further hospitalized cases, but it is clear that physical containment did not prevent occasional exposure of the workers in these laboratories.[27] A more sensitive test of exposure is seroconversion. Paul Berg, professor of Biochemistry at Stanford University, describes how in his P3 laboratory where Simian virus 40 (SV40) was being studied almost all personnel had developed substantial antibody titres to the virus after one half of a full year in the laboratory.[28] One would expect that exposure of laboratory workers to *E. coli* would be unavoidable in all but the most stringent containment facilities. Two of Wedum's recommendations at the end of his report are made in an attempt to insure that exposure is harmless:

For research on recombinant DNA, the most effective safety measure to prevent infection of laboratory personnel is to utilize a microorganism that will not infect humans. Otherwise, the greater the required human infective dose the safer it is.[27]

The other recommendation:

Microorganisms for which there is an effective vaccine, and also in some cases specific effective therapy, could be considered for use in research on recombinant DNA.[27]

does not currently apply to K12, for which there is currently no effective vaccine, and may be circumvented by both plasmid transfer and acquired or resident antibiotic resistance factors.

One underlying assumption in Wedum's report is that microepidemics may occur among those who come into contact with infectious microbes on dirty clothes, dead animals, or heavily exposed laboratory workers, but that the disease-causing agents will not spread to the general public. All known cases of laboratory-caused epidemics support this assertion, according to Wedum. Nevertheless, *E. coli*-borne recombinant DNA represents a special case. The containment achieved at Fort Detrick represents the results of constant medical surveillance — every illness reported by a laboratory worker was assumed to have occurred on the job until shown otherwise — and this was for diseases whose symptoms were recognizable and for which vaccines were known in most cases.

The bacterium harboring the recombinant DNA might not cause disease until it has passed through several hosts. Furthermore, a disease-causing bacterium may be difficult to trace from the laboratory or to recognize initially. Special hosts or environments inside or outside the laboratory, such as a person under antibiotic treatment, might allow the recombinant DNA and its host to flourish locally. The ways in which *E. coli* can disseminate and infect humans may surprise the builders of physical containment facilities. A recent study speculated that *E. coli* could be transferred by air from chickens to humans in much smaller quantities than is possible by ingestion, which usually required 10^6 to 10^8 organisms.[29] Another recent report on the spread of diarrhea-causing *E. coli* between infants in a special-care nursery suggested that the bacterium was transferred from patient to patient by the hands of hospital personnel.[30] Under the right conditions, only very low doses of *E. coli* must be ingested for infections to ensue.

B. Biological Containment

A lack of complete confidence in physical containment has led NIH to impose additional safeguards in the form of biological containment. Crippled bacterial hosts with a one in 10^8 chance of surviving outside the laboratory, designated EK2 and EK3, would be constructed for this work. The lowest level of biological containment, EK1, calls for *E. coli* to be used as a host in all experiments.[26] The two current host-vector systems recommended for EK2 status by the NIH Recombinant DNA Advisory Committee are the *E. coli* strain χ1776 of Curtiss[31] with the plasmid pMB9 and Leder's strain of λ bacteriophage with its own host derived from *E. coli* K12.[32] The approval of both host-vector systems has required some relaxation of the original criteria for EK2 status that were presented in the NIH Guidelines. The safety assessment of the χ1776 and λ systems has been limited to the laboratories where they were created. No outside laboratories have reported anything more than cursory tests for their adherence to the survival criteria presented in the Guidelines. This is not the best situation in view of the widespread use of χ1776. Some of the significant weaknesses of the Curtiss strain of *E. coli* K12 as it existed in March 1976[33] were outlined in a report by the Boston Area Recombinant DNA Group. Some of the major points of the criticism can be summarized and are serious safety considerations for any other planned host-vector systems.[34]

1. ". . . there is no available plasmid cloning vector which . . . is dependent on χ1776 or some other suppressor-containing strain for its replication."[33] In other words, the plasmid could survive well in some other more healthy *E. coli* if transferred there.
2. The data presented by Curtiss concerning the survival of χ1776 do not and cannot promise the reduced survival of the same strain carrying a cloned foreign DNA sequence. Experiments performed by Struhl et al.[35] have demonstrated that cloned DNA from a eukaryotic organism (*Saccharomyces cerevisiae*) can correct for genotypic deficiences of *E. coli*, such as auxotrophy for histidine biosynthetic pathway genes.
3. The strain survives better in tap water than in certain rich environments where the lack of certain specific nutrients which the cell requires, such as thymidine, causes the cells to commit suicide at a faster rate.
4. The ability of Curtiss' strain to transfer genetic information by known resistance factors is reduced only 100-fold at best for most transfer factors tested.
5. "Since markers are frequently lost during routine manipulations of a complex auxotroph" it is not certain how many of the 13 or more mutations in the strain will be lost during handling by other laboratories. Another criticism of the Curtiss strain is that of Adhya and Enquist who comment that "It was questionable that it meets the 10^{-8} [survival rate for] containment defined by the [NIH guidelines] committee."[30]

The laboratory environment presents opportunities for genetic exchange between crippled *E. coli* K12 and healthier bacterial strains. Although the NIH Guidelines ask that the phenotype of host *E. coli* K12 strains be checked regularly, wild-type bacterial and viral contaminations of laboratory cultures are very common occurrences in microbiological work despite the precautions taken against them. The contamination of routine, well-maintained *E. coli* cultures by weaker strains carrying foreign DNA is also possible. Any of these circumstances might completely frustrate biological containment.[7]

We have assumed up until this point that people would be the chief agents for the escape of *E. coli* K12 and most likely the first to suffer from the creation of a harmful organism. This human-centered perspective is shared by the Guidelines, where estimates of the dangers in the research are evaluated on the basis that DNA from a higher organism might cause harm to humankind. The Guidelines require considerably more strict containment for experiments using the DNA from primates than, for example, from insects not known to cause disease. "Shotgun" experiments, where the entire complement of the DNA of an organism is fragmented and inserted into a multitude of *E. coli* K12, are being carried out with insect DNA in P2 facilities where little more than careful microbiological technique and a sign on the door are required by the Guidelines. Shotgun experiments are the most unpredictable of recombinant DNA experiments and may be no less dangerous than experiments done with organisms closer to humans in evolution. There is no way to know what harmful genes the variety of insects and the variety of parasites accompanying them may carry.[37] Even though we unleash an organism by accident which does not have harmful consequences for humans directly, can we be sure that we might not cause great damage to the environment on which we depend?

Another danger is that recombinant DNA research is proliferating rapidly and with little control in major hospital and university laboratories often located in large cities around the world. How well regulated will these facilities be? In an academic or industrial environment, disrepair of equipment, carelessness, or pressure to work faster might make the exposure of laboratory personnel to recombinant organisms much more frequent than one would judge from the Detrick experience. Certainly one or a few national centers for this research would be preferable to the ever more widespread use of the technique, especially in regions of high population density.

With the many unquantifiable difficulties to physical containment, the irreversible spread of a recombinant plasmid from the laboratory is a possibility that must be taken seriously. Complete eradication of new, mutant, pathogenic strains of bacteria might be very difficult. In the EIS, it is stated that "reasonable means for minimizing further dispersal [of such a pathogen] could be undertaken." No such reasonable means are mentioned and indeed, they must be unknown to hospitals where the incidence of *E. coli* infections has been steadily rising.[38] Furthermore, the NIH has thus far failed to establish a systematic screening mechanism for monitoring the escape of recombinant organisms in laboratory personnel and the surrounding communities.

For research to be carried out in academic and industrial laboratories, much more public involvement in the decisions about the biohazards is necessary. Local biohazard safety committees mandated by the NIH Guidelines should include substantial membership from the people at risk from these experiments, not only lab technicians, custodial, and clerical workers, but also representatives from the local community. Some important recommendations are

1. A rigorous course on laboratory safety should be taught.
2. Biohazard reports should be included in the programs of scientific meetings, and more forums for the discussion of the social and biological impact of the research should be provided.
3. The Guidelines should be made binding as law to all laboratories, industrial and academic.
4. All shotgun experiments should be limited to a few laboratories where EK2 vectors and P3 containment are used.

In conclusion, we hope that in 20 years we can look back on the way in which this research was done and say that we may have guessed right or wrong in the assessment of the dangers, but right or wrong, we were not foolish.

C. Bridging the Evolutionary Gap

In the previous sections, some hard, technical questions were raised about recombinant DNA work; however, other broader questions deserve some examination. Some proponents of the research have asked if there is anything really serious or unique to contain. Bacteria may be constantly taking up DNA from higher organisms, e.g., from the cells which are shed from the lining of the intestines of mammals, or from any decaying animal or vegetable.[39] The assertion is not an easy one to prove, and there is no evidence to support it. No specific genes are known to be shared by *E. coli* and humans; therefore, there is no historical evidence of such a process. Nevertheless, many have stated that with approximately 10^{22} bacteria being excreted daily by the human population the transfer of human DNA to bacteria *must* occur to some extent. The rates cannot be defined, however, and the obstacles to the process are tremendous.

E. coli DNA can transfect *E. coli* under careful conditions in a test tube; this is, in fact, a key but inefficient step in recombinant DNA technology. Given the acidic and ionic conditions of the stomach, the number of bacteria competent to take up foreign DNA would be reduced drastically. More significantly, the survival of human DNA once inside the cell would be unpredictable given the unknown levels of internal bacterial restriction endonucleases, the lack of extensive regions of complementarity between bacterial and human DNA, and uncertainties of why recombination may or may not occur in the cell. A barrier to genetic exchange between higher and lower organisms may effectively exist with the purpose, as some have said, of keeping the two groups evolutionarily distinct: "The recombinant DNA techniques permit the creation and likely propagation of organisms (microorganisms at this time) that could not have derived from the ordinary processes of biological evolution."[39]

As simple-minded examples of novel and also harmful organisms, imagine the potency of an organism carrying the genes for a pharmacologically-active peptide or for a substance from plants or insects not previously known to be toxic to humans. Either may prove quite harmful were it lodged in a sensitive part of the human system. An entire biosynthetic pathway could be donated to *E. coli* since foreign DNA molecules containing enough space for as much as ten different genes are commonly part of recombinant plasmids.

The capability of an organism to cause disease and survive at the same time is not well understood. What is well understood, however, is the widespread effect of a new pathogen on an innocent population, as exemplified by the suffering of the American Indians from diseases imported from the Old World. Is our ecosphere as stable as we would like to believe with respect to the presence of an unanticipated infectious agent?

More subtle changes could arise from allowing *E. coli* access to human and viral genes. An autoimmune disease might be caused by the expression of human-like antigens on the surface of a bacterium, as in some cases of acute glomerulonephritis. Paul Berg, a leading researcher in the field, asks ". . . whether it would be harmful to have *E. coli* in our gut carrying genes from viruses that can change normal cells into cancer cells."[40] The concern these examples raise, along with many others of which we cannot conceive, will not be resolved in the foreseeable future. It should be evident how unquantifiable the risks of this research are, with or without the most earnest schemes for containment which we may put into effect.

D. Alternatives to Recombinant DNA Techniques

Whether one is allowed to use the recombinant DNA technique is not a question of academic freedom. As Richard Goldstein, Assistant Professor of Microbiology and Molecular Genetics at Harvard Medical School, has stated, "I don't think there is any question here about the freedom of inquiry. It is a matter of freedom of manufacture of a novel microorganism."[41] Many molecular biologists feel that recombinant DNA techniques are indispensible for their work. In studying every form of genetic control from differentiation and development in an embryo to the integration of a cancer-causing virus into the DNA of its host cell, scientists are searching hard for the individual stretches of DNA involved in these processes. Many of the techniques developed for these studies were left by the wayside when recombinant DNA technology arrived.

For example, the first eukaryotic gene to be isolated, the gene for hemoglobin, was obtained by making a copy of the messenger RNA for hemoglobin, without the use of a prokaryotic organism as a carrier at any point.[42] Many in vitro techniques, such as purifying DNA sequences which code for a particular messenger RNA species, are within the molecular biology community's grasp at the present and could replace recombinant DNA techniques in some instances. For example, with affinity column techniques using messenger RNA, one can look for homologous sequences in the cell's vast supply of heteronuclear RNA.[43] Similar techniques could be applied to the isolation of specific sequences of the cell's DNA as well. Other advantages presented by recombinant DNA cloning techniques, such as the translation of DNA into protein or the replication in large quantities of a purified DNA segment, can be currently performed with measurable efficiency in cell-free systems. None of these systems appears at present to be as convenient as recombinant DNA technology, but given the hazards of recombinant DNA research, it is the responsibility of the scientific community to actively pursue these other methods. Clearly, there are experiments with recombinant DNA techniques which do not have readily apparent in vitro parallels. This does not mean that the questions such experiments hope to answer cannot be approached through research which does not require the manufacture of novel microorganisms. The scientists wishing to conduct this research may simply have to be more patient.

It is certain that workers in the field of recombinant DNA research could develop a number of surprising, attractive, and unusual applications of the technique for medicine and industry. One such scheme would be to fabricate an *E. coli* which would either produce human insulin or the blood clotting factors which hemophiliacs lack. However, recombinant DNA techniques certainly cannot promise unique solutions to these problems. Insulin is currently obtained from bovine or porcine pancreas, and should it be in short supply, other methods, such as using cultured mammalian cells which produce larger amounts of the hormone, may be simpler to use (see Reference 44). In many other cases, recombinant DNA technology may not provide the only solution to problems for which it has been proposed as a technique of last resort.

III. ESTIMATING BENEFITS AND RISKS

As with all questions of technology policy, addressing recombinant DNA technology utimately requires an assessment of the benefits in relation to the above-mentioned risks. The catalog of possible risks has been claimed to span a vast realm from the precipitation of a cancer epidemic to new kinds of overt or subtle infection and environmental catastrophes. These risks are unpredictable in both character and likelihood. Accordingly, it becomes necessary to determine whether there are broad, profound, and definite benefits arising from recombinant DNA technology which would justify the risks taken in its development. Furthermore, the benefit to risk estimate must be placed in the context of the real world, taking account of the full impact of supposedly beneficial technologies as well as examining benefits and risks in terms of the social interests served.

Recent experience suggests that the introduction of new technologies often has complex and mixed effects which reflect the invaded economic and political milieu. Therefore, simple predictions of benefit, scaled up to global operations, are frequently not meaningful. Nevertheless, the proponents of recombinant DNA research make such simple predictions when arguing that fundamental advances will follow in food production, drug manufacturing, understanding disease (thus leading to "cures"), and genetic management generally, whether related to disease, behavior, or vulnerability to environmental insult. Paul Berg, one of the Asilomar participants, explained before an NIH conference on Guidelines for Recombinant DNA Research:

There are also important potential benefits to expanding the world food supply. The availability of natural and artificial fertilizers limits certain crop yields. But atmospheric nitrogen is an infinite source of ammonia, if only we could harness the microbial potential for nitrogen fixation. Experts in this field suggest that the introduction of the nitrogenase system from bacteria into plants or into symbiotic organisms is imminent and promises great rewards . . . I have no doubt that in time the opportunities will expand as the methodology becomes more sophisticated.[40]

Biomedical research has already begun to feel the attraction of working for genetic cures for cancer. An objective laid out in a 1974 report from the National Cancer Program Planning Conference was to "develop the means to modify individuals in order to minimize the risk of cancer development."[45] One approach to this objective was to "alter genetic make-up or genic expression to reduce the rate of cancer development." These ideas are shared by many in the field of recombinant DNA research. Even though Nobel Laureate David Baltimore concedes that 80% or more of human cancer is caused by environmental agents, he implies that recombinant DNA research is essential for dealing with cancer because prevention of it might be impractical:

Even if we identify the causes of breast cancer, cervical cancer, prostate cancer, colon cancer and bladder cancer, it is unlikely that we are going to be able to design a civilization that will be acceptable to the population and that will prevent the occurence of these terrible diseases. We should certainly make every effort we can to understand how our lifestyle causes cancer, but we must also push forward on a more basic attack on the problem, and I believe that attack is made much easier by the ability to manipulate recombinant DNA molecules.[46]

The Cetus Corporation, in Berkeley, California which includes on its board of advisors such eminent scientists as Joshua Lederberg, Donald A. Glaser, and Stanley N. Cohen, describes the promise of the future in glowing terms:

We propose to do no less than to stitch, into the DNA of industrial microorganisms, the genes to render them capable of producing vast quantities of vitally-needed human proteins. . . . This concept is so truly revolutionary to the biomedical sciences that we of Cetus predict that by the year 2000 virtually all the major human diseases will regularly succumb to treatment by disease-specific artificial proteins produced by specialized hybrid microorganisms.[41]

Those who look forward to a future ability to intervene in the genetic control of organisms, including humans, see recombinant DNA technology as an important step toward this goal. Presently, genetic engineering remains largely limited to screening and nongenetic remedial procedures.

There are a number of fallacies behind promises of great benefits from this technology. Foremost is the premise that solutions to major problems, such as insufficient world food production or human disease, are technology limited. The following section argues that fundamental solutions are not significantly limited by existing and available technology but rather by economic and political factors. Secondly, there is widespread conviction that new technologies can offer striking gains, with little appreciation that they frequently have associated with them a number of negative, and sometimes dominant, consequences. Finally, emphasis on "technological" solutions results in a diversion or distraction from other goals which are essential for real progress.

A. Producing Enough Food

In the case of world food production, it is claimed that existing productive capacity is insufficient to feed present and future populations. It is proposed that recombinant DNA research could be used, for example, to insert nitrogen-fixing ability into nonleguminous plants, thereby expanding output greatly. However, the limiting factor in food production is not capacity; rather, it is insufficient market demand. As explained in a recent study prepared under the auspices of the World Bank (but not endorsed by it),[47] the large number of people throughout the world without enough food have insufficient income to purchase food that would otherwise be readily available. The study found, for example, that the global calorie deficit below the recommended diet, while covering 75% of the population in underdeveloped countries, amounts to a mere 4% of the world cereal production.[48] Furthermore, even substantial reductions in food prices, as *might* happen with new technologies, would not bring this population into the food marketplace. Instead, the fundamental underlying problems would continue to be unemployment, income maldistribution, land ownership, and the retention of political power by elite, minority public interests. Recombinant DNA technology offers the prospect of increased yields and higher profits for the large owners of agricultural resources (the relatively few who own the best land in most underdeveloped countries) and vertically integrated corporations involved in all aspects of global agriculture (e.g., farm machinery, fertilizers, pesticides, processing, and marketing).

Recent advances in the technology of agriculture, applied globally, have had adverse effects. For example, the introduction of hybrid varieties during the "Green Revolution" tended to further stratify rural class structure, favoring large landowners who could afford the costly inputs required — irrigation, fertilization, mechanization — and encouraging the further consolidation of land holdings.[49-51] Landless and jobless peasants were forced into the cities at an even faster rate than before. There was destruction of traditional, local crops with further erosion of nutritional status, while the new crops had increased vulnerability to unexpected pests.[49,52] These changes exacerbated the maldistribution of income for the poorest sector in the Third World and consequently affected the ability of the people to buy, let alone grow, enough food. Presumably, new varieties created through recombinant DNA technologies will be introduced into a similar economic environment.

Accompanying the belief that advances in modern agriculture and population control will solve the problems of world food production (especially through the introduction of new plant varieties) is a political planning mode which renders essential changes more remote. These changes must include a full-employment rural economy and self-sufficient, small-scale industrialization in the context of rejecting the traditional, vested interests holding political leadership. "Advanced" agricultural technology requires heavy financing, which makes the credit vital for small farmers less available to them. It encourages government investment in irrigation, storage, and transport to suit the needs of export-oriented, large-scale farms. Furthermore, it creates dependence on agricultural inputs from foreign rather than local sources. Finally, it strengthens the economic and political position of the large operators and weakens the rural workers, thereby further delaying basic and required change in the rural economy.

B. Conquering Disease

It has been claimed that recombinant DNA techniques promise important advances in combating disease through lower drug prices and the creation of new drugs. The possibility of a rapid and inexpensive production of human insulin by genetically engineered bacteria has often been cited in this respect. Another frequent example is the rapid and efficient production of antibiotics by new fermentation processes; this might be achieved by implanting synthetic capabilities from slow-growing organisms such as yeasts and fungi into rapid-growing, easily cultured bacteria. However, although drug prices are exorbitant, the production costs of drugs are a small part of total costs in most cases; promotion, innovation for patent evasion or defence, regulatory requirements, and distribution comprise a far larger percentage of the price.[53]

Projections are made for more exotic medical benefits, such as the production of specific human proteins in large quantities (i.e., specific metabolic enzymes, blood clotting factors, immunoglobulins, and hormones), which would have a variety of uses. One direct application that has been suggested is to provide specific proteins which are absent or defective in persons having well-characterized genetic diseases. Rather than dwelling on technological solutions for medical problems, it must be recognized that the primary health problems facing people in the U.S. and the entire world derive directly from the conditions of their lives: working conditions, stresses, environmental quality, nutrition, and life style, none of which is primarily dependent on technological innovation for amelioration and all of which are the blunt consequences of the prevailing economic order. The resulting "civilization" or "technological" diseases include hypertension and other cardiovascular disease, cancer, anxiety-depression, alcoholism, obesity, and others.[54]

New drugs for the treatment of civilization diseases are increasingly sophisticated technologically, thus requiring more technical expertise for safe and effective use and providing ample opportunity for misuse. The proliferation and irresponsible use of antibiotics is a concrete illustration of this problem. Widespread use of antibiotics for prophylactic or improper purposes has created a selective environment favoring drug-resistant pathogens and compromising patients' immediate health. A number of drug-resistant strains of major pathogens are now in wide circulation and antibiotic-resistant epidemics have been both predicted and realized.[55] This trend toward technological medicine will be magnified multifold when the products of recombinant DNA techniques begin to appear. Awesome hazards are presented by the clinical development of these therapies, an activity which will continue and increase in Third World countries where the regulation of human experimentation is less stringent. Even the routine use of therapies derived from recombinant DNA techniques will probably be a positive advance only for those who have access to top quality medical care — for others, harm may be a likely outcome. Agressively marketed, difficult to regulate, and highly complex therapies will be available within a health care system incapable of assuring their uniform and correct use.

Emphasizing a new class of wonder drugs and therapies made possible by recombinant DNA research not only sustains the current excessive reliance on drugs as medical panaceas, devices to make medical management more efficient. It also diminishes the visibility of other alternatives for addressing contemporary problems, including those which look at the entire civilization and social order (e.g., pollution, working conditions) and those which increase the participation of patients in their own health management and responsibility. The present funding situation in cancer research is a graphic illustration of diversion by "high science." A small part (less than 10%) of the budget of the National Cancer Institute (NCI) of the NIH is budgeted for basic environmental epidemiology, i.e., identifying the agents responsible for most cancer and their modes of action.[56,57] The bulk of NCI funds support programs in viral research (based wholly on the *speculation* that viruses are involved in human cancers), immunology (hoping to utilize the resources of the immune system against preexisting cancer), and in chemotherapy (attempting to poison preexisting cancer). This is to say that cancer research is directed primarily toward treating the symptoms of cancer rather than eliminating the causes of cancer. Much of the basic science directed at cancer control has dealt with fundamental questions of genetic control, in part, because these questions are the more scientifically interesting ones. The danger remains that recombinant DNA-oriented research will sustain this bias instead of allowing emphasis to shift to vital epidemiological studies and programs aimed at eliminating exposure to carcinogens. Investment in the costly containment facilities and supplies required for recombinant DNA research only diverts funds from more directly health-oriented research, which would attempt to eliminate health problems at their sources, and may compound already existing health problems.

C. Mastering Genetic Control

There is currently a strong tendency in science and public policy to give special importance to genetic explanations of human behavior and disease. These positions are based on entirely speculative theories but derive their support from the expediencies that their conclusions allow. An example is the debate over the heritability of IQ[58] and the contention that innate learning potential is easily measured and catalogued using standard IQ tests. Screening for genetic factors in behavior provides a convenient method of dealing with problems whose causes are less willingly acknowledged by social planners. For example, the XYY karyotype in males was long believed (on the basis of poor and eventually discredited evidence) to predispose individuals to aggressive behavior.[59-61] Aggressive tendencies have often been explained in this way, and more specific and subtle components of human behavior are scheduled for "elucidation" through the new pseudoscience of sociobiology.[62]

In advancing recombinant DNA technologies as a means to understand and deal with a great variety of problems, the proponents of these technologies are contributing further to this emphasis on genetics in social policy. Consequently, programs for providing services that change the environment will be neglected in lieu of genetic assessments. Attempts at understanding the interaction of environmental and genetic factors in dealing with social problems with the intent of best serving people as they are will be deferred in favor of developing systems which track or channel individuals based on their genetic "limitations."

Civilization diseases may have some genetic components, but they remain largely undescribed. However, to focus on these undefined factors and pursue technological means for modifying or compensating for individual genetic variation amounts to defending the economic, social, and political order against the people, not vice versa.

Genetic screening is an example of a rapidly advancing technology which, under ideal circumstances, could provide social services of genuine value. Experience, however, does not demonstrate this value. Genetic screening techniques have created a new realm of stigmatization and discrimination which institutions have predictably used to their own ends. Insurance companies and employers in hazardous industries have used these techniques to avoid "susceptible" people,[63] while offering few clear-cut benefits to those being screened.

IV. SIGNIFICANCE OF THE RECOMBINANT DNA DEBATE TO SCIENCE

In its gravity and divisiveness in science, the debate surrounding recombinant DNA technology may be directly comparable to the debate surrounding the issues of nuclear power development. It has appeared at a time when technology in general is under suspicion and public scrutiny and when government at national and local levels are becoming increasingly uncomfortable with science policy decision making. While traditional academic freedom is being eroded by the necessity of the accountability of science to the public, some of its defenders threaten and coerce those who voice caution. A process, which began as an apparently pious act of self-regulation by scientists has ignited brush fires over the scientific landscape. Prominent members of the science community are in direct opposition over this issue — expert opinion is divided.[64]

How can objective observers be in such a state of disagreement? The answer is perhaps that there is no such thing as scientific objectivity, that perceptions of objective reality are wholly dependent on philosophical and ideological premises as well as on more immediate material factors in peoples' lives.

A. Nonobjectivity in Science

Because a large part of the benefit to risk estimate for recombinant DNA research is speculative, it is especially open to subjective valuation. As mentioned above, assessing the magnitude of benefits accruing from recombinant DNA technologies is contingent on one's view of the social role of technology, an inherently political and ideological consideration. In view of our ignorance of the ecology of prokaryotes, an estimation of the risks involved with this research is subject to the same criticism. An additional source of subjectivity arises at the personal level involving the self-perception of one's own contribution to science and technology and the views one accepts of the role of science and technology in society.

For many scientific workers, the value of their work depends to some extent on how they perceive their contribution directly or indirectly to human betterment. Societies where institutions do not operate *a priori* to serve desirable social ends create incentives toward believing that better technology tends to serve such ends and that new knowledge has intrinsic positive value. Consequently, many medical researchers pursue answers to problems for which other solutions, i.e. solutions that attempt to change social conditions, are entirely beyond their control or not even evident. There are some people who, for this reason, may have an unduly optimistic outlook in the debate over recombinant DNA research. Biases in perspective can also be observed in other scientific controversies, such as those surrounding techniques in the fields of psychosurgery, genetic screening, the funding of cancer research, and the role of computers in society.

Others in science view research issues largely from their individual, immediate vantage point and in terms of creative intellectual activity or employment. Many of these workers tend to envision limitations, constraints, and choices in scientific investigation in the same way that recognized scientific authority does. Many have careers whose success frankly depends on the rapid exploitation of opportunities in scientific discovery, sometimes involving developments with commercial implications. The resulting advantages include publications, appointments, the realization of creative potential, esteem with family and colleagues, and recognition by institutions and officials. In addition, the reward can be an entrance into business and government circles, including association with venture capitalists and invitations to corporate boardrooms. It is clear that in situations where advances are imminent, the personal benefits and risks of some scientists (as with investors) can understandably differ from those of the public. The influence of personal interests is no less likely a factor in determining their objective reality than in the case of dedicated medical researchers or others who view themselves as protagonists of the people's interests.

B. POPULAR DECISION MAKING IN SCIENCE POLICY

Research policy is inextricably tied to politics of the real world, even though many in science would prefer it otherwise. The development of knowledge reflects the dominant social forces in a society, which in the U.S. means the interests of those who hold wealth and property. Historically, this is apparent in theories of biological determinism which have been used to support the concept of social Darwinism and are currently employed to allege fundamental, innate sex-role and racial differences as well as the inheritability of intelligence. These theories have important ramifications for public policy in government, industry, and academia.[63]

Public policy for science must be determined by a process based on popular awareness, organization, and control. This determination would require a corresponding institutional environment which does not exist at the present time. The forms this process might take include unions with strong member participation and control, extensive internal education programs, and an active involvement in defining and enforcing both government legislation and corporate policy through collective bargaining and action. Another avenue for public control of science policy might be through community-based research and human experimentation boards, provided they are not under the control of government bodies. They must also attract a wide participation — not only residents of the local community but also people working in the affected institutions, including hospital and university workers, technicians, and other science workers.

Without this organization, public discussion, debate, and criticism must be employed as the major vehicle to effect the existing decision making apparatus. This process is not encouraged by most prominent members in science; rather, the consensus is that scientists should decide science policy. As with the issue of nuclear power, controversy surrounding recombinant DNA technology diffuses into the public domain despite the attempts of appointed experts to retain control of the decision making process.

How can good judgment on scientific issues be exercised by the masses? This is analogous to the question of how do top government leaders and policy experts decide questions of science and technology policy. They rely on experts whom they believe to be credible. The public should not only gain a better understanding of the science, but should also be able to evaluate the credibility of experts. What are these experts' views on the role of technology and specific issues bearing on the people's interests? How have they contributed to dealing with the real problems of the society, and what are their stakes in these matters? Evaluating experts is an important task for a properly functioning review board and was the primary concern of the Cambridge Experimentation Review Board established to assess recombinant DNA research policy. Through fortunate yet unusual circumstances, this board was selected to represent a broad sample of the population. Just as the rulers of the country can pick and choose between experts and the opinions that they espouse, so can the people.

SUMMARY AND CONCLUSIONS

To a large extent, the NIH *Guidelines for Research Involving Recombinant DNA Molecules* were authored by self-interested researchers in the field of molecular genetics who avoided the *a priori* question of whether recombinant DNA research should be pursued at all. This avoidance is revealed by the NIH Draft EIS, a document released 3 months subsequent to the release of the Guidelines, in direct violation of the NEPA of 1970. The Draft EIS discusses neither possible alternatives to the proposed action nor the global implications of the future conduct of the research, also in violation of the NEPA. The Guidelines are limited in applicability and establish little mechanism for enforcement, save the withdrawal of NIH-sponsored funding.

The choice of host organism for the receipt of foreign DNA, the bacterium *E. coli*, is a poor one. It is a poorly characterized pathogen which is ubiquitous in habitat. That foreign DNA implanted into *E. coli* K12 could eventually be transferred to another microorganism, resulting in the formation of a novel pathogen, cannot, at the present time, be ruled out by either experiment or by consensus within the scientific community.

Physical containment procedures established to reduce the likelihood of such an event cannot insure against the release of hybrid microorganisms generated by recombinant DNA techniques. Additional biological containment barriers do not measure up to expectations as worded in the Guidelines.

The benefits which are claimed to accrue from the use of these techniques are aimed at problems that are not limited by our technological ability. Rather, they are the blunt consequences of the prevailing social and economic order. Recombinant DNA research will do little more than divert funds from much-needed social and scientific programs aimed at correcting the sources of health-related problems. This effort will reinforce the societal predisposition to correct the symptoms of such problems as cancer, while effecting no substantial progress towards the elimination of these problems.

Alternatives to the present situation include abandonment of this research program, restriction of such recombinant research to regional facilities, or the tight regulation of recombinant DNA techniques at the community level as has been sought in Cambridge, Massachusetts. The use of alternative techniques which do not involve the creation of novel microorganisms should be explored by the major federal funding agencies. Presently, such techniques have the ability to supplant many of the proposed recombinant DNA experiments.

Self-regulation of science becomes untenable when that science begins to directly affect the lives of those who sponsor that societal luxury. Scientists must come to terms with the fact that they may be held accountable for the results of their work. Science must shed its guise of being "for the people" when it is not directed at eliminating the causes of social problems. At a time when the quality of the environment, and therefore life itself, is at a premium, science must redirect its goals in line with the social responsibility that demands public accountability of science.

REFERENCES

1. Cetus Corporation, Berkeley, California, Special Report (internal report for investors), October 1975.
2. **Slesin, L.,** Recombinant DNA research: a chronology, Occasional Paper No. 2, Massachusetts Institute of Technology, Laboratory of Architecture and Planning, Cambridge, Mass., November 10, 1976.
3. **Berg, P., Baltimore, D., Boyer, H. W., Cohen, S. N., Davis, R. W., Hogness, D. S., Nathans, D., Roblin, R. O., Watson, J. D., Weisman, S., and Zinder, N. D.,** Potential biohazards of recombinant DNA molecules, *Science,* 185, 303, 1974.
4. **Berg, P., Baltimore, D., Brenner, S., Roblin, R. O., and Singer, M. F.,** Summary statement of the Asilomar Conference on recombinant DNA molecules, *Proc. Natl. Acad. Sci. U.S.A.,* 72, 1981, 1975.
5. DNA committee has its critics, *Nature,* 257, 637, 1975.
6. **Wade, N.,** Recombinant DNA: NIH group stirs storm by drafting laxer rules, *Science,* 190, 767, 1975.
7. **Duncan, M,. Goldstein, R., Primakoff, P., and Orrego, C.,** Critique of Woods Hole Guidelines for recombinant DNA research, *Recombinant DNA Research,* Vol. 1, U.S. Department of Health, Education, and Welfare, National Institutes of Health, Bethesda, Md., 1976, 350.
8. **Frederickson, D. S.,** Decision of the director, NIH, to release guidelines for research on recombinant DNA research, *Fed. Regist.,* Vol. 41 (176), 38444, September 9, 1976.
9. **Gwinne, P., Michaud, S. G., and Cook, W. J.,** Politics and genes, *Newsweek,* 87, 51, January 12, 1976.
10. **Simring, F. R.,** The double helix of self-interest, *The Sciences,* Vol. 17(3), 10, 1977.
11. Cambridge Experimentation Review Board, Guidelines for the Use of Recombinant DNA Molecule Technology in the City of Cambridge, Mass., January 5, 1977, *Bull. At. Sci.,* 33(5), 22, 1977.
12. Department of Health, Education, and Welfare, National Institutes of Health, Recombinant DNA research guidelines, Draft Environmental Impact Statement, *Fed. Regist.,* 41 (176), 38426, September 9, 1976.
13. The National Environmental Policy Act, 42 U.S.Code 4341; Amended by Public Law 94—52, July 3, 1975; Public Law 94—83, August 9, 1975.
14. **Sack, R. B.,** Human diarrheal disease caused by enterotoxigenic *E. coli, Ann. Rev. Microbiol.,* 29, 333, 1975.
15. **Cooke, M. E.,** *Escherichia Coli and Man,* Churchill Livingstone, Edinburgh, Scotland, 1974, chap. 1.
16. **Falkow, S.,** *Infectious Multiple Drug Resistance,* Pion, London, 1975, 62.
17. **McGowan, J. E., Barnes, M. W., and Finland, M.,** Rates of bacteremic patients of Boston City Hospital, during 12 selected years 1935-1972, *J. Infect. Dis.,* 132, 316, 1975.

18. **Anderson, E. K.,** Viability of *E. coli* K12 in the human intestine, *Nature,* 255, 502, 1975.
19. **Smith, H. W.,** Survival of orally administered *E. coli* in the alimentary tract of man, *Nature,* 255, 500, 1975.
20. **Smith, M. G.,** R-factor transfer *in vivo* in sheep with *E. coli* K12, *Nature,* 261, 348, 1976.
21. **Avery, O. T., MacLeod, C. M., and McCarty, M.,** Studies on the chemical nature of the substance inducing transformation of pneumoccocal types, *J. Exp. Med.,* 79, 137, 1944.
22. **Fein, D., Burton, G., Tsutakawa, R., and Blenden, D.,** Matching of antibiotic resistance patterns of *Escherichia coli* of farm families and their animals, *J. Infect. Dis.,* 130, 274, 1974.
23. Department of Health, Education and Welfare, National Institutes of Health, Recombinant DNA research guidelines, Draft Environmental Impact Statement, *Fed. Regist.,* 41 (176), 38430, September 9, 1976.
24. **Edlin, G., Lin, L., and Kudrna, R.,** λ lysogens of *E. coli* reproduce more rapidly than non-lysogens, *Nature,* 255, 735, 1975.
25. **Levin, B. and Stewart, F.,** Probability of establishing chimeric plasmids in natural populations of bacteria, *Science,* 196, 218, 1977.
26. Department of Health, Education, and Welfare, National Institutes of Health, Guidelines for research involving recombinant DNA molecules, *Fed. Regist.,* 41 (176), 38443, September 9, 1976.
27. **Wedum, A. G.,** The Detrick Experience, in *Recombinant DNA Research,* Vol. 1, U.S. Department of Health, Education, and Welfare, National Institutes of Health, Bethesda, Md., 372, August 1976.
28. **Berg, P.,** Letter to DeWitt Stetten, Deputy Director for Science, National Institutes of Health, Bethesda, Md., September 2, 1975.
29. **Levy, S. B., FitzGerald, G. B., and Macone, A. B.,** Spread of antibiotic-resistant plasmids from chicken to chicken and from chicken to man, *Nature,* 260, 40, 1976.
30. **Ryder, R. W., Wachsmuth, I. K., Buxton, A. E., Evans, D. G., DuPont, H. L., Mason, E., and Barrett, F. F.,** Infantile diarrhea produced by heat-stable enterotoxigenic *Escherichia coli,* *N. Engl. J. Med.,* 295, 849, 1976.
31. **Curtiss, R., III, Pereira, D., Clark, J. E., Hsu, J. C., Goldschmidt, R., Hull, S. I., Moody, R., Maturin, L., and Inoue, M.,** Construction, properties and testing of χ1776, in *Recombinant DNA Research,* Vol. 2, U.S. Department of Health, Education, and Welfare, National Institutes of Health, Bethesda, Md., in press.
32. **Leder, P., Tiemier, D., and Enquist, L.,** EK2 derivatives of bacteriophage lambda, *Science,* 196, 175, 1977.
33. **Curtiss, R., III, Pereira, D., Clark, J. E., Hsu, J. C., Goldschmidt, R., Hull, S. I., Moody, R., Maturin, L,. and Inoue, M.,** Report to Recombinant DNA Advisory Committee, National Institute of Health, Bethesda, Md., March 30, 1976.
34. **Goldstein, R., Orrego, C., Youderian, P.,** Analysis and critique of the Curtiss report on the *Escherichia coli* strain intended for biological containment in DNA implantation research, in *Recombinant DNA Research,* Vol. 2, U.S. Department of Health, Education, and Welfare, National Institutes of Health, Bethesda, Md., in press.
35. **Struhl, K., Cameron, J. H., and Davis, R. W.,** Functional genetic expression of eukaryotic DNA in *Escherichia coli,* *Proc. Natl. Acad. Sci. U.S.A.,* 73, 1471, 1976.
36. **Adhya, S. and Enquist, L. W.,** Our Critique of Roy Curtiss' EK-2 Host Proposal, memo to Rowe, W. P., Chief, Laboratory of Viral Diseases, National Institutes of Health, Bethesda, Md., May 31, 1976, Report to Recombinant DNA Advisory Committee, National Institutes of Health, Bethesda, Md.
37. **Nightengale, E.,** Letter to DeWitt Stetten, Deputy Director for Science, National Institutes of Health, Bethesda, Md., in *Recombinant DNA Research,* Vol. 1, U.S. Department of Health, Education, and Welfare, National Institutes of Health, Bethesda, Md., August 1976, 489.
38. **Falkow, S.,** *Infectious Multiple Drug Resistance,* Pion, London, 1975, 239.
39. **Davis, B. and Sinsheimer, R. H.,** The hazards of recombinant DNA, in *Trends in Biochemical Sciences,* Vol. 1, International Union of Biochemistry, Elsevier, North-Holland, Amsterdam, N178, August 1976.
40. **Berg, P.,** Proceedings of a Conference on NIH Guidelines for Recombinant DNA Molecules, public hearings held at a meeting of the Advisory Committee to the Director, National Institutes of Health, Bethesda, Md., 23, 24, 1976.
41. **Goldstein, R.,** Proceedings of a Conference on NIH Guidelines for Recombinant DNA Molecules, public hearings held at a meeting of the Advisory Committee to the Director, National Institutes of Health, Bethesda, Md., 104, 1976.
42. **Rougeon, F., Kourlsky, P., and Mach, B.,** Insertion of a rabbit β-globulin gene sequence into a *E. coli* plasmid, *Nucleic Acids Res.,* 2, 2365, 1975.
43. **Williamson, B.,** Is HnRNA really pre-mRNA, *Nature,* 264, 397, 1976.
44. **Battelle Office of Coprorate Communications,** Membranes help insulin production. *Battelle Today,* No. 2, 2, November 1976.

45. **National Cancer Program, Planning Conference Report, National Institutes of Health, U.S. Department of Health, Education, and Welfare, No. (NIH) 76—952 Objective 2, 1974.**

46. **Baltimore, D.,** Proceedings of a Conference on NIH Guidelines for Recombinant DNA Molecules, public hearings held at a meeting of the Advisory Committee to the Director, National Institutes of Health, Bethesda, Md., 109, 1976.

47. **Reutlinger, S. and Selowsky, M,.** *Malnutrition and Poverty,* Johns Hopkins University Press, Baltimore, 1976, 1.

48. **Wade, N.,** Inequality the main cause of world hunger, *Science,* 194, 1142, 1976.

49. **Cleaver, H. M.,** The contradictions of the green revolution, *Monthly Rev.,* 24, (2), 80, 1972.

50. **Lappe, F. M., Collins, J., and Fowler, C.,** *Food First: Beyond the Myth of Scarcity,* Houghton Mifflin, Boston, 1977, 111ff.

51. **Wade, N.,** Green revolution I. A just technology, often unjust in use, *Science,* 186, 1093, 1974.

52. **Wade, N.** Green revolution II. Problems of adapting a western technology, *Science,* 186, 1186, 1974.

53. **Concerned Rush Students,** Turning perscriptions into profits, *Science for the People,* 9 (1), 6, 1977.

54. **Hall, R. H.,** *Food for Nought, The Decline of Nutrition,* Vintage, Random House, New York, 1976, 229.

55. **Culliton, B. J.,** Pencillin-resistant gonorrhea: new strain spreading worldwide, *Science,* 194, 1395, 1976.

56. **Epstein, S. E.,** Epidemic: the cancer producing society, *Science for the People,* 8 (4), 4, 1976.

57. **Epstein, S. E.,** The political and economic basis of cancer, *Technol. Rev.,* 78 (8) 35, 1976.

58. **Felman, M. W. and Lewontin, R. C.,** The heritability hang-up, *Science,* 190, 1163, 1975.

59. **King, J. and Beckwith, J.,** The XYY syndrome: a dangerous myth, *New Sci.,* 64, (921) 474, 1974.

60. **Genetic Engineering Group,** The XYY controversy, *Science for the People,* 7 (4), 28, 1975.

61. **Beckwith, J. and Miller, L.,** The XYY male: the making of a myth, *Harv. Mag.,* 79 (2), 30, 1976.

62. **Ann Arbor Science for the People Editorial Collective,** *Biology as a Social Weapon,* Burgess Publishing Co., Minneapolis, Minn., 1977, 133.

63. **Miller, L. and Beckwith, J.,** Genetic Screening: benefits and limitations, in *The Biosciences and Society,* Munroe, R. E., Ed., Liverpool University Press, England, in press.

64. **Bennett, W. and Gurin, J.,** Science that frightens scientists: the great debate over DNA, *Atlantic,* 239 (2), 43, 1976.

Chapter 10

MICROBIAL GENETIC ENGINEERING BY NATURAL PLASMID TRANSFER — Some Representative Benefits and Biohazards

A.M. Chakrabarty and J.F. Brown, Jr.

TABLE OF CONTENTS

I. INTRODUCTION

Microbial genetic engineering — that is, the deliberate modification of the gene structure of procaryotic microorganisms so as to achieve a human benefit — is not a new concept. For many years, microbiologists and fermentation engineers have been "engineering" microorganisms to produce strains that would permit increased production rates and culture stabilities in industrial fermentations, enhance the yields of valuable microbial gene products (e.g., amino acids, alcohols and antibiotics), or improve the quality of foods or beverages conventionally prepared by fermentation. The technique used to create these conventional "genetically engineered" strains is to expose the microbes to mutagenic agents and then select mutants having the desired performance characteristics. This technique permits only limited, localized changes in the gene structure of the microbes used, typically, those affecting the expression of genes already present. Hence, it results in genetically engineered mutants with metabolic capabilities that differ quantitatively rather than qualitatively from those of their progenitors. Consequently, these conventional mutant strains can neither offer any human benefit that is not offered to at least some small degree by an existing wild microbe nor present any hazards that are not already present to some degree.

Since 1974, a powerful new technique for microbial genetic engineering, *in vitro* recombinant DNA formation, (described in detail in other chapters of this book) has permitted the development of a new type of mutant strain. These recombinant DNA strains must still be regarded as mutants of the original parent, since their overall genetic composition is still more than 99% that of their progenitor. They differ from the earlier man-made mutants, however, in that the last 0.1 to 1.0% of the gene structure may contain additional types of genes, including some that might never be introduced by natural gene transfer processes. As a result, they may (1) exhibit metabolic capabilities that are qualitatively different from those of the unmutated parent; (2) offer qualitatively new types of human benefits, such as the production of eukaryotic gene products by prokaryotic microbial fermentations; and (3) present new types of hazards as well. The potential benefits and hazards of this recombinant DNA technique have been discussed previously.[1]

The purpose of this chapter is to consider the potentialities of a third genetic engineering methodology, namely the transfer of natural plasmids between prokaryotes. This technique, like the recombinant DNA technology, results in mutant strains that have some active genes and hence some metabolic capabilities that differ from those of the parent strain. They differ, however, in that they contain no genetic material or metabolic capabilities other than those already present collectively in the naturally interbreeding microbial populations from which the parent strains were derived.

To explore the potential benefits and hazards that might result from the use of the plasmid transfer technique, this chapter will examine two areas where the use of this technique has been considered in attacking practical problems relating to environmental pollution and new energy sources. These areas are oil spill cleanups and the enhanced production of methane from cellulosic waste. For each of these areas, the chapter shall examine, in turn, the reasons why microbial genetic improvements are needed, the potentialities of the natural plasmid transfer technique for providing the necessary degree of genetic modification, and finally the biohazards that might be associated with the mutant strains thereby created.

II. CONSTRUCTION OF MICROBIAL STRAINS FOR OIL SPILL CLEANUP

The problems inherent in using microorganisms for oil spill cleanup have been discussed in Reference 2. Although various microorganisms can degrade and thereby grow efficiently upon crude oil, each individual strain is usually characterized by an ability to oxidize only a few kinds of hydrocarbons. Yeasts, for example, although used for the manufacture of protein from alkanes, can oxidize only the aliphatic hydrocarbons. Bacterial genera such as *Pseudomonas, Arthrobacter,* and *Acinetobacter* contain species that together can degrade most constituents of crude oil, including the aliphatic, aromatic, cyclic aliphatic, and polynuclear aromatic hydrocarbons. However, pure cultures of the individual species have only limited substrate ranges and are of little help in consuming the complex hydrocarbon mixtures found in crude oil.[3] Most effort on microbial oil spill cleanup has been directed towards the identification and use of mixed cultures wherein the individual species would have the abilities to consume different hydrocarbon components of the oil. Studies of the biodegradability of different kinds of crude oil with such mixed cultures indicate that while some types of oil are rather recalcitrant to microbial action, as much as 85% of others can be consumed within a few weeks.[4]

Nevertheless, there are serious problems in using mixed cultures for oil spill cleanup. First, the process is slow. Because of interactions among themselves, only a few strains grow rapidly, consuming the assimilable hydrocarbons. Once these fractions are exhausted and the culture growth slows, other strains begin growing and consume residual hydrocarbon fractions. The oil utilization process is thus sequential and associated with a declining rate of hydrocarbon degradation. Second, it is almost impossible to achieve mutational improvement in a mixed culture, since growth of the individual strains cannot be regulated. A third problem in microbial oil spill cleanup is supplying nitrogenous nutrients to the cultures enabling their proliferation. Since sea, river, or lake water have little nitrogen and phosphorus, rapid oil utilization will require an external supply of these elements for the oil-consuming microorganisms. At any event, no attempts to use mixed cultures for large-scale oil spill cleanup have been successful.[3,5] Thus, the challenge to the geneticist is that the use of mixed cultures for oil spills presents problems that can be overcome only by the use of pure cultures or simple mixtures of pure cultures, while that of pure cultures has the drawback that they can degrade only limited fractions of the oil.

A few years ago, Chakrabarty and Brown began investigations to determine whether this challenge could be met if the genetic competences for various types of natural hydrocarbon degradation could be inserted into a single strain. Since *Pseudomonas* species are often involved in the biodegradation of various aliphatic, aromatic, cyclo-paraffinic, and polynuclear aromatic hydrocarbons and since many of these degradative abilities are known to be specified by plasmid-borne genes,[6] one of our earlier objectives was to discover if the natural hydrocarbon-utilization genes from different *Pseudomonas* species could be transferred from the parent cells to a single *Pseudomonas* culture. An example of this sort of genetic transfer is shown in Table 9.1. The recipient was a strain of *P. aeruginosa*, which grows very rapidly with hexadecane, octadecane, and their higher homologs, but fails to utilize tetradecane or its lower homologs. It is, however, possible to isolate certain *P. putida* or *P. aeruginosa* strains that can utilize these lower aliphatic hydrocarbons at a rapid rate. For example, strain *P. putida* AC59 harbors a recombinant CAM-OCT plasmid and can grow rapidly with hexane, octane, and decane but very poorly with dodecane and tetradecane. Strain AC63, on the other hand, grows poorly with hexane and octane, but grows well with dodecane and tetradecane. The ability of strain 63 to grow rapidly with dodecane and tetradecane is due to the presence of a plasmid termed DOD. Both DOD and OCT are nontransmissible plasmids; however, OCT can be transferred to *P. aeruginosa* as part of the recombinant transmissible plasmid CAM-OCT, while DOD can be transferred to *P. aeruginosa* with the help of a transfer plasmid termed factor K.[6] The introduction of both OCT and DOD plasmids from two different strains to a single culture of *P. aeruginosa* allowed this strain to decompose all linear aliphatic hydrocarbons of chain lengths C_6 to C_{18} at a rapid rate (Table 1).

The introduction of XYL degradative plasmid[7] permits utilization of toluene and xylenes in a similar manner. In addition, the genetic competences required for the biodegradation of the alicyclic hydrocarbons, such as terpenes and polynuclear aromatic hydrocarbons, can also be introduced into the appropriate recipient to increase its oil-consuming potentialities.[2]

Another improvement that could possibly be made to such strains would be to render them capable of fixing their own nitrogen. As mentioned before, one of the problems of microbial oil spill cleanup involves supplying enough nitrogenous nutrients to the cells to sustain their metabolic activities at the maximal rate. Although use of oil-soluble N and P sources has been suggested,[8] it would be very expensive to supply such

TABLE 1

Transfer of OCT and DOD Plasmids to *Pseudomonas aeruginosa*

Donor	Recipient	Transfer frequency	Exconjugant phenotype
AC59 Trp⁻(CAM-OCT)⁺	*P. aeruginosa*	10^{-4}	$C_6^+ C_8^+ C_{10}^+ C_{16}^+ C_{18}^+$
AC63 Met⁻ DOD⁺	CAM-OCT⁺ *P. aeruginosa*	10^{-7}	$C_6^+ C_8^+ C_{10}^+ C_{12}^+ C_{14}^+ C_{16}^+ C_{18}^+$

nitrogenous materials to the vast areas of lake or ocean waters where an oil spill had occurred.

A better solution would be to transfer the nitrogen-fixation capabilities from nitrogen-fixing bacteria to the multiplasmid, oil-consuming strain. The construction of a *Pseudomonas* drug-resistance plasmid harboring the *Klebsiella pneumoniae his nif* genes has previously been achieved.[9] However, this plasmid could not be transferred to His⁻ mutants of *P. aeruginosa* strain PAO.[9] Since a transformation system with plasmid DNA is now available in *P. putida*,[10] Chakrabarty and co-workers introduced by transformation *Klebsiella his nif* genes as part of the plasmid RP41 to a number of His⁻ mutants of *P. putida* (Table 2). RP41 appears to be compatible with a number of degradative plasmids and may coexist stably in the multiplasmid strain. However, further mutational improvement is necessary before such strains actually fix their nitrogen, since the *Klebsiella* nitrogenase is sensitive to oxygen and does not function in *P. putida* or *P. aeruginosa* cells growing aerobically upon crude oil. These authors are presently attempting to isolate mutants in RP41-positive *Pseudomonas* cultures that would produce an oxygen-tolerant nitrogenase and hence be able to fix N_2 under aerobic conditions.

At present, it would appear that a suitably engineered multiplasmid pure culture could present acceptable solutions to many of the problems limiting the use of mixed cultures for oil spill cleanup. Since the plasmids are expressed simultaneously to form the degradative enzymes when the hydrocarbon inducers are present in the crude oil, the rate of oil consumption by such strains is rapid, and most of the major component fractions are used up quickly. This is partly because of a relaxed substrate and inducer specificity of the enzymes coded by these plasmids. Introduction of the CAM plasmid leads to the utilization of a number of terpenes, bornanes, and steroids,[11] while introduction of the plasmid XYL leads to that of toluene, xylenes, and other trimethyl benzene derivatives. It is likely that a large number of analogous hydrocarbons present in crude oil, but never individually tested in the laboratory, are also utilized. An obvious benefit is the ease of documenting pathogenicity, toxinogenicity, or the lack thereof in a single culture. The possibility of introducing nitrogen fixation capabilities into a single culture poses yet another potential benefit. It would be virtually impossible to introduce the nitrogen fixation capability into all of the sequentially blooming mixed cultures. Similar arguments can be applied to genetic improvements leading to enhanced rate of degradation of oil components or extended range of substrates.

Another potential benefit of the use of the multiplasmid bacteria should be improved ecological compatibility. Oil spills can occur on lakes, rivers, or oceans. It is clear that a single type of mixed culture, e.g., one that could grow on river or lake water, would

TABLE 2

Transformation of His⁻ Mutants of *Pseudomonas putida* with RP41 Plasmid DNA

Bacteria	RP41 DNA (μg/ml)	Treatment	His⁺ Transformant/ μg DNA[a]
AC13[b]	8	None	0
His⁻ *P. putida*	8	$CaCl_2$, 0.1 *M*	1×10^2
AC724[b]	8	None	0
His⁻ *P. putida*	8	$CaCl_2$, 0.1 *M*	3×10^2

[a]   All His⁺ transformants were Cb′Tc′Km′.

[b]   AC13 and AC724 are two different His⁻ mutants, where the mutations map at 12 and 30 min, respectively, on the chromosome.

be unable to function significantly in the high salinity of the sea water. It would, therefore, be desirable to transfer the degradative plasmids to marine pseudomonads for their use on ocean water, while pseudomonads isolated from nonmarine sources could be used for oil spills on rivers or lakes. Similarly, transfer of such plasmids to microorganisms isolated from Arctic areas such as the North Alaskan slopes might be one way to construct microorganisms for cleanups in that area. These authors assume that the degradative plasmids would be functional in marine or arctic environments, although no definitive studies have been conducted in this regard. The possibility of using air-dried powders of single cultures (which can survive for long periods of time) as opposed to mixed cultures (which tend to lose their potency because of susceptability of some strains to air drying) also needs exploration.

There are still many technical problems in using multiplasmid microorganisms for oil spill cleanup. While more than 70 or 80% of some types of crude oil are known to be reasonably subject to microbial decomposition,[4] other types are much more recalcitrant to microbial attack and probably could not be treated by the organisms. Not all species of microbes are convertible into multiplasmid strains. Many of the microorganisms that oxidize oil components may turn out to have the genetic competence on their chromosome, so that such capabilities could not be multiply transferred. There is an almost total lack of knowledge regarding the status of hydrocarbon-degradative genes in hydrocarbon-utilizing bacteria other than pseudomonads. Therefore, further genetic studies have to be conducted on such species to find or construct specific hydrocarbon-degradative plasmids. Even in multiplasmid pseudomonads, incompatibility among the degradative plasmids is a serious technical problem. While it has been possible to recombine some of them so that they may become part of the same replicon,[12] the other alternative has been to group compatible plasmids together and introduce them as packets of three or four plasmids into the same cell. Large numbers of plasmids in various combinations can thus be introduced into different cells of the same parentage, which can be grown separately and mixed together before use. Since isogenic cells derived from the same strain do not interact with one another, there is little effect on the growth of the individual multiplasmid strains. One of the persistent problems in constructing these strains is not only continued and enhanced loss of many plasmids from the cell, but that introduction of several plasmids greatly overwhelms the genetic economy of the cell, causing them to grow much more slowly. Introduction of packets of three or four plasmids greatly alleviates this problem, although no doubt it contributes to a decrease in the overall rate of oil consumption. Finally, oil spill cleanup by microbial means is only a supplementary step to clean up the layer of oil which is otherwise difficult to remove by floating booms or other mechanical means. The oil consumed is lost, of course, and cannot be recovered; although it might still be argued that the bacterial proteins obtained at the expense of the oil can still enter the food chain and help nourish the aquatic animals whose lives were threatened by the spill.

A. The Biohazard Potential of Multiplasmid Oil-degrading Microorganisms

Before introducing new technology, it is important to identify and evaluate any possible deleterious side effects. In the case of the proposed development and use of multiplasmid oil-degrading microbes, three conceivable side effects have been suggested. These are

1. The possible transfer of the plasmid clusters from the harmless pseudomonads to pathogenic or toxin-producing prokaryotes, thereby giving rise to new toxic species that would be likely to proliferate wherever an oil spill occurred
2. The possible escape of the multiplasmid organism into the general environment,

followed by a widespread destruction of petroleum-based hydrocarbon everywhere
— in oil wells, gasoline tanks of cars, and even in asphalt roads and roofing
3.   Possible change of the local ecological balance due to the massive seeding with the
multiplasmid organism that might be required to rapidly digest the oil spill

The first two hazards can be disposed of by noting that the conditions used in the
laboratory for carrying out interspecies transfers of natural hydrocarbon-degrading
plasmids are no different from those that occur naturally in the vicinity of any oil
seepage through soil. The transfers of such plasmids from one hydrocarbon-degrading
species to another to produce multiplasmid organisms or to toxin-producing prokaryotes to produce pathogenic hydrocarbon degraders, must both represent events that
are occurring constantly in nature. The failure of the resulting hybrid species to exist
in wild populations must indicate their inability to compete and survive. In this sense,
the development of the multiplasmid bacteria is analogous to that of domesticated
plants and animals by cross-breeding between wild populations and selecting the progeny. The result is a species that can perform a valuable economic function in a protected, humanly controlled environment, but not compete in the wild. In the case of
the multiplasmid hydrocarbon degraders, this would be particularly true for wild environments that are generally marginal for microbial growth anyway, e.g., the hot,
anaerobic environment within an oil well, the water-deficient environment of the inside
of a gasoline or oil storage tank, or the nutrient-poor one of an asphalt road. Any
microbial growth that occurred under conditions approaching any of these would have
to be dominated by the existing, wild, single-plasmid hydrocarbon degraders, since
these are the only ones that would have the requisite survival capacity.

The third conceivable hazard, namely that of acute local environmental side effects,
change of the ecological balance, etc., due to the sudden large release of a single microbial strain into the oil spill environment, is less simply disposed of. Such effects
have not been reported in connection with other releases of pure cultures (e.g., from
large-scale fermentation operations); however, these have involved species other than
those of interest here. Obviously, tests for acute toxicity (which can be done by a
variety of simple, unambiguous procedures) should be carried out before any large-
scale manufacture or release of multiplasmid hydrocarbon-degrading microorganisms.

III. CONSTRUCTION OF MICROBIAL STRAINS FOR METHANE PRODUCTION

The anaerobic microbial conversion of cellulose to methane (natural gas) is a process
that results in an almost complete retention of the fuel value of the original solid carbohydrate. If it could be applied to all of the Nation's cellulosic wastes, it would generate enough gas to supply all current needs. The only problem is that the process
occurs too slowly to be practical for large-scale use. Accordingly, it appeared of considerable importance to examine both the mechanism of microbial methane production
and the possibilities of genetically engineering the microbes involved to improve their
production rates.

The anaerobic conversion of dead plant tissue to methane requires the combined
action of at least two different groups of microorganisms. One group acts on polymeric
carbohydrate substrates, including cellulose, to form low molecular weight organic
acids such as formate, acetate, propionate, etc. These are subsequently acted upon by
the second group to form methane. The microorganisms that produce methane from
formate, acetate, or CO_2 and H_2, when isolated as pure cultures, have been found to
exhibit extreme fastidiousness. They are very sensitive to small amounts of oxygen,
metal ions, or slight changes in pH and have a very narrow substrate range. For ex-

TABLE 3

Transfer of Genes Specifying Cellulose Degradation from *Pseudomonas fluorescens* to *Escherichia coli*

Donor	Recipient	Select	Frequency
AC 143	AC 501	Xyl$^+$	10^{-5}
(XYL-K)$^+$ Met$^-$	Cls$^+$Trp$^-$ *P. fluorescens*		
AC 506	AC 100	Cls$^+$	3×10^{-8}
Cls$^+$Trp$^-$(XYL-K)$^+$	*E. coli* C		
	AC 80	Cls$^+$	1×10^{-8}
	Res$^-$Thr$^-$Leu$^-$Met$^-$ *E. coli*		

Note: Cls, Cellulose; Xyl, xylene; Trp, tryptophan; Res, restriction; Thr, threonine; Leu, leucine; Met, methionine.

ample, they can utilize ammonia as a source of nitrogen, but not amino acids, peptides, or other nitrogeneous compounds.[13] As carbon sources, they can utilize CO_2, formate, and acetate, but none of the sugars, alcohols, or amino acids. These limited nutritional capabilities, coupled with their extreme sensitivity to oxygen and other physical and chemical agents, make the conversion of cellulose to methane an extremely slow process.

A conceivable way to avoid these limitations and improve methane production yields would be to create microbial strains in which the capacities to produce cellulase and form methane had been transferred to a single nonfastidious microbe. To begin assessing the possibility of doing this by natural plasmid transfer, Chakrabarty and Brown examined the feasibility of constructing a cellulolytic *E. coli.*

The results presented in Table 3 show that it was possible to transfer the cellulolytic genes, specifying cellulase enzymes, from a cellulolytic *P. fluorescens* strain[14] to a strain of *E. coli* C or a restriction-negative mutant of *E. coli* K-12 without using in vitro recombinant DNA techniques. To accomplish gene transfer, a recombinant XYL-K plasmid was first transferred to the *P. fluorescens* strain. These authors have earlier shown that the transfer plasmid, factor K, is capable of mobilizing nontransmissible plasmids and chromosomal genes at a high frequency from K$^+$ donor cells to K$^-$ recipients. Since K itself does not have a selectable character, a recombinant XYL-K plasmid was used so that K$^+$ *P. fluorescens* strains could be isolated on xylene plates. Such XYL-K$^+$ *P. fluorescens* cells are perfectly capable of transferring the cellulolytic properties to *E. coli* cells, although at a very low frequency (Table 3). The cellulolytic *E. coli* produces cellulases that appear to have essentially the same properties as those of the original *P. fluorescens* strain.

The feasibility of cloning methane-producing genes in cellulolytic *E. coli* using vectors such as pCR1 or pMB9 has not been assessed. Chakrabarty and Brown's work with the cellulolytic *E. coli* has been suspended and the cultures destroyed pending an appraisal of its biohazards.

A. Biohazard Potential of Cellulolytic *E. Coli*

To date, these authors have identified three possible biohazards that might result from the use of *E. coli* strains containing cellulase genes on plasmids. The first would be the escape of the organism to the biosphere, followed by an increase in the cellulolytic activity of the soil microflora to the point where living plant tissues (e.g., roots) as well as dead tissues might come under attack. The increased cellulolytic activity might result from either the survival and propagation of the cellulolytic *E. coli* or the transfer of the cellulase plasmid to some other soil microbe. Neither of these routes to increased cellulase activity seems particularly plausible, however. *E. coli* does not live in soil, and the natural soil microorganisms already have ample access to cellulase genes, including those of the *Pseudomonas* that was the original source of the cellulase plasmid used in Table 3.

The second conceivable biohazard is the colonization of the human gut by accidently released *E. coli* and is disposed of less simply. Since other *E. coli* strains are natural inhabitants of the human colon, the possibility of its establishing residence there is not implausible. This might subsequently present three side effects: (1) a problem of enhanced gas formation; (2) a disturbance of the balance among the normal intestinal microflora, which constitute one of the body's primary defenses against microbial infections; and (3) a reduction in the roughage passing through the gut, a concern in that recent data suggest a correlation between lack of roughage and increased bowel cancer.

The third conceivable biohazard would be the accidental exposure of the *E. coli* to the enteric bacteria of the human gut, followed by the transmission of the cellulase genes to one of the toxin-producing strains that might be present. If the resulting cellulose-digesting capacity leads to a competitive growth advantage, an increased population of a normally minor toxinogenic species might follow, with increased toxic effects on the individual.

It should be pointed out, however, that powerful arguments exist denying the plausibility of either of the last two types of biohazards. Anaerobic cellulolytic prokaryotes are already widely distributed in natural environments, including the intestinal tracts of warm-blooded animals. Many nonruminant mammals, including equids, rodents, and even some species of monkeys, harbor sufficient levels of such microbes to permit them to live on high cellulose diets. It would be almost inconceivable that low levels of cellulolytic bacteria do not already exist in the human intestinal tract as well and that these have been participating in gene exchange processes with the other intestinal prokaryotes for as long as the primate intestine has existed. The failure of such microbes to bloom forth into prominence must be related not to their absence, but to the competitive disadvantage provided by cellulolytic activity. In a nutrient-rich, microbially crowded environment like the human colon, any microorganism that wasted its metabolic capacities producing an extracellular enzyme such as cellulase, which would then produce soluble sugars accessible to all the neighboring microbes, would have difficulty outgrowing its lazier neighbors. It is difficult to imagine that a sizable unfilled niche for cellulolytic anaerobes may exist in the human intestinal tract.

Nevertheless, there is currently no solid proof that this inference is valid, and hence the last two biohazards noted above cannot be completely discounted. Accordingly, these authors decided to discontinue work with cellulase genes in *E. coli* or microorganisms that can exchange genes with enteric bacteria until additional evidence regarding the survivability of such genes in the human intestinal microflora becomes available.

CONCLUSION

The authors have chosen to discuss two problem areas where advances in genetic engineering would appear to offer sizable human benefit in combating environmental pollution and energy shortages. In each of these areas, there appears to be a plausible opportunity for achieving the required advance by transferring into the recipient microorganisms new combinations of existing natural plasmids rather than synthetic plasmids prepared by the in vitro recombinant DNA technique. In each case, it was also possible to identify conceivable biohazards that might be associated with the new multiplasmid organisms produced. Nevertheless, these hazards generally appeared considerably less plausible than those attributed to organisms produced by in vitro recombinant DNA techniques. Since the multiplasmid strains are created by assembling transmissible genetic fragments from species that already exchange genetic information naturally, the authors must presume these strains to represent life forms already tried and found unable to compete in a wild environment. Like domesticated plants and animals ("genetically improved" by analogous forms of crossbreeding and selection), they may be able to perform valuable economic functions in a humanly protected environment. However, they are incapable of activities that are qualitatively different from those already present in the natural mixed microbial populations from which they were derived and hence would seem unlikely to present new types of biohazards.

REFERENCES

1. **Chakrabarty, A. M.,** Recombinant DNA: areas of potential applications, in *The Recombinant DNA Debate,* Jackson, D. A. and Stich, S. P., Eds., Prentice-Hall, Englewood Cliffs, N. J., 1978, in press.
2. **Friello, D. A., Mylroie, J. R., and Chakrabarty, A. M.,** Use of genetically engineered multi-plasmid microorganisms for rapid degradation of fuel hydrocarbons, in *Proceedings of the Third International Biodegradation Symposium,* Sharpley, J. M. and Kaplan, A. M., Eds., Applied Science, London, 1976, 205.
3. **Soli, G. and Bens, E. M.,** Selective substrate utilization by marine hydrocarbonoclastic bacteria, *Biotechnol. Bioeng.,* 15, 285, 1973.
4. **Walker, J. D., Petrakis, L., and Colwell, R. R.,** Comparison of the biodegradability of crude and fuel oils, *Can. J. Microbiol.,* 22, 598, 1976.
5. **Miget, R.,** Bacterial seeding to enhance biodegradation of oil slicks, in *The Microbial Degradation of Oil Pollutants,* Ahern, D. G. and Meyers, S. P., Eds., Center for Wetland Resources, Louisiana State University, Baton Rouge, Publ. No. LSU-SG-73-01, 1973.
6. **Chakrabarty, A. M.,** Plasmids in *Pseudomonas, Annu. Rev. Genet.,* 10, 7, 1976.
7. **Friello, D. A., Mylroie, J. R., Gibson, D. T., Rogers, J. E., and Chakrabarty, A. M.,** XYL, a nonconjugative xylene-degradative plasmid in *Pseudomonas* Pxy., *J. Bacteriol.,* 127, 1217, 1976.
8. **Bartha, R. and Atlas, R. M.,** Biodegradation of oil in sea water: limiting factors and artificial stimulation, in *The Microbial Degradation of Oil Pollutants,* Ahern, D. G. and Meyers, S. P., Eds., Center for Wetland Resources, Louisiana State University, Baton Rouge, Publ. No. LSU-SG-73-01, 1973.
9. **Dixon, R., Cannon, F., and Kondorosi, A.,** Construction of a P. plasmid carrying nitrogen fixation genes from *Klebsiella pneumoniae, Nature (London),* 260, 268, 1976.
10. **Chakrabarty, A. M., Mylroie, J. R., Friello, D. A., and Vacca, J. G.,** Transformation of *Pseudomonas putida* and *Escherchia coli* with plasmid-linked drug-resistance factor DNA, *Proc. Natl. Acad. Sci. U.S.A.,* 72, 3647, 1975.
11. **Gunsalus, I. C., Conrad, H. E., Trudgil, P. W., and Jacobson, L. A.,** Regulations of catabolic metabolism, *Isr. J. Med. Sci.,* 1, 1099, 1965.
12. **Chakrabarty, A. M.,** Genetic fusion of incompatible plasmids in *Pseudomonas, Proc. Natl. Acad. Sci. U.S.A.,* 70, 1641, 1973.
13. **Pohland, F. G.,** *Anaerobic Biological Treatment Processes,* American Chemical Society, Washington, D.C., 1971.
14. **Yamane, K., Suzuki, H., and Nisizawa, K.,** Purification and properties of extracellular and cell-bound cellulase components of *Pseudomonas fluorescens* var. *cellulosa, J. Biochem.* (Tokyo), 67, 19, 1970.

INDEX

A